IB DIPLOMA PROGRAMME

Environmental Systems and Societies

Course Companion

Jill Rutherford

OXFORD
UNIVERSITY PRESS

UNIVERSITY PRESS

Great Clarendon Street, Oxford OX2 6DP

Oxford University Press is a department of the University of Oxford.
It furthers the University's objective of excellence in research, scholarship,
and education by publishing worldwide in

Oxford New York

Auckland Cape Town Dar es Salaam Hong Kong Karachi
Kuala Lumpur Madrid Melbourne Mexico City Nairobi
New Delhi Shanghai Taipei Toronto

With offices in

Argentina Austria Brazil Chile Czech Republic France Greece
Guatemala Hungary Italy Japan Poland Portugal Singapore
South Korea Switzerland Thailand Turkey Ukraine Vietnam

© Jill Rutherford 2009

The moral rights of the authors have been asserted

Database right Oxford University Press (maker)

First published 2009

British Library Cataloguing in Publication Data

Data available

ISBN- 978-0199152278

10 9 8 7 6

Printed in Singapore by KHL Printing Co. Pte Ltd

Paper used in the production of this book is a natural, recyclable product
made from wood grown in sustainable forests. The manufacturing process
conforms to the environmental regulations of the country of origin.

Acknowledgments

Cover image: Visions of America/Joe Sohm/Getty Images;
p. 7 Sammis, 1864 Museum of History and Industry, Seattle; p. 11 Ross White,
AFP/Getty Images; pp. 13, 69 NASA Images; pp. 28, 32, 35, 43, 44, 45, 48, 49,
53, 59, 71, 72, 75, 78, 79, 80, 81, 83, 84, 85, 138, 160, 162 Nigel Gardiner,
www.sciencebitz.com; p. 34 with permission of Pearson Education; p. 36 The
Whittaker Bome Diagram from RH Whittaker, *Communities and Ecosystems*,
Macmillan, New York, 1975. With permission of Pearson Education; p. 37
© Crown Copyright 2008, the Met Office; p. 38 from TC Whitmore, *An
Introduction to Tropical Rainforests*, Clarendon Press, Oxford University Press,
1990, with permission of the publisher; pp. 40, 46, 142 David Harrison; pp.
47, 58, 67, 75 ILEA *Advanced Biology Alternative Learning Project, Unit 9: Ecology*,
Cambridge University Press, 1985, with permission of the publisher; p. 52
from *Biology: Nuffield Coordinated Science*, Longman, 1992, by permission of
The Nuffield Foundation; pp. 60, 62 from JJ Chapman and MJ Reiss, *Ecology
Principles and Applications*, Cambridge University Press, 1992, by permission of
the publishers; p. 82 from TH Odum, *Fundamentals of Ecology*, Saunders, 1953;
pp. 87–8 from EP and HT Odum, *Ecology and Our Endangered Life Support Systems*,
Stanford, CT, 1989, by permission of the authors; p. 94 © 2005–8 Geology.
com; p. 105 JM Watson, United States Geological Survey; p. 106 Andrew
Sarnecki, Jupiter Images; p. 107 Austin Post, US Geological Survey; p. 108
Bettmann/Corbis; p. 110 T-Rex Art, iStockphoto; p. 111 Michael Nicholson,
Corbis; Tibor Bognar, Art Directors & Trip Photo Library; Jeremy Sutton-
Hibbert, Alamy; p. 112 Brandon D Cole, Corbis; p. 113 John Carnemolla,
Corbis; p. 115 iStockphoto; Lordprice Collection, Alamy; p. 118 Martin Shields,
Alamy;p. 126 David Reed, Corbis; p.129 © 2002 J Michael Reed; p. 131 Pete
Oxford, Nature Picture Library/Rex Features; p. 132 Natalia Bratslavsky,
Shutterstock; p. 137 source data from Climate Change 2007: The Physics
Science Basis, a report presented to the Intergovernmental Panel on Climate
Change; p. 138 Ashley Cooper/Alamy; p. 141 from Mark Maslin: *Global Warming:
A Very Short Introduction*, Oxford University Press, 2004, with the permission of
the publisher; p. 142 plate Tectonic and Paleogeographic maps and animations
by CR Scotese, © 2005, PALEOMAP Project. CO2 after RA Berner © 2001
(GEOCARB III); p. 143 source data from (US) National Oceanic and Atmospheric
Administration (NOAA), Earth System Research Laboratory (ESRL); p. 144
source data from Met Office Hadley Research Centre and University of East
Anglia Climatic Research Unit; p. 150 from Moore, Frances C, *Carbon Dioxide
Emissions Accelerating Rapidly*, published at www.earthpolicy.org. Copyright
© 2008 Earth Policy Institute; p. 154 Sheila Terry, Science Photo Library;
p. 160 source data from US Census Bureau, International Programs; p. 170
source data Population Connnections (USA) and UNdata; p. 174 source data ©
EarthTrends 2003; pp. 177–8, 180, 182 source data US Census Bureau; p. 185 JP
Nel, Big Stock Photo; p. 205 Eyepress, Associated Press; pp. 193, 194 from Rees,
William, Wackernagel, Mathis and Testemale, Phil (illustrator), *Our Ecological

Footprint: reducing human impact on Earth*, New Society Publishers, 1996, by
permission of the illustrator; p. 203 source data from US Energy Information
Administration (EIA), Oct 2008; p. 203 source data from BP Worldwide; p. 204
source data from REN21 2006 Global Status Report on Renewables and the
BP 2006 Statistical Review; p. 206 source data from Re-Energy.ca; p. 209 US
Geological Survey; p. 210 The Oceanic Conveyor Belt of Heat: A conceptual
model of global ocean circulation © WS Broeker, Columbia University; p.
233 source data from GRID-Arendal/United Nations Environment Programme
(UNEP); p. 234 Corbis; p. 235 Shirley Barenholz, Photographers Direct; p. 238
Frans Lanting, Corbis; p. 241 Saskatchewan Archives Board; p. 242 Jeremy
Hartley, Panos; p. 244 Keith Weller, USDA; p. 246, source data from The
Food and Agricultural Organization of the United Nations (FAO) Food Prices
Indices, February 2008; p. 248 Aurelia Zizo; source data from The Food and
Agricultural Organization of the United Nations (FAO); p. 249 GP Bowater,
Alamy; p. 252 source data from The Food and Agricultural Organization of
the United Nations (FAO); p. 252 from K Byrne: *Environmental Science*, 2nd edn,
Nelson Thornes, 2001, by permission of the publisher; p. 253 from Janet
Larsen, "The Other Fish in the Sea, But For How Long?", *Eco-Economy Update*,
Washington, DC Earth Policy Institute, 16 July 2003; p. 262–3 source data from
The Food and Agricultural Organization of the United Nations (FAO); p. 274
Jim Vallance, US Geological Survey; p. 285 Andrew Paterson, Alamy; p. 287 Jeff
Schmaltz, MODIS Rapid Response Team, NASA/GSFC; p. 294 Yannis Behrakis,
Reuters; p. 296 Stan Kujawa, Alamy; p. 309 from TJ King, *Ecology*, Nelson
Thorne, 1989. Published with the permission of TJ King; p. 310 from Singsby,
D and Cook, C, *Practical Ecology*, Macmillan, 1986, by permission of Palgrave
Macmillan; pp. 315–16 from Carrie et. al., *Biology in Practice*, Oxford University
Press, 1990, with the permission of the publisher; p. 322 source data from
US Census Bureau; p. 343 The Cartoonists and Writers Syndicate, New York;
p. from JA Vucetich and RO Peterson, Michigan Technological University;
p. 346 from CB Cox and PD Moore, *Biogeography*, Blackwell, 2000, with the
permission of Wiley-Blackwell; p. 353 from Michael Carrr, *Patterns, Process and
Change in Human Geography*, Basingstoke, Macmillan, 1987, with permission of
the author.

[Extracts]
pp. 24–5 Viktor Vovk, from the introduction to *Sustainable Development for the
Second World*, Worldwatch Paper 167, September 2003, with permission of
Worldwatch Institute; p. 74, Bill Bryson, *A Short History of Nearly Everything*,
Doubleday, 2003, with permission of The Random House Group; p. 74
Marcus Chown, "Science: Quantum theory and relativity explained", *The
Daily Telegraph*, 20 Nov 2007, with permission of Telegraph Media Group;
pp. 89–90 James Lovelock "The Earth is about to catch a morbid fever that
may last as long as 100,000 years", *The Independent*,16 Jan 2006, copyright
© The Independent 2006, with permission of The Independent; p. 156 'The
Science' on the website for *An Inconvenient Truth* by Al Gore, with permission
of Participant Media for Climatecrisis.net; pp. 253–5 from Janet Larsen, "Other
Fish in the Sea, But for How Long?", *Eco-Economy Update*, Washington, DC,
Earth Polity Institute, 16 July 2003, copyright © 2003 Earth Policy Institute,
with permission of the Earth Policy Institute; pp. George Monbiot, *Heat: How
to Stop the Planet Burning*, Penguin, 2007, copyright © George Monbiot 2006,
with permission of Penguin Books; pp. 331–2 Chris Wilmott, *Journal of Biological
Education* 37 (3), 2003, with permission of the Institute of Biology.

We have tried to trace and contact all copyright holders before publication. If
notified the publishers will be pleased to rectify any errors or omissions at the
earliest opportunity.

Course Companion definition

The IB Diploma Programme Course Companions are resource materials designed to provide students with extra support through their two-year course of study. These books will help students gain an understanding of what is expected from the study of an IB Diploma Programme subject.

The Course Companions reflect the philosophy and approach of the IB Diploma Programme and present content in a way that illustrates the purpose and aims of the IB. They encourage a deep understanding of each subject by making connections to wider issues and providing opportunities for critical thinking.

These Course Companions, therefore, may or may not contain all of the curriculum content required in each IB Diploma Programme subject, and so are not designed to be complete and prescriptive textbooks. Each book will try to ensure that areas of curriculum that are unique to the IB or to a new course revision are thoroughly covered. These books mirror the IB philosophy of viewing the curriculum in terms of a whole-course approach; the use of a wide range of resources; international-mindedness; the IB learner profile and the IB Diploma Programme core requirements; theory of knowledge; the extended essay; and creativity, action, service (CAS).

In addition, the Course Companions provide advice and guidance on the specific course assessment requirements and also on academic honesty protocol.

The Course Companions are not designed to be:

- study/revision guides or a one-stop solution for students to pass the subjects
- prescriptive or essential subject textbooks.

IB mission statement

The International Baccalaureate aims to develop inquiring, knowledgeable, and caring young people who help to create a better and more peaceful world through intercultural understanding and respect.

To this end the IB works with schools, governments, and international organizations to develop challenging programmes of international education and rigorous assessment.

These programmes encourage students across the world to become active, compassionate, and lifelong learners who understand that other people, with their differences, can also be right.

The IB learner profile

The International Baccalaureate aims to develop internationally minded people who, recognizing their common humanity and shared guardianship of the planet, help to create a better and more peaceful world. IB learners strive to be:

Inquirers   They develop their natural curiosity. They acquire the skills necessary to conduct inquiry and research and show independence in learning. They actively enjoy learning and this love of learning will be sustained throughout their lives.

Knowledgeable   They explore concepts, ideas, and issues that have local and global significance. In so doing, they acquire in-depth knowledge and develop understanding across a broad and balanced range of disciplines.

Thinkers   They exercise initiative in applying thinking skills critically and creatively to recognize and approach complex problems, and make reasoned, ethical decisions.

Communicators   They understand and express ideas and information confidently and creatively in more than one language and in a variety of modes of communication. They work effectively and willingly in collaboration with others.

Principled   They act with integrity and honesty, with a strong sense of fairness, justice, and respect for the dignity of the individual, groups, and communities. They take responsibility for their own actions and the consequences that accompany them.

Open-minded   They understand and appreciate their own cultures and personal histories, and are open to the perspectives, values, and traditions of other individuals and communities. They are accustomed to seeking and evaluating a range of points of view, and are willing to grow from the experience.

Caring   They show empathy, compassion, and respect towards the needs and feelings of others. They have a personal commitment to service, and act to make a positive difference to the lives of others and to the environment.

Risk-takers   They approach unfamiliar situations and uncertainty with courage and forethought, and have the independence of spirit to explore new roles, ideas, and strategies. They are brave and articulate in defending their beliefs.

Balanced   They understand the importance of intellectual, physical, and emotional balance to achieve personal well-being for themselves and others.

Reflective   They give thoughtful consideration to their own learning and experience. They are able to assess and understand their strengths and limitations in order to support their learning and personal development.

A note on academic honesty

It is of vital importance to acknowledge and appropriately credit the owners of information when that information is used in your work. After all, owners of ideas (intellectual property) have property rights. To have an authentic piece of work, it must be based on your individual and original ideas with the work of others fully acknowledged. Therefore, all assignments, written or oral, completed for assessment must use your own language and expression. Where sources are used or referred to, whether in the form of direct quotation or paraphrase, such sources must be appropriately acknowledged.

How do I acknowledge the work of others?
The way that you acknowledge that you have used the ideas of other people is through the use of footnotes and bibliographies.

Footnotes (placed at the bottom of a page) or endnotes (placed at the end of a document) are to be provided when you quote or paraphrase from another document, or closely summarize the information provided in another document. You do not need to provide a footnote for information that is part of a 'body of knowledge'. That is, definitions do not need to be footnoted as they are part of the assumed knowledge.

Bibliographies should include a formal list of the resources that you used in your work. 'Formal' means that you should use one of the several accepted forms of presentation. This usually involves separating the resources that you use into different categories (e.g. books, magazines, newspaper articles, Internet-based resources, CDs and works of art) and providing full information as to how a reader or viewer of your work can find the same information. A bibliography is compulsory in the extended essay.

What constitutes malpractice?

Malpractice is behaviour that results in, or may result in, you or any student gaining an unfair advantage in one or more assessment component. Malpractice includes plagiarism and collusion.

Plagiarism is defined as the representation of the ideas or work of another person as your own. The following are some of the ways to avoid plagiarism:

- Words and ideas of another person used to support one's arguments must be acknowledged.

- Passages that are quoted verbatim must be enclosed within quotation marks and acknowledged.
- CD-ROMs, email messages, web sites on the Internet, and any other electronic media must be treated in the same way as books and journals.
- The sources of all photographs, maps, illustrations, computer programs, data, graphs, audio-visual, and similar material must be acknowledged if they are not your own work.
- Works of art, whether music, film, dance, theatre arts, or visual arts, and where the creative use of a part of a work takes place, must be acknowledged.

Collusion is defined as supporting malpractice by another student. This includes:

- allowing your work to be copied or submitted for assessment by another student
- duplicating work for different assessment components and/or diploma requirements.

Other forms of malpractice include any action that gives you an unfair advantage or affects the results of another student. Examples include, taking unauthorized material into an examination room, misconduct during an examination, and falsifying a CAS record.

Contents

Introduction

This book is a Course Companion to the first transdisciplinary IB Diploma Programme course – Environmental Systems and Societies. Although this course has had a history of some decades in several forms, the one that you are studying now is the culmination of the ideas and work of many teachers and their students. It introduces you to some big environmental issues facing humans and the world that we inhabit.

The IB mission statement, which is also expressed in the IB learner profile characteristics, is at the heart of this course. As you read and refer to this Course Companion, consider the examples, case studies and questions with reference to your characteristics as a learner and the characteristics of the learner profile. We try to communicate clearly the issues facing the Earth which are described in this book. We must continue to enquire into and think about the environment and our actions within it so that we can build up knowledge across disciplines in order to solve problems. Governments, groups and individuals taking decisions on environmental issues must evaluate the different viewpoints with an open mind and balance the risks and benefits of their actions. We would not be adequate guardians of the planet unless we care about it, have principles by which we live, and accept accountability for our actions after due reflection. The maxim "Think globally, act locally" is a driver of this course and of the IB Diploma Programme CAS requirement. If you are carrying out CAS activities, many of these could also involve protecting or repairing your environment, and we hope that you may gain some ideas for this from this book.

Writing this Course Companion would not have been possible without a team approach. Many IB Diploma Programme teachers have contributed in varying ways and the author is most grateful to these busy people, living in different countries and biomes, ecosystems and environments. We believe that this team approach gives the book a truly international flavour, with case studies from many countries and viewpoints. Thank you to all teachers (and students) of this course and particularly to those who have given their words and ideas for this text: Alison Underwood, Alistair Robertson, Brian Sweeney, Cheryl Moulder,

David Harrison, Douglas Glasenapp, Ellie Alchin, Gary Seston, Henno van Beek, Jan Kent, Kelvin Moon, Nigel Gardner, Patrick Armstrong and Tejas Ewing. The book would not have been written at all were it not for past students who, with goodwill, enthusiasm and interest, were willing to get cold, wet and muddy for the purpose of collecting data and gaining understanding.

Any errors and omissions are entirely those of the author and we welcome communication from you to point out where these are and to suggest improvements and updates for the next edition.

How to use this Course Companion

Each chapter starts with the key ideas that it covers and ends with a list of key words that you should know having read that chapter. Although we believe that the sequence of chapters 1–15 is a logical route through the course, you may wish to dip into various sections in a different order. Chapters 16–19 cover assessment, both internal and external, and the extended essay.

Special features included in the first 15 chapters are:

- "To do" – suggestions for self-assessment or reinforcement
- "To think about" – interesting ideas to link with your Theory of Knowledge course
- "Test yourself" – some questions (usually data-based) to test application of knowledge
- "Review" – to review your understanding of concepts
- "To research" – something to find out for yourself
- Case studies – longer examples with some questions
- – more information on a topic.

Chapter 20 contains selected further reading and website addresses of organizations that have been mentioned in the text. Here too are some thoughts by Tejas Ewing, a graduate of the IB Diploma Programme, on careers in sustainable development and related fields.

Jill Rutherford
November 2008

1 Historical overview of the modern environmental movement

Key points

- This chapter looks at key historical influences on the development of the modern environmental movement. We start in a different way to what you may expect from a book on this course but this sets the scene for understanding why the environmental movement is what it is now.

Setting the scene

To think about

Look up Chief Seattle on the Internet. His famous speech was in the Lushootseed language, translated into Chinook Indian trade language, and then into English. He may not have said these exact words, but what do you think of their message? Do you agree?

As you can see, although the modern environmental movement originated in the 1960s, a concern for the effect of humans on their environment was evident long before. Problems such as air and water pollution were reported on back to at least Roman times. Between the late 14th century and the mid 16th century, waste produced by humans was associated with the spread of epidemic disease in Europe, and soil conservation was practised in China, India and Peru as early as 2 000 years ago. In general, however, such concerns did not give rise to widespread public activism. To understand modern environmentalism we must look back historically at the events which caused concern over environmental impacts and at the responses of individuals, groups of individuals, governments and the United Nations to these impacts. Notice how influential individuals and independent pressure groups, with their use of media, have catalysed the movement, making it a people's or "grass roots" movement. Also notice that there has been a continuing divide in philosophy between those who see the reason to conserve nature being to continue to supply goods and services to humankind in a sustainable way (environmental managers) and those who believe that we should conserve nature unconditionally, for its spiritual value (deep and self-reliance ecologists); i.e. do we save it for *our* sake or for *its* sake?

Who is involved in the environmental movement?

It is probably fair to say that the majority of people in the world do not spend much time focusing on environmental issues unless it is brought to their attention or affects them directly. However, norms of behaviour (e.g. purchasing choices such as dolphin-friendly tuna and recycling) and even political choices (e.g. the successes of the

"Whatever befalls the Earth, befalls the sons of the Earth. Humankind has not woven the web of life. We are but one thread within it. Whatever we do to the web, we do to ourselves. All things are bound together. All things connect."

attributed to Chief Seattle, 1855

Fig. 1.1 The only known photo of Chief Seattle, taken in the 1860s.

"We abuse land because we regard it as a commodity belonging to us. When we see land as a community to which we belong, we may begin to use it with love and respect."

Aldo Leopold, *A Sand County Almanac*

"Green Party") have been influenced. This has been through the activities of a number of groups.

Influential individuals, often through media publications (e.g. Aldo Leopold's *A Sand County Almanac* or Rachel Carson's *Silent Spring*), commonly raise issues and start the debate.

Independent pressure groups, often through awareness campaigns (e.g. Greenpeace, "Save the Whale") influence the public and use this to influence government and corporate business organizations. These groups are called non-governmental organizations (NGOs). "Friends of the Earth" is another example.

Corporate businesses (especially multinational corporations – MNCs – and transnational corporations – TNCs) are involved since they are supplying consumer demand and in doing so using resources and creating environmental impact (e.g. mining for minerals or burning of fossil fuels).

Governments make policy decisions including environmental ones (e.g. planning permission for land use), and apply legislation (laws) to manage the country (e.g. emissions controls over factories). They also meet with other governments to consider international agreements (e.g. United Nations Environment Programme, UNEP). Different countries are at different stages of environmental awareness, as are different individuals. Legislating about emissions is important but so is making sure there is enough food for the population. While different countries may put environmental awareness at different levels of priority, all are aware of the issues facing the Earth and that all must be involved in finding solutions.

Intergovernmental bodies such as the **United Nations** have become highly influential in more recent times by holding Earth Summits to bring together governments, NGOs and corporations to consider global environmental and world development issues.

The growth of the modern environmental movement in outline

The human population began to rise after the **Neolithic Agricultural Revolution** which started 10 000 years ago when humans settled to become farmers instead of nomadic hunter-gatherers. This meant managing local resources (food, water, fuel) sustainably from around the settlement.

However, over the last 200 years, population growth and resource usage has escalated at an unprecedented rate. In Europe, the **Industrial Revolution** of the early 1800s produced goods and services for all but this demanded the burning of large amounts of fuel in the form of trees and coal, and the mining of minerals from the earth to produce metals to make machines and lime for cement. Land was cleared, natural waterways polluted, cities became crowded and smoky. Our urban consumer society arose.

A few early **conservationists** spoke out about the loss of "wilderness" areas but, more generally at this time, the attitude of the day was that the world contained an infinite supply of resources, a land of plenty.

Parts of the New World such as North America, Asia and Australasia were being colonized by "the West" and their resources must have seemed endless. Sustainability became a forgotten and unnecessary concept and was replaced by exploitation economics.

The **Green Revolution** of the 1940s to 1960s mechanized agriculture and boosted food production massively but required the building of machinery and burning of enormous amounts of fossil fuels such as oil. Technology was applied to agriculture. New crop varieties were developed and fertilizer and pesticide use rose sharply. The world population grew to about 3 billion in this time. Our resource use and waste production rocketed.

The impacts followed and became more global: collapsing fish stocks, endangered species, pesticide poisoning, deforestation, nuclear waste, ozone layer depletion, global warming, acid precipitation, etc. These were all highlighted to the public by the new breed of **environmentalists** who had scientific backgrounds and spearheaded the modern environmental movement. Influential individuals wrote books, NGOs campaigned, the media reported, governments formed nature reserves and put environmental issues on their agenda, some businesses marketed themselves as environmentally friendly and UNEP organized Earth Summits on the environment. The **modern environmental movement** became public and gained momentum.

Landmarks in the growth of the modern environmental movement

1800–1900: Early days

The environmental movement, as we know it today, is founded on three basic principles. These are that:

- natural resources are not infinite – if we do not conserve them, we shall run out
- our natural environment and the life within it must be preserved for future generations to enjoy and exploit
- the pollution that we generate and our activities endanger life on Earth.

Two hundred years ago in North America, James Fenimore Cooper, author of *The Last of the Mohicans*, articulated this in his novels of the 1820s, *The Pioneers* and *The Prairie*. Following this, Ralph Waldo Emerson and Henry David Thoreau wrote about a reverence for the natural world and an almost mystical experience when surrounded by nature that could not be rationalized – transcendentalism. But they were not scientists nor did they formalize the effects humans were having on the environment. This was first done in America by George Perkins Marsh who wrote *Man and Nature* in 1864 and was the first to write that deforestation could cause desertification, loss of species and degradation of the ecosystem.

Theodore Roosevelt, US president 1901–9, was the first US president to act on environmental concerns and was instrumental in founding the US Forest Service and setting aside lands for national parks. He and his colleagues argued that the health of the nation depended on

> "*If all mankind were to disappear, the world would regenerate back to the rich state of equilibrium that existed ten thousand years ago. If insects were to vanish, the environment would collapse into chaos.*"
>
> Edward O. Wilson

To think about

New and Old Worlds

The term New World was first used in the 15th century when Christopher Columbus returned to Europe from his first voyage to the Americas in 1493. Africa, Europe and Asia were then the Old World as they were known to these explorers but, of course, the New World was the Old World to those who lived in it.

New World is now used for the non-Eurasia and non-African continents of Americas and Australasia and the term has stuck to describe species that are found in one or the other, e.g. llamas in the New World (South America), camels in the Old; marsupials in Australasia but not the Old World. It is still also used in a historical context.

Question

Is New World purely a descriptive term or does it carry evaluative meaning?

the health of the land and that use of resources could not grow indefinitely. "Teddy" Roosevelt worked with naturalist, John Muir, perhaps the first **preservationist**, who founded the Sierra Club in 1892 and fought for the preservation of Yosemite National Park for its own sake, not as a resource for humans to use. The first director of the US Forest Service, Gifford Pinchot, was a **conservationist** who believed that forests should be used and renewed both to maintain a flow of resources for human use and to conserve the environment. Muir and Pinchot disagreed over the Hetch Hetchy Dam project which was to flood a valley within the Yosemite National Park to provide water and electricity to San Francisco and surrounding areas. The Sierra Club fought this for 10 years but the dam was sanctioned in 1913, after Roosevelt left office, and completed in 1923 and the valley flooded. The debate still goes on.

With the move towards urbanization and the massive growth in cities, the impact of humans on the local environment became obvious with problems of fuel supply (trees cleared), waste disposal, clean drinking water, air pollution and smogs. Early government focus was on making healthier cities. However, a few writers of the day lamented the loss of wilderness places. Thoreau lived alone for two years by Walden Pond in a wood in Massachusetts, USA and wrote his book, *Walden*, in 1848. His work inspired people to appreciate the harmony of living with nature and this harmony sentiment was championed by visionary John Muir.

"I would rather sit on a pumpkin and have it all to myself, than be crowded on a velvet cushion."

Henry David Thoreau, *Walden* (1848)

The reserves formed in countries such as USA, Canada, Australia and New Zealand at this time were allocated as a future resource of timber, oil, coal and water. However the idea of nature conservation grew and NGOs began to form. In Britain, the Royal Society for the Protection of Birds (RSPB) was formed in 1893 and the National Trust in 1894.

1900–1960: Environmental awareness grows

The impacts of humans on a more global scale were first highlighted when a number of popular species such as the tiger and the North American buffalo became endangered. William Hornaby wrote the influential book *Our Vanishing Wildlife* in 1913, and in 1914 the passenger pigeon, which was once the most numerous bird in the world, became extinct when Martha, the last of its species, died in Cincinnati Zoo. The conservation movement grew and in 1949 Aldo Leopold wrote what some people believe is the most important book on conservation ever written, *A Sand County Almanac*. In it he argues that nature is not our servant to be used and abused as we wish but that we have an ethical responsibility to maintain the balance of nature. He pioneered the concept of environmental stewardship, our duty being to benefit from and protect nature. In 1951, Britain formed its 10 National Parks, protected by legislation.

"Let us a little permit Nature to take her own way; she better understands her own affairs than we."

Michel de Montaigne, translated

In the 1930s, in the grass plains of North America, drought and winds turned the farms into the famous "Dust Bowl" and ruined their agriculture. At the time it was said to be a natural occurrence but with the increased agricultural mechanization, overgrazing, overploughing and overplanting that had become commonplace, some suspected it was human-induced.

Mercury was used in the hat-making industry into the 20th century. Hat makers were known to often suffer mental illnesses although the source of such illnesses was unknown. This is the basis of the name of the "Mad Hatter" character in Lewis Carroll's *Alice in Wonderland* and the phrase "as mad as a hatter".

In 1956 a large number of people in a fishing community in Japan in an area called Minamata Bay developed a strange disease that affected their nervous systems, deformed many and eventually killed over 900 people. After long investigations it was found that the cause was heavy metal mercury poisoning. The mercury had found its way from a chemical factory discharging waste into the bay, through the food chain, and into the shellfish that people ate. The chemical factory continued to discharge the poisons until 1968. The effects are still evident in the area today.

Fig. 1.2 The crippled hand of a Minamata disease victim.

1960s: The modern environmental movement gains momentum

There is general agreement that the modern environmental movement was catalysed by Rachel Carson's book, *Silent Spring*, published in 1962. Carson warned of the effects of pesticides on insects, both pest and others and how this was being passed along the food chain to kill others, including birds (hence a silent spring). What really gained people's attention was her belief that pesticides such as DDT (**d**ichloro**d**iphenyl**t**richloroethane, a persistent, synthetic insecticide) were finding their way into people and accumulating in fatty tissues, causing higher risks of cancer. Chemical industries tried to ban the book but many scientists shared her concerns and when an investigation, ordered by US president John F. Kennedy, confirmed her fears, DDT was banned.

Silent Spring had a powerful impact on the environmental movement and inspired activism which led eventually to the deep ecology movement and the rise of ecofeminism.

Then began a new awareness among the public that the effect of humans on the environment was having direct effects on them personally, rather than just on other species and on the countryside.

Through the 1960s the movement became organized and pressure groups came together to drive campaigns. The leading NGO was the World Wildlife Fund (WWF, the global conservation organization) which formed in 1961.

"For the first time in the history of the world, every human being is now subjected to contact with dangerous chemicals, from the moment of conception until death."

Rachel Carson, *Silent Spring*, 1962

"There are no passengers on Spaceship Earth. We are all crew."

Marshall McLuhan, 1964

1970s: The movement goes global

In the early 1970s a number of other important NGOs formed. Friends of the Earth and Greenpeace were both established in 1971. Initially they led campaigns against exploitation of endangered species such as tigers, pandas, whales, rhino, elephants and fur seals.

Media coverage of environmental issues became much more common and public awareness grew dramatically.

Fig. 1.3 The *Rainbow Warrior* after an attack by the French secret service in 1985.

In 1972 the first Earth Summit on the environment (United Nations Conference on the Human Environment) was held in Stockholm, Sweden, and attended by 113 nations. It heralded the beginning of an era of international environmentalism. The outcomes were the Declaration of the United Nations Conference, an Action Plan for the Human Environment and an Environment Fund. At this summit UNEP was formed and summits were planned to take place every 10 years. The summit also showed a world divide, with the developed nations overusing resources and perpetuating the unequal distribution of wealth between them and developing nations (see page 167 for more on terms "developed" and "developing" nations).

In 1975, the Convention on International Trade in Endangered Species (CITES) came into force for the 173 countries that agreed to take part. Produced by the IUCN (World Conservation Union), it aimed to identify endangered species and limit trade to protect them from extinction.

In the mid 1970s, philosophers also formed a new area of environmental philosophy and argued that nature has its own intrinsic value beside its value to humans.

> *"We do not inherit the earth from our ancestors, we borrow it from our children."*
>
> Native American proverb

A major shift in philosophy was the move from regarding the environment as an inexhaustible and resilient provider to regarding humankind as a steward responsible for caring for the environment. This was called **stewardship**.

In 1979, the author James Lovelock published *Gaia – A new look at life on Earth*. In his controversial Gaia Theory he describes the Earth as a single integrated system able to maintain homeostatic equilibrium ideal for life through feedback controls and able to protect itself from attack.

> *"The rose has thorns only for those who would gather it."*
>
> Chinese proverb

Lovelock's work encouraged environmentalists to take a systems approach to the environment. More recently he has suggested that Gaia's control capacities are being stretched by human impacts, especially related to global warming, and that the Earth will become uninhabitable for humans in many regions within 100 years.

1980s: Sustainability, the way forward

The 1982 Earth Summit in Nairobi, Kenya was generally thought to be ineffective.

In 1983, the UN General Assembly formed the UN World Commission on Environment and Development (notice the link to development now) which published the Brundtland report. This report officially introduced the concept of sustainability defined as: "Development that meets the needs of the present without compromising the ability of future generations to meet their own needs". Sustainable development was seen as the way forward.

> *"There is a sufficiency in the world for man's need but not for man's greed."*
>
> Mohandas K. Gandhi

In the early hours of the morning of 3 December 1984, in the centre of the city of Bhopal, India, in the state of Madhya Pradesh, a Union Carbide pesticide plant released 40 tonnes of methyl isocyanate (MIC) gas, immediately killing nearly 3000 people and ultimately causing at least 15 000 to 22 000 total deaths. This has been called the Bhopal Disaster and is considered to be the world's worst industrial disaster. The world was in shock.

In 1986, a further disaster struck as the Chernobyl nuclear power plant in Ukraine exploded, sending radioactive fallout over the Soviet Union, northern Europe and even as far as north east America. In Ukraine and Belarus, 336 000 people were evacuated but it was estimated that 6.6 million people had been highly exposed. The effects of contamination caused 4000 deaths from thyroid cancer but the World Health Organization predict from 30 000 to 60 000 further deaths from cancer. The ecological damage was huge and areas are considered still unsafe to enter.

In the mid 1980s, alarm bells were also rung by pressure groups with particular concerns over the findings of the British Antarctic Survey team in 1985 that the ice sheets were thinning and that the ozone layer was being depleted by CFCs and other gases. The Ozone Hole over Antarctica made big news. Again the direct threat of skin cancer to the public generated a powerful reaction and ozone-friendly products hit the supermarket shelves.

In 1987, the Montreal Protocol was signed. This agreement required nations who signed to reduce emissions of CFCs and other ozone-depleting gases. It has been said that this is the most successful international agreement so far.

Political parties go green

Since the early 1980s, green political parties around the world have formed. These share common goals of increasing awareness about environmental issues and reducing our use of resources through participatory democracy. Founded in the early 1970s, the world's first green political party was the Values Party in New Zealand, followed shortly by the United Tasmania Group in Australia. The first green party to be officially elected was in Switzerland in 1979. Later, in 1981, four green candidates won legislative seats in Belgium. Green parties also have been formed in the former Soviet bloc, where they were instrumental in the collapse of some communist regimes. There have been politically green parties in developing countries in Asia, South America and Africa, but they have not achieved success in elections. The most successful environmental party has been the **German Green Party** (die Grünen), founded in 1980. It entered the Bundestag (parliament) by winning 5.6% of the national vote in 1983 and 8.4% in 1987.

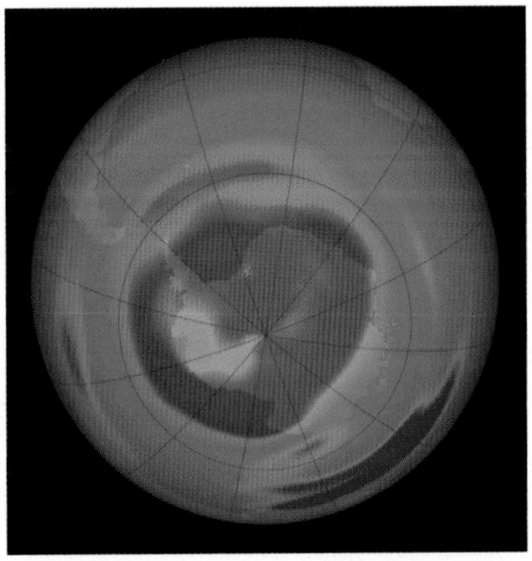

Fig. 1.4 The ozone hole in the atmosphere over Antarctica in October 2004.

In 1988, UNEP formed an advisory panel called the Intergovernmental Panel on Climate Change (IPCC) which looks at up-to-date scientific information and assesses the risks of climate change. It has been informing the Earth Summits and policy makers ever since.

1990s: Global warming and global economics

Rio de Janeiro, Brazil, was the location for the Earth Summit in 1992. It demonstrated clearly how environmental problems are linked to global economics and development. The major issue was carbon emissions (CO_2) from the burning of fossil fuels such as coal and oil causing global warming due to the enhanced greenhouse effect. Based on information from the IPCC, the Kyoto Protocol was introduced and nations that signed agreed to reduce carbon emissions by 5% between 2008 and 2012 (later revised to 12.5%

"Climate change is for real. We have just a small window of opportunity and it is closing rather rapidly. There is not a moment to lose."

Dr Rajendra Pachauri, Chairman, Intergovernmental Panel on Climate Change

13

below 1990 levels following further IPCC findings). However, major economies based on oil such as the USA and Saudi Arabia were unhappy and the USA refused to commit to anything at the time.

Green awareness became widespread and powerful in the developed nations. Recycling bins appeared (partly because landfill space was critically low), environmentally friendly products boomed, the Green party was winning election seats in countries over Europe and a favourite children's cartoon superhero, guaranteed to save the Earth every week, was "Captain Planet and his Planeteers".

The 1990s were also the start of big business in eco-tourism. While its environmental soundness might be debatable, it made good economic sense. The whale-watching industry is now valued at over US$1 billion, much more lucrative than whale hunting.

Early 2000s: Climate change, a time for action?

The 2002 summit in Johannesburg, South Africa, was the biggest so far with 65 000 politicians, huge attendance by NGOs and heavy media coverage. Developing countries had much more presence and were much more vocal. Five areas identified as needing attention globally were: water and sanitation, energy, health, agriculture and biodiversity. The summit agreed to halve the number of people in the world who suffer a lack of basic sanitation by 2015. It also agreed to reduce agricultural and energy usage in developed countries and stop degradation of fish and forest resources, without setting a date. Many environmentalists believe that the event was hijacked by corporate interests and that the USA, Japan and the oil companies resisted the development of renewable energy sources. Meanwhile many scientists are becoming ever more convinced that climate change is occurring and that this is, at least in part, due to fossil fuel use. The Kyoto Protocol became a legal requirement in 2005. This is being assessed by the IPCC as part of an ongoing study.

"Don't blow it – good planets are hard to find."

quoted in *Time magazine*

A review of major landmarks in environmentalism

Years	Events	Significance
10 000 years BP	Neolithic agricultural revolution	Settlements, population increase, local resource management began
Early 1800s	Industrial revolution in Europe	Increased urbanization, resource usage and pollution
Late 1800s	Influential individuals such as Thoreau and Muir write books on conservation	First conservation groups form and nature reserves established. NGOs form (RSPB for protection of birds and National Trust for conservation of nature in UK)
1914	Once the most prolific bird, the passenger pigeon, becomes the extinct	Conservation movement grows. Concern for tigers, buffalo, etc.
1940s	Green Revolution – intensive technological agriculture	Resource use (especially fossil fuel use) and pollution increased. Human population rises sharply
1930s and 1940s	Dustbowl in North America	Recognition that agricultural practices may affects soils and climate
1949	Leopold writes *A Sand County Almanac*	Concept of "stewardship" is applied to nature
1951	UK's ten National Parks are established	Recognition of need to conserve natural areas
1956 to 1968	Minamata Bay Disaster	Emphasizes the ability of food chains to accumulate toxins into higher trophic levels, including into humans
1962	Rachel Carson publishes *Silent Spring*	General acceptance of dangers of chemical toxins affecting humans. The pesticide DDT is banned

1960s and early 1970s	NGOs gain greater following	Public awareness grows. WWF, Greenpeace, Friends of the Earth all formed
1972	First Earth Summit – UN Conference on the Human Environment	Declaration of UN conference. Action Plan for the Human Environment. Environment Fund established. Formation of UN Environment Programme (UNEP). Earth Summits planned at 10-year intervals
1975	CITES formed by IUCN	Endangered species protected from international trade
Mid 1970s	Environmental philosophy established	Recognition that nature has intrinsic value. Stewardship ethic grows
1979	James Lovelock publishes *Gaia: A new look at life on Earth* and presents the "Gaia hypothesis"	Systems approach to studying the environment begins. Nature seen as self-regulating
1982	Nairobi Earth Summit	Ineffective
1983	UN World Commission on Environment and Development publishes the Brundtland Report	Sustainability established as the way forward
1984	Bhopal Disaster	World's worst industrial disaster
1986	Chernobyl Disaster	Nuclear fallout affects millions
Mid 1980s	British Antarctic Survey Team detects ice sheets thinning and ozone hole	Public awareness of ozone depletion and risks of skin cancer
1987	Montreal Protocol	Nations agree to reduce CFC use
1980s	Green political parties form around the world	Political pressure placed on governments
1988	IPCC formed by UNEP	Advises governments on the risks of climate change
1992	Rio Earth Summit and Kyoto Protocol	Agreement to reduce carbon (CO_2) emissions to counter enhanced greenhouse effect and global warming
1990s	Green awareness strengthens	Environmentally friendly products, recycling and eco-tourism become popular
2002	Johannesburg Earth Summit	Plans to globally improve: water and sanitation, energy supply issues, health, agricultural abuse, biodiversity reduction
2005	Kyoto Protocol becomes a legal requirement	174 countries signed and are expected to reduce carbon emissions to some 15% below expected emissions in 2008. It expires in 2012
2006	Film *An Inconvenient Truth* released	Documentary by Al Gore, former US vice president, on global warming
2007	Nobel Peace Prize	Awarded jointly to Al Gore and IPCC for their work on climate change
	IPCC release 4th assessment report in Nov 2007	Report states that "Warming of the climate system is unequivocal" and "Most of the observed increase in globally averaged temperatures since the mid-20th century is *very likely* due to the observed increase in anthropogenic greenhouse gas concentrations"
	UN Bali meeting Dec 07	187 countries meet and agree to open negotiations on an international climate change deal
2008	EU Climate summit, Brussels, Dec 08	Reaffirmed 20-20-20 commitment – cut emissions by 20%, make 20% energy savings and generate 20% of energy from renewable sources by 2020
	UN Climate Change conference, Poznan, Poland, Dec 08	Negotiations continued towards an agreement in Copenhagen in 2009

To research

There are many quotations on the environment in this chapter. Look up the names of some of the authors and write three sentences on each person. Find out:

1 Who they are/were and where and when they live(d).

2 What they did.

3 What their influence has been.

We suggest you consider: Mahatma Mohandas Gandhi, Henry David Thoreau, Aldo Leopold, John Muir and E. O. Wilson; but you could look up any other authors of the quotations as well.

To do

Carbon emissions to double by 2050

Japan removes humpback whales from Antarctic hunt

Sugarcane ethanol: Brazil's biofuel success

Yangtze dam risks disaster

Amazon deforestation seen surging

Decide on polar bears first, then oil

Water becomes the new oil as world runs dry

Mercury poses serious health hazard

Coral reef loss at unprecedented levels

a Look at newspaper headlines for one week. Copy out the headlines that refer to environmental issues. Put these in a table or on a notice board.

Good news	Bad news

b Discuss with your fellow students what the environmental headlines may be in 2020 and 2050.

To do

Find a local environmental issue where a pressure group is fighting for a cause.

1 Describe the issue and state the argument of the pressure group.

2 What are the opposing arguments to their case? These may be economic, aesthetic, socio-political or cultural.

3 State your own position on this issue and defend your argument.

Earth Days

In the late 1960s, after *Silent Spring*, environmentalism turned to action, particularly in North America. "Earth Days" were founded to encourage us all to be aware of the wonder of life and the need to protect it. There are two different Earth Days. The UN Earth Day each year is on the Spring equinox (so in March in the Northern hemisphere and September in the Southern when the sun is directly above the equator). John McConnell, an activist for peace, drove this concept and designed its flag.

The other Earth Day is on 22 April each year, and was founded by US politician Gaylord Nelson as an educational tool on the environment. Up to 500 million people now take part in its activities worldwide each year. It has a different flag.

Question

Some people are critical of Earth Days as marginalized activities that do not change the actions of politicians. Do you think they have an effect?

Key words

environmentalism
environmental value systems
deep ecology
TNCs
Industrial Revolution
Green Revolution
conservationists
preservationists

stewardship
modern environmental
   movement
intergovernmental bodies/
   organizations
pressure groups
NGOs
anthropogenic

Rachel Carson and *Silent Spring*
Minamata
Bhopal
Chernobyl
Greenpeace
Earth Summits
The Gaia theory

2 Environmental perspectives

"Nothing in this world is so powerful as an idea whose time has come."

Victor Hugo 1802–1885

Key points

- There are a range of environmental philosophies that people hold.
- Your perspective on the environment may also be called your value system, paradigm or viewpoint. (We shall use these words interchangeably.)
- Your environmental philosophy influences your decision-making process on environmental and other issues.
- You should know your own environmental philosophy and understand what influences this and how it may change.
- Different societies hold different environmental philosophies and comparing these helps explain why societies make different choices.
- The environment or any organism can have its own intrinsic value regardless of its value to humans. How we measure this value is a key to understanding the value we place on our environment.

Setting the scene

For much of history, our viewpoint has been that the Earth's resources are unlimited and that we can exploit them with no fear of them running out. And for much of history that has been true. A much smaller human population in the past has been just one species among many. The words and phrases we use describe how we have seen the environment: "fighting for survival", "battle against nature", "man or beast", "conquering Everest", "beating the elements". It has only been in very recent times that humans have been able to control our environment and even think about terraforming (altering conditions to make it habitable for humans) on Mars. The Industrial Revolution heralded the arrival of the "unbound Prometheus[1]" of technological development when we were driven to explore, conquer and subdue the planet to the will of industrial growth. This ideology has reigned in the industrial world with the worldview that economic growth improves the lot of us all. But now it is clearer that the Earth's resources are not limitless as the Earth is not limitless. Humans may be the first species to change the conditions on Earth and so make it unfit for human life.

To think about

Our relationship with the Earth

1 Can you think of other phrases that describe our relationship with nature and the Earth?

 The words we use are often evaluative and not purely descriptive.

2 How do you think our language has influenced human perspectives on the environment?

3 How does the language we use influence your viewpoint?

What is your environmental worldview?

You have a view of the world that is formed through your experiences of life – your background, culture, education and the society in which you live. This is your paradigm or worldview. You may be optimistic or pessimistic in outlook – see the glass as half full or half empty.

[1] Prometheus, a Titan, stole fire from Zeus and gave it to mortal humans to use. As punishment, he was chained to a rock and a vulture ate his liver – every day, as it grew back each night.

Fig. 2.1 Is this half full or half empty?

The glass is half full | The glass is half empty

Your paradigm affects how you view environmental issues and you will have made various assumptions based on your values and attitudes. This is your environmental value system.

But what are your assumptions and values (and where do they come from)?

To do

Environmental attitudes questionnaire

Consider these statements and decide if you agree strongly, agree, don't know, disagree or disagree strongly with each.

1 Humans are part of nature.
2 Humans are to blame for all the world's environmental problems.
3 We depend on the environment for our resources (food, water, fuel).
4 Nomadic and indigenous peoples live in balance with their environment.
5 Traditional farming methods do not damage the environment.
6 Nature will make good any damage that humans do to the Earth.
7 Humans have every right to use all resources on the planet Earth.
8 Technology will solve our energy crisis.
9 We have passed the tipping point on climate change and the Earth is warming up and we cannot stop it.
10 Animals and plants have as much right to live on Earth as humans.
11 Looking at a beautiful view is not as important as economic progress.
12 Species have always become extinct on Earth and so it does not matter that humans are causing extinctions.

Discuss your responses with your colleagues. Do they have different ones? Why do you think this is?

To think about

Consider these words:

- environment
- natural
- nature.

Think about what they mean to you. Write down your responses.

Now discuss what you wrote with two of your classmates. Do you agree?

What have you written that is similar or different?

Why do you think your responses may be different?

How different do you think the responses of someone from a different century or culture may be? Discuss some examples.

Guns, Germs and Steel: The fates of human societies and *Collapse*, both by Jared Diamond

In 1997, Jared Diamond, professor at UCLA, published *Guns, Germs and Steel* and won a Pulitzer Prize for it. He tries to explain that it is because of their environment that many European civilizations have survived, formed empires and conquered others. But the Europeans are not intellectually or genetically superior to other races. So how did they do it? His premise is that it is because Eurasia is aligned with a long east–west axis in its landmass while the Americas and Africa have a long north–south axis. Eurasia is larger than the Americas or Africa as well and these two factors mean that there are more continuous ecological regions in Eurasia than in the other continents. This has meant that domesticated plants and animals and technology spread further and faster across Eurasia, societies mixed more easily and food surpluses meant there was more time to develop sedentary civilizations, inventions and specialization of labour.

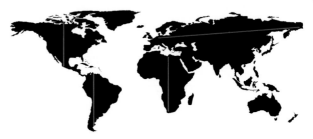

Fig. 2.2 The world showing continents and their longest axes

Eurasian animals – cattle and sheep – which are fairly placid herd animals, were easier to domesticate, whereas American (buffalo, llamas) and African (elephants, zebra) large herbivores are harder to domesticate and keep. Eurasian plant species were also easier to domesticate (wheat, rice) and could grow across the continent which is mostly in the northern temperate zone. Because the other continents spread across several biomes (tropical rainforest, desert, savanna, tundra), species could not grow throughout the continent.

With larger populations, bureaucracies and armies were needed to control them, and armies can conquer lands of other peoples. Because of the density of population of Eurasia and living next to domesticated animals, a good degree of resistance to disease built up. So a small number of Europeans could conquer larger indigenous populations (in South America, for example) using more advanced technology and, accidentally, the secret weapons of diseases.

There has been controversy over this book as some people understood it to be racially biased towards the superiority of the Eurasian peoples. Diamond refutes this totally.

In *Collapse*, Diamond's next book, he lists eight factors which have contributed to the collapse of past societies and gives examples including the Greenland Norse, Easter Islanders and the Mayans. The factors are:

1 Deforestation and habitat destruction
2 Soil problems (erosion, salinization, and soil fertility losses)
3 Water management problems
4 Overhunting
5 Overfishing
6 Effects of introduced species on native species
7 Human population growth
8 Increased per capita impact of people

He also looks at present-day societies (Montana, China, Australia) and says four new factors may contribute to the weakening and collapse of these:

1 Human-caused climate change
2 Build-up of toxic chemicals in the environment
3 Energy shortages
4 Full human utilization of the Earth's photosynthetic capacity

In nearly all examples, the cause of the problem is overpopulation relative to the carrying capacity of the environment. He does not say that the collapse of all societies is due to environmental factors as war and political conflict can also destroy – Carthage destroyed by Rome, communist states falling in the second half of the 20th century. But he says that overpopulation has been a cause for many collapses of civilizations.

Both books are worth a further look.

A classification of different environmental philosophies

Humans like to classify and categorize, and environmental philosophies are no exception to this. The major categories are:

● the **ecocentric** worldview – life-centred – which respects the rights of nature and the dependence of humans on nature so has a holistic view of life which is earth-centred and
● the **anthropocentric or technocentric** worldview – human-centred – in which humans are not dependent on nature but nature is there to benefit humankind. This worldview puts the individual organism or species at the centre.

Most people living in the more economically developed countries (MEDCs) in the industrial world have an **anthropocentric (human-centred) or technocentric (planetary management) worldview**. In this humans are seen as the dominant species on Earth and we can manage the environment to suit our needs. Other species only then have value if they are useful to us. This can be summarized as:

● We are the Earth's most important species, we are in charge.
● There will always be more resources to exploit.
● We will control and manage these resources and be successful.
● We can solve any pollution problem that we cause.
● Economic growth is a good thing and we can always keep the economy growing.
● In summary – whatever we do, we can solve it.

Technocentric worldviews include the cornucopians and the environmental managers.

Cornucopians include those people who see the world as having infinite resources to benefit humanity. Cornucopians think that through technology and our inventiveness, we can solve any environmental problem and continually improve our living standards. For them, it is growth that will provide the answers and wealth to improve the lot of all and nothing should stand in the way of this. Through a free-market economy – capitalism with minimal government control or interference – will be the best way to manage markets and the planet. Some see the Earth as a spaceship and we are its captain and crew. If we understand the machine, we can steer it.

Environmental managers see the Earth as a garden that needs tending – the **stewardship** worldview. We have an ethical duty to protect and nurture the Earth. Environmental managers hold the view that there are problems and we need governments to legislate to protect the environment and resources from overexploitation and to make sustainable economies. We may need to compensate those who suffer from environmental degradation and the state has a duty to intervene. Environmental managers believe that if we look after the planet, it looks after us.

The **ecocentric worldview** believes that the views above are too simplistic. We do not even know what species are alive on Earth at the moment and certainly do not know how they interact so it is arrogant of us to think that we can manage it all. To think that we can continue economic growth until every person alive has as high a standard of living as the most affluent, is just not possible and so we shall either fall off the treadmill of growth or find it stops beneath us. **Biocentric** (life-centred) thinkers see all life as having an inherent value – a value for its own sake, not just for humans. So animals are not just for hunting and eating, trees for logging, lakes for fishing. We should not cause the premature extinction of any other species, whether it does us harm or good or neither. An extreme view of this is that we should not cause the harm of any individual of a species, which is what animal rights activists believe. Others who also call themselves ecocentric (earth-centred) broaden this out to the protection of ecosystems and habitats in which the species live. If we can preserve the ecological integrity and complexity of systems, then life will thrive. To broaden this further, some emphasize the holistic nature of our ethical obligation to the Earth. We are just one species, no more important than the others. Because we are sentient beings and can alter our environment, it is our duty to restore degraded ecosystems, remove pollution and deal with global environmental problems.

To summarize the ecocentric view:

- The Earth is here for all species.
- Resources are limited.
- We should manage growth so that only beneficial forms occur.
- We must work with the Earth, not against it.
- We need the Earth more than it needs us.

Ecocentrists include the **self-reliant or soft technologists**. They believe in the importance of small-scale, local community action and the actions of individuals making a difference. They view materialism and our need for more as wrong and do not like centralized decision-making.

At the end of the continuum are the **deep ecologists** who put more value on nature than humanity. They believe in **biorights** – universal rights where all species and ecosystems have an inherent value and humans have no right to interfere with this. Deep ecologists would like policies to be altered to reduce our impact on the environment which includes a decrease in the human population and consuming less. Deep ecology is not an ecoreligion but a set of guidelines and values to help us think about our relationship with the Earth and our obligations towards it.

Another way of looking at these two environmental value systems is to consider them as **nurturing** (ecocentric) and **intervening or manipulative** (technocentric/anthropocentric). These are two extremes of the spectrum on environmental values but most of us also think in both ecocentric and technocentric ways about issues and we may change our minds depending on various factors and as we get older. It is too simplistic to say that we fit into one or the other group all the time.

As we can only experience the world through our human perceptions, our views of the environment are biased by this. We talk of animal rights but can only discuss these using our anthropocentric viewpoint. Most of us will take an accommodating view of the environment ("light-green") – faith in the ability of our institutions to adapt to environmental demands and changes and in communities to work together to reduce resource use (e.g. bottle banks, recycling aluminium cans) – and so be classified as environmental managers in Table 2.1. A large minority are cornucopians ("bright-green") with faith in the appliance of science to solve environmental problems and very few are deep ecologists ("deep-green" or "dark-green") who believe in green rights and the survival of the Earth above the survival of the human species.

> *"A thing is right when it tends to preserve the integrity, stability and beauty of the biotic community. It is wrong when it tends otherwise."*
>
> Aldo Leopold

Cost-benefit analysis and the environment

Environmental economists working in industry may be asked how much pollution should be removed from a smokestack of a chimney before the waste is released to the atmosphere. All the pollutant could be removed but at a high cost financially and, in doing so, the company may not be able to afford cleaning up the outflow of heavy metals into a nearby ditch. The opportunity cost of the action is high. There are limited funds and unlimited demands on those funds. Usually costs are passed on to the consumer. So decisions may have to be made that mean some pollution escapes but both demands are met to some extent. Often a cost-benefit analysis is carried out to trade off the costs and benefits.

But valuing the environmental cost is very difficult and it can be argued that cost-benefit analysis cannot apply to these nonmarket effects. How do you value an undisturbed ecosystem or a wild animal or human health? Cost-benefit analysis is still used in decision-making for industry as it is transparent but it may not be the best way. Later in this book, we talk more about how to value the environment, but do be aware that an environmentalist may not always promote the total clean-up or eliminate solution if the opportunity cost is too high. When you add in questions of ethical practice and what is fair to do, you can see how complex this can become.

To do

1 Draw a table with two columns labelled "Ecocentric" and "Anthropocentric/Technocentric".

2 Put each of the words or phrases below in one of these columns. Don't think for too long about each one. Go with your instinct now you have read about environmental value systems.

Aesthetic

Animal rights

Authoritarian

Belief in technology

Capitalism

Centralist

Competitive

Consumerism

Cooperative

Earth-centred

Ecology

Economy

Feminist

Global co-existence

Holistic

Human-centred

Individual

Intervening

Managerial

Manipulative

Nurturing

Participatory

Preservation

Reductionist

Seeking progress

Seeking stability

Utilitarian

Then put a tick next to the words that best describe your environmental viewpoint.

Draw a line with ecocentric on the left-hand side and technocentric/anthropocentric on the right.

Put a cross which you think gives your position and get all your classmates to mark their own as well.

Review this at the end of the course and see if you have moved along the line – to left or right – or moved relative to your classmates.

Review

Copy and complete this table to show the main points of the different environmental philosophies.

Environmental value system	Ecocentric	Anthropocentric/Technocentric
Environmental management strategies		
Environmental philosophies		
Labels and characteristics		
Social movements		
Politics		

Table 2.1

Various environmental worldviews

Communism and capitalism in Germany

After the Iron Curtain and Berlin Wall fell in Germany in 1989, western journalists rushed to see East Germany and report upon it. Communism was seen as the antidote to capitalist greed and communists claimed that their system could produce more wealth than capitalism and distribute it more evenly, in the process curing social ills including environmental degradation. But journalists reported on a polluted country in East Germany with the Buna

"The earth shall rise on new foundations: We have been nought, we shall be all!"

from *The Internationale*, anthem of international socialists and communists chemical works dumping ten times more mercury into its neighbouring river in a day than a comparable West German plant did in a year. And the smoky two-stroke Trabant cars emitting one hundred times as much carbon monoxide as a western car with a catalytic converter. The message was that capitalism would clean up the industry – but it was not such a non-polluter itself. In some ways the paternalistic communist state had protected the interests of primary producers like farmers and fishermen and so the environment. There was a law that made smelters shut down and so not pollute in spring when crops were growing.

Native American environmental worldview

While there are many native American views, a broad generalization of their views is that they tend to hold property in common (communal), have a subsistence economy, barter for goods rather than use money, and use low-impact technologies. Politically, they come to consensus agreements by participation in a democratic process. The laws are handed down by oral tradition. Most communities have a matrilineal line (descent follows the female side) as opposed to patriarchal, with extended families and low population density. In terms of religion, they are polytheistic (worshipping many gods) and hold that animals and plants as well as natural objects have a spirituality.

The modern Western worldview

The two religions on Earth with the most adherents are Christianity and Islam, together numbering some 3.6 billion. They share the belief in a separation of spirit and matter or body and soul and a notion of "dominion" or mastery over the Earth. But the ancient Greek view of citizenship and democracy, the Judaic notion of the covenant and the Christian view of unconditional love are examples which have perhaps been distorted in the anthropocentric views of the West. In the biblical book of *Genesis*, God commands humans to "replenish the earth, and subdue it; and have dominion over it" (*Genesis* 1:28). But what does this mean? Are humans to be masters or stewards of the Earth? Do stewards own something or just look after it? Another layer comes from ecofeminism as an environmental movement in which ecofeminists argue that it is the rise of male-dominated societies since the advent of agriculture that has led to our view of nature as a foe to be conquered rather than a nurturing Earth mother.

Buddhism's environmental worldview – a religious ecology

Buddhism has evolved over 2500 years to see the world as conjoined in four ways – morally, existentially, cosmologically and ontologically. Buddhists believe that all sentient beings share the conditions of birth, old age, suffering and death and that every living thing in the world is co-dependent. Buddhist belief teaches that as we are all dependent on each other, whether plant or animal, we are not autonomous and humans cannot be more important than other living things and must extend loving-kindness and compassion not just to life but to the Earth itself.

Adapted from Introduction to *Sustainable Development for the Second World* by Viktor Vovk, Worldwatch paper 167, Sept 2003

In the summer of 1969, oil wastes dumped into the Cuyahoga River in Cleveland, Ohio—part of the old industrial heartland of the free-market, capitalist US economy—caught fire. It had happened before and was not considered especially remarkable at the time. But circumstances of local politics and fortuitous media coverage turned the image of the blazing river into an icon, created a powerful "teachable moment" for millions of Americans, and helped galvanize support for legislation to clean up surface waters all over the country. The Clean Water Act was passed three years later.

In 1986 in Ukraine, before that country achieved independence following the breakup of the Soviet Union, a reactor at the Chernobyl nuclear power station blew up and released a cloud of radioactive gas that spread hundreds of miles. This disaster, an iconic event of a wholly greater magnitude than the incident on the Cuyahoga, was the product of an industrial *communist* economy. It too stimulated powerful political forces for environmental reforms.

Today, the economies of Ukraine, the former Soviet Union, and many other post-communist nations of the Second World are once again largely focused on resource extraction and industrial production, with little regard for the environment. (Russian oil companies, for instance, spill more oil every day than was lost when the *Exxon Valdez* oil tanker ran aground off Alaska in 1989.) The environmental performance of the capitalist West is better in many respects, but still far from sustainable. The United States, the winner of the Cold War and the avatar of triumphant capitalism, in particular has much to answer for. It continues, for example, to emit greenhouse gases in ever-larger quantities even as the evidence piles up that such gases are warming the globe. And despite the European Union's relatively progressive environmental policies, its own environmental assessment body has concluded that conditions there are worsening. As for the rest of the planet, the record is mostly one of accelerating pollution, biodiversity loss, habitat destruction and deforestation, and resource depletion, regardless of the economic model employed.

Capitalism (based on neoclassical economics) and communism differ in obvious and important ways, but both are expressions of the industrialism that was born in the 19th century. Both succeeded in achieving traditional industrial goals by marshalling large amounts of energy and other forms of natural capital and turning them into manufactured capital (buildings, cars, ships, planes, weapons, appliances, etc.). Both view the natural world as merely a stock of resources without intrinsic value, a source of inputs to the economic process. Both see that process, in turn, as linear rather than cyclical, and in general have tended to abstract from reality and paint incomplete and dangerously misleading pictures of the way the world works as a physical system.

In short, both capitalist and communist economies operate in violent disharmony with the global ecosystem. The theories that underlie them—neoclassical economics and Marxism—ignore or disregard fundamental truths about the relationship between ecosystems and economies and thus sow the seeds of their own destruction.

The emergence of the "post-industrial" economy in the West—with its theoretical emphasis on services, dematerialization, industrial ecology and information—may point the way toward long-term sustainability. But the case is not yet proven, and recent political developments in the United States remind us that progress is intermittent and painful. A more interesting, and possibly more hopeful, situation has arisen in the post-communist nations, where the collapse of the old communist order has triggered a difficult period of change and adjustment—and possibly opened a window of opportunity for a more radical vision of sustainability to blossom.

That window, however, is in danger of closing. The collapse offered an unexpected chance for the post-communist nations to reorganize their economies and societies according to new principles. However, Western institutions stepped in with a ready-made blueprint of the post-collapse transition to democracy and market economies. Specifically, the World Bank, the International Monetary Fund and other international institutions imposed the so-called "Washington consensus," a suite of structural adjustment measures that have been routinely applied, with deeply mixed results, to many developing countries. This is unlikely to work well in the nations in transition. For one thing, decades of repressive and paternalistic government have generally left them without the civic traditions (faith in government and respect for civil rights and the rule of law, for instance) that are necessary for democracy and rational markets. It is naive to simply impose democratic structures and market mechanisms in place of the old structures and expect the result to work as they do in Western nations.

More importantly, the imposition of the Washington consensus reforms represents a grave failure of imagination. Swapping formulaic communism for formulaic capitalism is simply the substitution of one kind of flawed industrialism for another. And industrialism in *any* form is wildly out of touch with the ecological demands and limits of the real world.

What's needed is a better theory: an interdisciplinary synthesis of ecological ideas, new economic theory, social science and politics. The paradigm of political economy that emerges from this synthesis would seek to harmonize human economic activity with ecological reality. It would also strive to balance the drive for personal freedom and expression of self-interest with society's necessary role as trustee for the interests of the public at large, future generations and the global ecosystem. And just as this paradigm explicitly acknowledges the value of biodiversity in preserving the ecological functions of the Earth's systems, it also recognizes the importance of preserving ethnic and cultural diversity and national identity. There can be no universally applicable plan for sustainability.

Questions

1 What do you think is the environmental value system of the author?

2 Do you agree with him? If not, why not?

Read the following articles and decide whether each is describing an ecocentric or technocentric viewpoint.

Case one

In 1972, an innovative plan was introduced on an experimental basis to curb deforestation in the fringe forest villages of Arabari, West Bengal. Local villagers were involved in protecting coppices of Sal (*Shorea robusta*) trees in return for rights on all non-timber forest products, additional employment, and a promise of 25% share of the net cash benefits from the sale of Sal poles. About 1270 hectares of degraded Sal forests were taken up for revival. Initially a population of 3600 were involved through "forest protection committees." The non-timber forest products like Sal leaves and seeds, mushrooms, silk cocoons, medicinal plants, edible roots and tubers, etc. motivated the villagers. Encouraged by the experience of the Arabari experiment, the State Government decided in 1987 to encourage fringe forest populations to actively participate in managing and rehabilitating degraded forests all over southwest Bengal. This movement spread like a wild fire. In 1990, the State Government officially recognized the forest protection committees (FPC) in southwest Bengal.

Case two

The 6000 indigenous people, living along the fringes of Nagarhole National Park, in South India, face eviction.

Reason? To save the forests and wildlife. Never mind that only 30% of primary forests remain; the rest reduced by logging and timber plantations not done by them. Though they have lived in the area for decades, authorities consider them illegal occupants. They are forbidden from entering the forest to gather minor forest products. They have no rights for cultivation, keeping domestic animals, collecting food from the forest, hunting small game, building houses, using roads and transporting materials and, most importantly, for cultural practices and religious rituals.

Nagarhole is one of the seven protected areas for which the World Bank is financing US$68 million to the Government of India for the "Eco-Development Project." The project covers a total area of 6714 sq. km comprising other Protected Areas and tiger reserves in several states and affecting an overall population of 48800 tribal people. According to the World Bank, "local people, living in these protected areas, do not use natural resources in a sustainable way." Ironically the World Bank talks of "voluntary displacement" instead of forced displacement. The 6000 inhabitants of Nagarhole are protesting and asking for self-rule. The tribal people who have been living here from time immemorial now have to sneak into the forests to collect berries, honey and other food.

To do

Find two contrasting newspaper/internet/magazine articles about any environmental issue. The articles you select may be about different issues, but should take an ecocentric, technocentric or anthropocentric environmental view.

On a large piece of (coloured) paper make a well-presented poster with both articles enlarged and pasted opposite each other. In any way you think attractive (lists or annotations), use phrases from this chapter to show the different viewpoints.

Shades of green: Where are we now?

In any political movement, there will be changes and developments. It is now difficult to avoid marketing that is based on environmental well-being, often related to human well-being. Organic, biotic, low emissions, energy-saving, sustainable, free-range, green credentials are all terms used in green marketing of products although exactly what they mean and how we perceive them is questionable. Greenwash and Green sheen are terms that describe activities that are not as good for the environment as the producer would like us to believe.

A way of classifying environmentalists today is as dark greens, light greens and bright greens.

Dark greens are dissenters seeking political change in a radical way as they believe that economic development and industrial growth are not the answer. They see a change in the status quo and a reduction in the size of the human population as the way to go. Light greens are individuals who do not want to work politically for change but change their own lifestyles to use fewer resources. For them, it is an individual choice. Bright greens want to use technological developments and social manipulation to make us live sustainably and believe that this can be done by innovation. For bright greens, economic growth may be beneficial if it means more of us live in efficient cities, use more renewable energy and reduce the size of our ecological footprints while increasing our standard of living. For them, we can have it all. The viridian design movement is a spin-off from the bright greens and is about global citizenship and improved design of green products.

Question

What shade of green are you?

Key words

environmental worldview	environmental manager
environmental value systems	cornucopian
ecocentric	intrinsic
technocentric	holistic
anthropocentric	greenwash
biocentric	deep green
biorights	dark green
deep ecologist	light green
self-reliant ecologist	bright green

3 Ecosystems

Key points

- Ecosystems have biotic and abiotic components.
- Photosynthesis, respiration and productivity are key concepts in understanding ecology.
- A biome is a collection of ecosystems with similar climatic conditions.
- The location of a biome depends on climate and limiting factors.
- The most important climatic factors determining biome distribution are precipitation and temperature.
- Trophic (feeding) levels, food chains, food webs and pyramids of number, biomass and productivity all help us in understanding how ecosystems function.
- Populations of organisms interact in different ways.
- Different biotic and abiotic factors influence population size.
- Species show different characteristics with respect to the ecological niche they fill.
- Energy flows through ecosystems and most is lost to the system as heat.
- Humans manipulate ecosystems to get the greatest productivity from them.

"And this, our life, exempt from public haunt, finds tongues in trees, books in the running brooks, sermons in stones, and good in everything."

William Shakespeare

Setting the scene

This is a big chapter because it contains most of the ecology of the course. It is divided into three main sections: biomes, ecosystem structure and ecosystem function.

Look in Chapter 8 on population dynamics for ecosystem changes as this links with human population change. Succession has its own chapter later on and measuring components of systems and how they change is covered in Chapter 16.

Within this section are several ecological concepts but there are three that are vital to your understanding of how everything else works. These are photosynthesis, respiration and productivity. If you have a grasp of the basics of these, everything else makes more sense. We shall discuss them in outline and come back to them as the picture builds.

Biotic and abiotic components of ecosystems

Ecosystems are made up of the organisms and physical environment and the interactions between the living and non-living components within them. It is impossible to think of an ecosystem without including these interactions. The living components of an ecosystem are known as the **biotic** factors – organisms or their products that directly or indirectly affect an organism in its environment. This includes organisms, their interaction, and waste. Predation (one animal eating another), parasitism (one organism living on or in and gaining its food from another) and disease are also classified as biotic factors. Biotic factors interact as: producer, consumer, detritivore (detrivore), decomposer, parasite, host, predator, prey, competitor, herbivore, mutualist and pathogen.

The non-living, physical and chemical components of an ecosystem are called the **abiotic factors** and include:

- the atmosphere
- climate – light intensity, temperature range, precipitation
- soil structure and chemistry
- water chemistry
- seasonality
- level of pollutants.

Factors which prevent a community, population or organism growing larger are called **limiting factors**. Many limiting factors restrict the growth of populations in nature. Examples of this are phosphate being in limited supply (limiting) in most aquatic systems, and low temperature in the tundra which freezes the soil and limits water availability to plants.

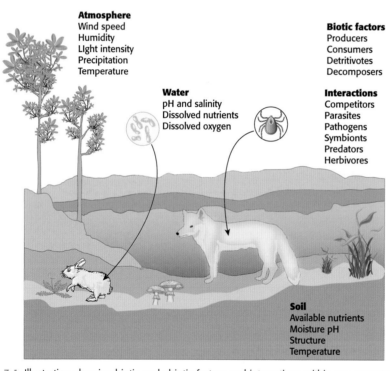

Atmosphere
Wind speed
Humidity
LIght intensity
Precipitation
Temperature

Biotic factors
Producers
Consumers
Detritivotes
Decomposers

Water
pH and salinity
Dissolved nutrients
Dissolved oxygen

Interactions
Competitors
Parasites
Pathogens
Symbionts
Predators
Herbivores

Soil
Available nutrients
Moisture pH
Structure
Temperature

Fig. 3.1 Illustration showing biotic and abiotic factors and interactions within an ecosystem

Two examples of abiotic and biotic factors affecting ecosystems are described.

An oak tree in a temperate deciduous woodland is a producer providing the basic unit of energy for the rest of the ecosystem. But at the same time it competes for light with other trees and may be the host to parasitic plants such as mistletoe or decomposing fungi. During the annual cycle in the wood, the tree will at times take water and mineral nutrients from the soil and at others return nutrients as fallen wood and leaves. Within the centre of the canopy of an oak tree the relative humidity is higher than in the woodland as a whole. This provides the physical and chemical conditions needed for a community of mosses, lichens and ferns to develop. In a very simplistic form, it is the availability of a suitable abiotic environment that provides the conditions for a distinct biotic community to exist. Importantly though, the biotic community can greatly influence and even change the abiotic one.

In commercial forestry in parts of Scotland, until the 1970s, large areas of Scottish blanket bog were viewed as beyond the reach of commercial forestry operations. It was too wet for Sitka spruce, the predominant cash wood crop, to grow and too expensive to drain. Then it was discovered that if a "nurse" crop of Lodgepole pine was planted ahead of the Sitka, even though the pines would eventually die in the very wet conditions, they would dry the soil enough to allow the Sitka to take hold. Along with this, the drying of the area and closing in of the canopy with trees planted tightly in rows would prevent continued growth and accumulation of sphagnum moss. This in turn aided the drying process. This is an example of positive feedback when the trend is enhanced – here drying leads to more drying.

Many different abiotic factors and animal or plant species also interact and change with time themselves, e.g. temperature is dependent on solar radiation, wind speed, time of year, time of day, altitude and aspect. Temperature affects water loss from organisms and respiration, and for plants, the rate of photosynthesis. Changes in temperature affect relative humidity and evaporation from water bodies and soils.

Getting the terminology

A **species** is a particular type of organism. Individuals of this type can interbreed and produce fertile offspring. Examples of species are humans, giraffes and pine trees. Each species is given a scientific name composed of two parts: the genus name and then the species name. Scientific names are always underlined or in italics and the genus name is given first with a capital letter:

Common name	Scientific or binomial name
Human	*Homo sapiens*
Giraffe	*Giraffa camelopardalis*
Scots pine	*Pinus sylvestris*
Aardvark	*Orycteropus afer*

A **population** is a group of individuals of the same species living in the same area at the same time. They are able to interbreed. Snails of one species in a pond form a population but the snails in another pond are a different population. A road or river may separate two populations from each other and stop them interbreeding.

A **habitat** is the environment where a species normally lives. It is the natural environment in which a species or a population of a species lives and includes the physical (abiotic) environment. Many populations of different species (a community) may share the same habitat.

An **ecological niche** is how an organism makes a living. This includes every relationship that organism may have – where it lives, how it responds to resources available, to predators, to competitors and how it alters these biotic factors. It also includes abiotic factors – how much space there is, and availability of light and water. No two species can inhabit the same ecological niche in the same place at the same time. However, many species may live together – because they have slightly different needs and responses they are not in the same niche.

A **community** is a group of populations living and interacting with each other in a common habitat (the same place). A community contains all the biotic components of a habitat. A tropical rainforest is a community of plants and animals, bacteria and fungi, as is an aquarium.

An **ecosystem** is a community of interdependent organisms and the physical (abiotic) environment which they inhabit. The term was coined in 1930 and modified by Arthur Tansley, a British ecologist, to describe the complex relationships between organisms and their abiotic environment. Ecosystems may be of varying sizes from a drop of rainwater to a forest. Human ecosystems include a household or a school or a nation state. Ecosystems do not exist independently but interact to make up the biosphere. As virtually all parts of the Earth have an impact from humans upon them, all ecosystems may be considered to be affected by humans.

A **biome** is a collection of ecosystems sharing common climatic conditions, e.g tundra, desert, tropical rainforest.

The **biosphere** is that part of the Earth inhabited by organisms. It is a thin layer that extends from the upper part of the atmosphere down to the deepest parts of the oceans which support life – 11 km below sea level to 10 km above.

To do

1 Copy the table and fill in the gaps.

Component	Local example	International/global example
Species		
Population		
Community		
Habitat		
Ecosystem		
Biome		

2 Make a model of the biosphere (in two or three dimensions) to include as many terms as you can from those above.

Respiration

All living things must respire to get energy to stay alive. If they do not do this, they die. Respiration involves breaking down food, often in the form of glucose, to release energy which is used in living processes. These processes are movement, respiration, sensitivity, growth, reproduction, excretion, nutrition, and some people remember these by their first letters which spell MRS GREN. Respiration can use oxygen (aerobic) or not (anaerobic). In aerobic respiration, energy is released and used and the waste products are carbon dioxide and water. Whether plants or animals, bacteria or fungi, all living things respire all the time, in the light and dark, when asleep or awake.

Aerobic respiration can be summarized as:

glucose + oxygen → energy + water + carbon dioxide

$$C_6H_{12}O_6 + 6\,O_2 \rightarrow \text{energy} + 6\,H_2O + 6\,CO_2$$

Photosynthesis

Green plants (also called autotrophs) photosynthesize in sunlight. This is the process by which they make their own food from water and carbon dioxide. In their chloroplasts, they use sunlight energy to split water and combine it with carbon dioxide to make food in the form of the sugar, glucose. Glucose is then used as the starting point for the plant to make every other molecule that it needs. In complex chemical pathways in cells, plants add nitrogen and sulfur to make amino acids and then protein; they rearrange carbon, hydrogen and oxygen and add phosphorus to make fatty acids and lipoproteins which make up cell membranes. Animals are totally dependent on the chemicals produced by plants. Although we can make most of the ones we need, we can only obtain essential amino acids from plants.

The waste product of photosynthesis is oxygen. This is essential to life as oxygen is used in respiration. In photosynthesis, light energy is changed into chemical energy in the form of glucose so this is a transformation of energy from one form to another.

Photosynthesis can be summarized as:

carbon dioxide + water $\xrightarrow[\text{chlorophyll}]{\text{light energy}}$ glucose + oxygen

$$6\,CO_2 + 6\,H_2O \xrightarrow[\text{chlorophyll}]{\text{light energy}} C_6H_{12}O_6 + 6\,O_2$$

But remember that glucose is then used to make many other chemicals.

Green plants respire in the dark and photosynthesize and respire in the light. Water reaches the leaves from the roots by transpiration. When all the carbon dioxide that plants produce in respiration is used up in photosynthesis, the rates of the two processes are equal and there is no net release of either oxygen or carbon dioxide. This usually occurs at dawn and dusk when the light intensity is not too high. This point is called the **compensation point** of a plant and it is neither adding biomass nor using it up to stay alive at this point. It is just maintaining itself. This is important to remember when we come to think about succession and biomes.

Productivity

First some definitions.

Production is making something. **Productivity** is making something per unit area per unit time, e.g. per metre squared per year. Productivity is the rate of growth or biomass increase in plants and animals.

Gross refers to the total amount of something made as a result of an activity, e.g. profit from a business or salary from a job.

Net refers to the amount left after deductions are made, e.g. costs of production or subtracting tax and insurance from a salary. The net amount is what you have left and is always lower than the gross amount.

Primary production in ecology means to do with plants. **Secondary** production is to do with animals.

Biomass is the living mass of an organism or organisms but sometimes refers to dry mass.

Now we know these terms, we can put some together.

We usually talk about productivity and not production in ecology as then we have defined the area or volume and time period.

Gross productivity (GP) is the total gain in energy or biomass per unit area per unit time. It is the biomass that could be gained by an organism before any deductions.

But all organisms have to respire to stay alive so some of this energy is used up in staying alive instead of being used to grow.

So **net productivity** (NP) is the gain in energy or biomass per unit area per unit time that remains after deductions due to respiration.

Autotrophs (organisms that can make their own food) are the base unit of all stored energy in any ecosystem. Light energy is converted into chemical energy by photosynthesis using chlorophyll within the cells of plants.

Because all of the light energy fixed by plants is converted to sugars it is in theory possible to calculate a plant's energy uptake by measuring the amount of sugar produced. This is **gross primary productivity** (GPP), because plants are the first organisms in the production chain. However, measuring the sugar produced is extremely difficult as much of it is used up by plants almost as soon as it is produced. A more useful way of looking at production of plants is the measurement of biomass to calculate **net primary productivity**. An ecosystem's NPP is the rate at which plants accumulate dry mass (actual plant material) usually measured in $g\ m^{-2}$. This store of energy is potential food for consumers within the ecosystem. NPP represents the difference between the rate at which plants photosynthesize, GPP, and the rate at which they respire. This is because the glucose produced in photosynthesis has two main fates. Some provides for growth, maintenance and reproduction (life processes) with energy being lost as heat during processes of respiration. The remainder is deposited in and around cells as new material and represents the stored dry mass (NPP).

$$NPP = GPP - R$$

where R is respiratory loss.

This accumulation of dry mass is usually termed biomass and provides a useful measure of the both production and utilisation of resources. Primary productivity is the available energy for the animals of an ecosystem.

Biological communities include plants, herbivores carnivores, decomposers and detritivores (organisms which feed on dead material). Plant tissue may be eaten or it may die and decay. Herbivores gain energy by consuming plants and carnivores gain their energy by consuming other animals. We can follow the flow of energy and materials through the community as food chains or food webs.

The plant material eaten by herbivores not only represents biomass that has been removed from primary productivity, it is also the theoretical maximum amount of energy that is available to all the animals, both the herbivores themselves and the carnivores that feed on them.

In reality the amount of energy available from the consumption of primary productivity is not the total consumed by herbivores. Some of the ingested plant material will pass straight through the herbivore and be released as faeces (this is **egestion** or removal of undigested food from the gut). Food is only absorbed when it crosses the wall of the alimentary canal (gut wall) of animals. The food that passes through an animal and is egested is not absorbed and provides the animal with no energy. Only the food eaten that is actually absorbed, the **assimilated food energy**, is of any use to power life processes. Of the assimilated energy, the part that passes through the gut wall, some is used in cellular respiration to provide energy for life processes. Some is removed as nitrogenous waste, in most animals as urine. The rest is stored in the dry mass of new tissue as **net secondary productivity** (NSP).

So the net productivity of herbivores (net secondary productivity) is the energy in the food ingested less the energy lost in egestion and the energy used in respiration. Total food ingested including the food that is egested is the measure of **gross secondary productivity** (GSP). Therefore net secondary productivity can be thought of in the same way as net primary productivity:

$$NSP = GSP - R$$

where R = respiratory loss and
GSP = (Food eaten − Energy in waste)

NPP = GPP − R

Glucose produced during photosynthesis (GPP)

Some glucose used to supply energy to drive cellular processes (Respiration)

Remaining glucose available to be laid down as new material – biomass (NPP)

Fig. 3.2 Diagram of glucose productivity using Sun's energy

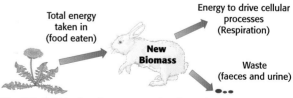

Total energy taken in (food eaten)

Energy to drive cellular processes (Respiration)

New Biomass

Waste (faeces and urine)

Fig. 3.3 Rabbit showing energy taken in and out

Only a very small percentage of the original NPP of plants is turned into secondary production by herbivores and it is this secondary production which is available to consumers at the next trophic level.

Carnivores, animals that eat other animals, are the next up the trophic ladder. Secondary consumers are those that eat herbivores and tertiary consumers those whose main source of energy is other carnivores. Omnivores feed at all trophic levels. The ability of carnivores to assimilate energy follows the same basic path as that of herbivores, though secondary and tertiary consumers have higher protein diets consisting of meat, which is more easily digested and assimilated. On average carnivores assimilate 80% of the energy in their diets and egest less than 20%. Most primary consumers assimilate only about 40% and egest 60% of the energy in their diet. This higher energy intake is offset by increased respiration during hunting: herbivores graze static plants, and carnivores usually have to chase moving animals. Also more biomass is locked up in the prey foods of carnivores in the form of non-digestible skeletal parts, such as bone, horn and antler. So carnivores have to assimilate the maximum amount of energy that they can from any digestible food.

Summary of the terms

Gross primary productivity (GPP) is the total gain in energy or biomass per unit area per unit time by green plants. It is the energy fixed (or converted from light to chemical energy) by green plants by photosynthesis. But, some of this is used in respiration so...

Net primary productivity (NPP) is the total gain in energy or biomass per unit area per unit time by green plants after allowing for losses to respiration. This is the increase in biomass of the plant – how much it grows – and is the biomass that is potentially available to consumers (animals) that eat the plant.

Primary productivity also varies with time. When plants first colonize bare ground, primary productivity is low as there are not many plants and they are starting from a seed. It rises quickly as more plants germinate and the biomass accumulates. When a climax community (see Chapter 14) is reached (stable community of plant and animal species), productivity levels off as energy being fixed by the producers is approximately equal to the rate at which energy is being used in respiration, and emitted as heat.

Gross secondary productivity (GSP) is the total gain in energy or biomass per unit area per unit time by consumers (animals) through absorption. Animals are known as heterotrophs or heterotrophic organisms to distinguish them from plants (autotrophs). Troph is derived from the ancient Greek word for food, so plants are auto-feeding and animals other-feeding (hetero = other) or feed on others.

Net secondary productivity (NSP) is the total gain in energy or biomass per unit area per unit time by consumers after allowing for losses to respiration. There are other losses in animals as well as to respiration but we shall come back to these later.

Productivity is normally measured as dry mass/g or energy/kJ produced in unit area/m^2 in unit time/per day or year. Typical units might be kJ m^{-2} yr^{-1}.

Biomes

A **biome** is a collection of ecosystems that share similar climatic conditions and so give rise to similar vegetation patterns, e.g. tundra, tropical rainforest. An **ecosystem** is a community of interdependent organisms and the physical environment that they inhabit.

How many biomes are there?

Opinion differs slightly on the number of biomes, because they are not a natural classification but one devised by humans, but it is possible to group biomes into six major types with subdivisions in each type:

Freshwater: swamp forests, lakes and ponds, streams and rivers, bogs

Marine: rocky shore, mud flats, coral reef, mangrove swamp, continental shelf, deep ocean

Deserts: hot and cold

Forests: tropical, temperate and boreal (taiga)

Grasslands: tropical or savanna and temperate

Tundra: arctic and alpine

We shall look in more detail at these biomes: tropical rainforest, hot desert, tundra, temperate forest and temperate grassland.

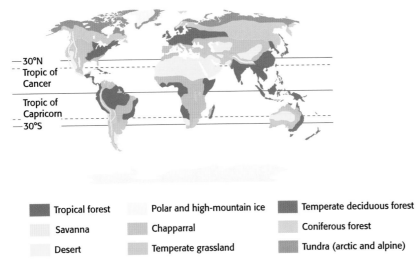

- ■ Tropical forest
- ░ Savanna
- ░ Desert
- ░ Polar and high-mountain ice
- ▓ Chapparral
- ▓ Temperate grassland
- ■ Temperate deciduous forest
- ░ Coniferous forest
- ▓ Tundra (arctic and alpine)

Fig. 3.4 Terrestrial biome distribution map

Why biomes are where they are

The climate is the major factor that determines what grows where and so what lives where. The other important factor is the terrain or geography – slope, aspect and altitude. Climate is made up of general weather patterns, seasons, extremes of weather and other factors but two factors are most important – temperature and precipitation (rain and snowfall).

It is stating the obvious to say that the climate is hotter nearer the equator and generally gets cooler as we go towards the poles (increasing latitude). At the equator, solar energy is most intense as it hits the Earth at a 90° angle.

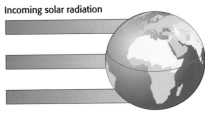

Incoming solar radiation

Fig. 3.5 Solar radiation hitting the Earth

Elsewhere on Earth, the rays hit the Earth at a more acute angle so are spread over more surface area than at the equator. It is like shining a torch beam directly at an object which is flat in front of the torch or shining it at an angle.

Latitude (distance from the equator) and altitude (height above sea level) both influence climate and biomes. It generally gets colder as you increase latitude or increase altitude. So there is snow on Mount Kilimanjaro and the Himalayas and Andes and they have alpine or polar biomes even though they are at lower latitudes.

The surplus heat energy at the equator is distributed towards the poles by winds and ocean currents. Air moving horizontally at the surface of the Earth is called wind. Winds blow from high to low pressure areas. Winds cause the ocean currents. The important thing to know is that it is water that is responsible for transferring the heat. Water can exist in three states – solid (ice and snow), liquid (water) and gas (water vapour). As it changes from state to state it either gives out or takes in heat. This is called **latent heat**. As water changes from solid to liquid (melts) to gas (evaporates), it takes in heat as more energy is needed to break the intermolecular bonds holding the molecules together. Therefore the temperature of the surroundings falls. As water changes from gas to liquid (condenses) to solid (freezes), it gives out heat to its surroundings. It is this change that distributes heat around the Earth. Water is the only substance that occurs naturally in the atmosphere that can exist in the three states within the normal climatic conditions on Earth.

As well as orbiting around the sun, the Earth rotates and is tilted at 23.5° on its axis. It takes 365 days (and a quarter) for the Earth to go once round the sun and this gives us a year and our seasons.

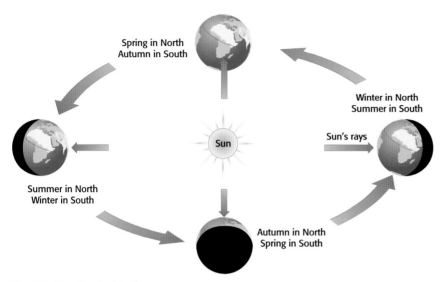

Fig. 3.6 How the Earth's tilt causes seasons

We stated that precipitation and temperature are the most important abiotic (physical) factors influencing biomes or what grows where. Increasing temperature causes increasing evaporation so the relationship between precipitation and evaporation is also important. Plants may be short of water even if it rains or snows a lot if the water evaporates straight away (deserts) or is frozen as ice (tundra). So we must also consider the P/E ratio (precipitation to evaporation ratio). This is easy to calculate. For example:

If 75 cm of snow falls in Norway per year and 50 cm are lost by evaporation, then P>E and the P/E ratio is 75/50 or 1.5. If 5 cm of rain falls in a desert and 50 cm are lost by evaporation, then P<E and the P/E ratio is 5/50 or 0.1. If the **P/E ratio is much greater than 1**, i.e. it rains or snows a lot and evaporation rates are low, then there is leaching in the soil when soluble minerals are washed downwards. If the **P/E ratio is approximately 1**, when precipitation is about the same as evaporation, the soils tend to be rich and fertile. If the **P/E ratio is far less than 1**, water moves upwards through the soil and then evaporates from the surface. This leaves salts behind and the soil salinity increases to the point that plants cannot grow (salinization).

Net primary productivity (NPP) is the total gain in energy or biomass per unit area per unit time by green plants after allowing for losses to respiration. Different biomes have differing levels of productivity because some raw materials or the energy source (light) for photosynthesis may be in short supply – are limiting. Solar radiation and heat may be limited at the South Pole in winter; water is in limited supply in a desert. All food webs depend on photosynthesis by green plants to provide the initial energy store so, if they cannot photosynthesize to their maximum capacity, other organisms will have a problem getting enough food.

Productivity is greater in low latitudes (nearer the equator), where temperatures are high all through the year, sunlight input is high and precipitation is also high. These conditions are ideal for photosynthesis. Moving towards the poles, where temperatures and amount of sunlight decline, the rate at which plants can photosynthesize is lower, and thus both GPP and NPP values are lower. In the terrestrial areas of the Arctic, Antarctic and adjacent regions (i.e. in high latitudes), low temperatures, permanently frozen ground (permafrost), long periods in winter when there is perpetual darkness, and low precipitation (cold air cannot hold as much moisture as warm air) all tend to cause a reduction in photosynthesis and lower productivity values. Obviously, in desert areas (such as the Sahara and much of Saudi Arabia) and semi-arid areas (e.g. central Australia and the southwest USA), the absence of moisture for long periods lowers productivity values severely, even though temperatures may be high and sunlight is abundant. Temperate deciduous forests would become temperate rainforest if precipitation were higher and temperate grassland if it were lower. However, these are generalizations, and variations are considerable. In a few, sheltered, favourable places in Greenland and South Georgia (in the Arctic and sub-Antarctic, respectively) productivity values close to those of mid-latitude forest have been recorded.

Whittaker, an American ecologist, first plotted biomes against temperature and precipitation. Notice that the temperature axis shows decreasing temperatures.

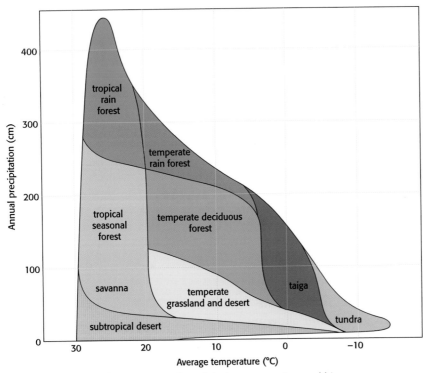

Fig. 3.7 Relationship between annual precipitation, temperature and biomes

How the atmosphere circulates: the tricellular model of atmospheric circulation

(Although you are not required to know this in the *Environmental Systems and Societies Guide*, understanding it will help you understand the distribution of biomes.) The equator receives most insolation (solar radiation) per unit area on Earth. This heats up the air which rises (hot air rises because it is less dense). As it rises, it cools and the water vapour in the air condenses as rain. This causes the afternoon thunderstorms and low pressure areas of the tropical rainforests. There is so much energy in this air that it continues to rise until it is pushed away from the equator to north and south. As it moves away, it cools and then sinks (cooler air is more dense) at about 30° N and S of the equator forming high pressure areas (where air density is higher). This air is dry and this is where the desert biome lies. Some of the air then returns to the equator and some blows to higher latitudes. The air that blows towards the equator completes a circle or cell to end up where it started. This movement of air is given the name "trade winds" which always blow towards the equator. At the equator, the north and south trade winds converge and rise again at the ITCZ (inter-tropical convergence zone). Here are the doldrums or area of little wind. The cell is called the Hadley cell.

The air that blows to higher latitudes at 30° N and S forms the winds known as the westerlies and they collect water vapour from the oceans as they blow towards the poles. At about 60° N and S, these winds meet cold polar air and so rise as they are less dense. As they rise, the water in them condenses and falls as precipitation where the temperate forest and grassland biomes are found. This is another low pressure area and is associated with depressions and heavy cyclonic rainfall.

The air then continues to flow, some to the poles and some back towards the equator. This air forms another cell known as the Ferrel cell. The air that continues towards the poles then descends as it gets cooler and more dense and forms a high pressure area at the poles, completing the polar cell. This then returns to lower latitudes as winds known as the easterlies.

Three cells form between the equator and each pole, hence the name tricellular model.

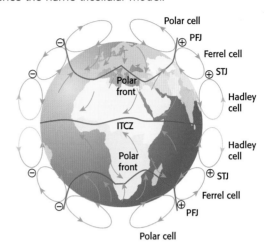

Fig. 3.8 Idealized representation of the general circulation of the atmosphere showing the positions of polar front; ITCZ (inter-tropical convergence zone); subtropical jets (STJ), polar front jets (PFJ)

The winds do not blow directly north or south because the Earth is rotating towards the east, i.e. if you viewed it from above the North Pole, it would be turning counter-clockwise.

Because of this, anything not fixed on the Earth (the oceans and the atmosphere) move in a straight line due to low friction with the surface but this causes them to veer to the right in the northern hemisphere and left in the southern as the Earth rotates. This apparent deflection is called the **Coriolis effect** and it means that the trade winds, westerlies and other winds are deflected to east or west.

North of the equator, the trade winds blow from the northeast and westerlies from the southwest: south of the equator, trade winds blow from the southeast, westerlies from the northwest. This allowed sailing ships to cross the major oceans on trade routes as they could find the prevailing winds by altering their latitude.

Review

To test yourself, try these questions.

1 What is latitude?
2 Do temperatures increase, or decrease with increasing (a) latitude, (b) altitude? Explain why.
3 Describe how the tilt of the Earth's rotational axis causes differences in the amount of heat received at the Earth's surface.
4 What are trade winds?
5 What causes high and low pressure?

We shall now look at some biomes in more detail.

Tropical rainforest

Fig. 3.9 A tropical rainforest in Borneo

Fig. 3.10 Tropical rainforest structure showing layers

What	Hot and wet areas with broadleaved evergreen forest.
Where (distribution)	Within 5° North and South of the equator.
Climate and limiting factors	High rainfall 2000-5000 mm yr^{-1}. High temperatures 26–28°C and little seasonal variation. High insolation as near equator. P and E are not limiting but rain washes nutrients out of the soil (leaching) so nutrients may be limiting plant growth. P>E.
What's there (structure)	Amazingly high levels of biodiversity – many species and many individuals of each species. Plants compete for light and so grow tall to absorb it so there is a multi-storey profile to the forests – called stratification – with very tall emergent trees, a canopy of others, understorey of smaller trees and shrub layer under this. Vines, climbers and orchids (epiphytes) live on the larger trees and use them for support. In primary forest (not logged by humans), so little light reaches the forest floor that few plants can live here. Nearly all the sunlight has been intercepted before it can reach the ground. Because there are so many plant species and a stratification of them, there are many niches and habitats for animals and large mammals can get enough food. Plants have shallow roots as most nutrients are near the surface so they have buttress roots to support them.
Net productivity	Estimated to produce 40% of NPP of terrestrial ecosystems. Growing season all year round, fast rates of decomposition, respiration and photosynthesis. Plants grow faster. But respiration is also high and for a large mature tree in the rainforest, all the glucose made in photosynthesis is used in respiration so there is no net gain. However, when rainforest plants are immature, their growth rates are huge and biomass gain very high. Rapid recycling of nutrients.
Human activity	The problem is that more than 50% of the world's human population live in the tropics and subtropics and one in eight of us live in or near a tropical rainforest. With fewer humans, the forest could provide enough resources for the population but there are now too many exploiting the forest and it does not have time to recover. Low levels of soil nutrients are quickly exhausted by agriculture and washed away as soil is eroded when forest cover is removed. This is not sustainable. In addition, commercial logging of valuable timber, e.g. mahogany, and clear felling to convert the land to grazing cattle/plantations all destroy the forest.
Issues	Logging, clear felling, conversion to grazing and to plantations, e.g. oil palm. Tropical rainforests are mostly in LEDCs and have been exploited for economic development.
Examples	Amazon rainforest, Congo in Africa, Borneo rainforest.

Table 3.1 Characteristics of tropical rainforest

Desert

Fig. 3.11 A desert in South-West America

What	Dry areas which are usually hot in the day and cold at night as skies are clear and there is little vegetation to insulate the ground. There are tropical, temperate and cold deserts.
Where (distribution)	Cover 20–30% of the Earth's land surface about 30° North and South of the equator where dry air descends. Most are in the middle of continents. (Some deserts are cold deserts, e.g. the Gobi desert.) The Atacama desert in Chile can have no rain for 20 years or more. It is the driest place on Earth.
Climate and limiting factors	Water is limiting. Precipitation less than 250 mm per year and irregular. Usually evaporation exceeds precipitation. E>P.
What's there (structure)	Few species and low biodiversity but what can survive in deserts is well adapted to the conditions. Soils can be rich in nutrients as they are not washed away. Plants are drought-resistant and mostly cacti and succulents with adaptations to store water and reduce transpiration, e.g. leaves reduced to spines, thick cuticles to reduce transpiration. Animals too are adapted to drought conditions. Reptiles are dominant, e.g. snakes, lizards. Small mammals can survive by adapting to be nocturnal (come out at night and stay in a burrow in the heat of the day, e.g. kangaroo rat) or reduce water loss by having no sweat glands and absorbing water from their food. There are few large mammals in deserts. Slow rate of decomposition.
Net productivity	Both primary (plants) and secondary (animals) are low because water is limiting and plant biomass cannot build up to large amounts. Food chains tend to be short because of this.
Human activity	Traditionally, nomadic tribes herd animals such as camels and goats in deserts as agriculture has not been possible except around oases or waterholes. Population density has been low as the environment cannot support large numbers. Oil has been found under deserts in the Gulf States and many deserts are rich in minerals including gold and silver. Irrigation is possible by tapping underground water stores or aquifers so, in some deserts, crops are grown. But there is a high rate of evaporation of this water and, as it evaporates, it leaves salts behind. Eventually these reach such high concentrations that crops will not grow (salinization).
Issues	Desertification – when an area becomes a desert through overgrazing, overcultivation or drought or all of these, e.g. the Sahel.
Examples	Sahara and Namib in Africa, Gobi in China.

Table 3.2 Characteristics of desert

Temperate grassland

Fig. 3.12 Temperate grassland

What	Fairly flat areas dominated by grasses and herbaceous (non-woody) plants.
Where (distribution)	In centres of continents 40–60° North of equator.
Climate and limiting factors	P = E or P slightly > E. Temperature range high as not near the sea, which moderates temperatures. Clear skies. Low rainfall, threat of drought.
What's there (structure)	Grasses, wide diversity. Probably not a climax community as arrested by grazing animals. Grasses die back in winter but roots survive. Decomposed vegetation forms a mat, high levels of nutrients in this. Kangaroo, bison, antelope and burrowing animals (rabbits, gophers). Carnivores – wolves, coyotes. No trees.
Net productivity	Not very high.
Human activity	Used for cereal crops. Cereals are annual grasses. Black earth soils of the steppes rich in organic matter and deep so ideal for agriculture. Prairies in North America are less fertile soils so have to add fertilizers. Called world's bread basket as so much of the world's wheat and maize is grown here. Plus livestock – cattle and sheep that feed on the grasses.
Issues	Dust bowl in 1930s in America when overcropping and drought led to soil being blown away on the Great Plains – ecological disaster. Overgrazing reduces temperate grasslands to desert or semidesert.
Examples	North American prairies, Russian steppes in northern hemisphere, pampas in Argentina, veldt in South Africa (30–40° South).

Table 3.3 Characteristics of temperate grassland

Temperate forest

Fig. 3.13 Temperate forest

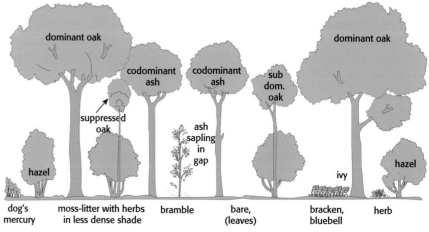

Fig. 3.14 Temperate forest structure in Britain

What	Mild climate, deciduous forest.
Where (distribution)	Between 40° and 60° North and South of the equator.
Climate and limiting factors	P> E. Rainfall is 500–1500 mm per year. Winters freezing in some (Eastern China and NE USA), milder in western Europe due to the Gulf Stream. Temp range –30°C to + 30°C. Summers cool.
What's there (structure)	Fewer species than tropical rainforests, e.g. in Britain, oaks, which can reach heights of 30–40 m, become the dominant species of the climax vegetation. Other trees, such as the elm, beech, sycamore, ash and chestnut, grow a little less high. Relatively few species and many woodlands are dominated by one species, e.g. beech. In USA there can be over 30 species per sq. km. Trees have a growing season of 6–8 months, may only grow by about 50 cm a year. Woodlands show stratification. Beneath the canopy is a lower shrub layer varying between 5 m (holly, hazel and hawthorn) and 20 m (ash and birch). The forest floor, if the shrub layer is not too dense, is often covered in a thick undergrowth of brambles, grass, bracken and ferns. Many flowering plants (bluebells) bloom early in the year before the taller trees have developed their full foliage. Epiphytes, e.g. mistletoe, mosses, lichens and algae, grow on the branches. The forest floor has a reasonably thick leaf litter that is readily broken down. Rapid recycling of nutrients, although some are lost through leaching. The leaching of humus and nutrients and the mixing by biota produce a brown-coloured soil. Well-developed food chains in these forests with many autotrophs, herbivores (rabbits, deer and mice) and carnivores (foxes). Deciduous trees give way to coniferous towards polar latitudes and where there is an increase in either altitude or steepness of slope. P>E sufficiently to cause some leaching.
Net productivity	2nd highest NPP after tropical rainforests but much lower than these because of leaf fall in winter so reduced photosynthesis and transpiration and frozen soils when water is limiting. Temperatures and insolation lower in winters too as further from the equator.
Human activity	Much temperate forest has been cleared for agriculture or urban developments. Large predators (wolves, bears) virtually wiped out.
Issues	Most of Europe's natural primary deciduous woodland has been cleared for farming, for use as fuel and in building, and for urban development. Some that is left is under threat, e.g. US Pacific Northwest old-growth temperate and coniferous forests. Often mineral wealth under forests is mined.
Examples	US Pacific Northwest.

Table 3.4 Characteristics of temperate forest

Arctic tundra

Fig. 3.15 Arctic tundra

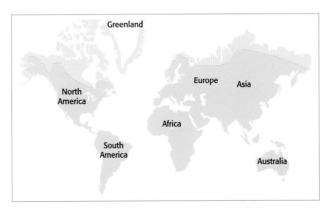

Fig. 3.16 Distribution of Arctic tundra (shown in yellow)

What	Treeless plain where subsoil is permanently frozen (permafrost). Cold, low precipitation, long, dark winters. 10% of Earth's land surface. Youngest of all the biomes as it was formed after the retreat of the continental glacier only 10 000 years ago. P>E.
Where (distribution)	Just south of the Arctic ice cap and small amounts in southern hemisphere. (Alpine tundra is found as isolated patches on high mountains from the poles to the tropics.)
Climate and limiting factors	Water is limiting but fire can also stop the climax community forming. Frozen ground (permafrost), cold, high winds and little precipitation mean the growing season is only 6 weeks per year. Permafrost reaches to the surface in winter but in summer the top layers of soil defrost and plants can grow. Low temperatures so rates of respiration, photosynthesis and decomposition are low. Slow growth and slow recycling of nutrients. Water, temperature, insolation and nutrients can be limiting. During spring and summer, animals are active, and plants begin to grow rapidly. Sometimes temperatures reach 30°C. Much of this energy is absorbed as the latent heat of melting of ice to water. In the winter, the northern hemisphere, where the arctic tundra is located, tilts away from the sun. In the spring equinox, the northern hemisphere is in constant sunlight. For nearly three months, from late May to August, the sun never sets. This is because the arctic regions of the earth are tilted toward the sun. With this continuous sun, the ice from the winter season begins to melt quickly. In southern hemisphere, where a small amount of tundra is also located, the seasons are reversed.
What's there (structure)	No trees but thick mat of low-growing plants – grasses, mosses, small shrubs. Adapted to withstand drying out with leathery leaves or underground storage organs. Animals also adapted with thick fur and small ears to reduce heat loss. Mostly small mammals, e.g. lemmings, hares, voles. Predators – Arctic fox, lynx, snowy owl. Most hibernate and make burrows. Simple ecosystems with few species. Low biodiversity – 900 species of plants compared with 40 000 or more in the Amazon rainforest. Soil poor, low inorganic matter and minerals.
Net productivity	Very low. Slow decomposition so many peat bogs where most of the carbon is stored.
Human activity	Few humans but mining and oil – see oil sands, page 207. Nomadic groups herding reindeer.
Issues	Fragile ecosystems that take a very long time to recover from disruption. May take decades to recover after being walked across. Mining and oil extraction in Siberia and Canada destroy tundra. Many scientists feel that global warming caused by greenhouse gases may eliminate Arctic regions, including the tundra, forever. The global rise in temperature may damage the Arctic and Antarctic more than any other biome because the arctic tundra's winter will be shortened, melting snow cover and parts of the permafrost, leading to flooding of some coastal areas. Plants will die, animal migrating patterns will change, and the tundra biome as we know it will be gone. The effect is uncertain but we do know the tundra, being the most fragile biome, will be the first to reflect any change in the Earth. Very large amounts of methane are locked up in tundra ice in clathrates (molecular cages of water that trap methane inside them). If these are released into the atmosphere (they contain 3000 times as much methane as is in the atmosphere now and methane is more than 20 times as strong a greenhouse gas as carbon dioxide) then huge increase in greenhouse gases.
Examples	Siberia, Alaska.

Table 3.5 Characteristics of Arctic tundra

Comparison of biomes

Biome	Net primary productivity g m⁻² yr⁻¹	Annual precipitation mm y⁻¹	Area 10⁶ km²	Plant biomass 10⁹ t	Mean biomass kg m⁻²	Animal biomass 10⁶ t	Solar radiation W m⁻² yr⁻¹
Tropical rainforests*	2200	2000–5000	17.0	765	45	330	175
Temperate forests	1200	600–2500	12.0	385	32.5	160	125
Boreal forests	800	300–500	12.0	240	20	57	100
Tropical grasslands (savannas)	900	500–1300	15.0	60	4	220	225
Temperate grassland	600	250–1000	9.0	14	1.6	60	150
Tundra and alpine*	140	<250	8.0	5	0.6	3.5	90
Desert* (rock, sand, ice)	90	<250	24.0	0.5	0.02	0.02	75

Table 3.6

*The *Environmental Systems and Societies Guide* requires you to study these biomes and any one other.

Review

To test yourself, try these questions.

1 What does deciduous mean?

2 What is net primary productivity (NPP)?

3 Which biome has the highest NPP per m² per year and why?

4 Which biome has the largest NPP in total?

5 Why is NPP low in tundra and deserts?

6 Why are there no large deciduous forests or tundra in the southern hemisphere?

7 Why are some trees in temperate biomes deciduous? Why may it be an advantage to a tree to lose its leaves in winter?

8 How do the number of tree species and their distribution differ in temperate and tropical rainforests?

9 What are the main factors causing the distribution of biomes?

10 Which biome(s) are most threatened and why?

11 Which biome(s) have been most changed by human activity?

12 What effect may climate change have on biome distribution?

Ecosystem structure

Food chains and trophic levels

All energy on Earth comes from the sun so solar energy (solar radiation) is the start of every food chain. (Well, very nearly all; some deep ocean vents give out heat from the core of the planet and some organisms get their energy from this. But most organisms get energy from the sun's energy.)

A **food chain** shows the flow of energy from one organism to the next. It shows the feeding relationships between species in an ecosystem. Arrows connect the species, usually pointing towards the species that consumes the other: so in the direction of transfer of biomass.

"In all things of nature, there is something of the marvellous."

Aristotle

Fig. 3.17 A food chain   **Sun**   **Grass**   **Zebra**   **Lion**

A **trophic level** is the position that an organism or a group of organisms in a community occupies in a food chain. Organisms are grouped into trophic (or feeding) levels. Trophic levels usually start with a primary producer and end with a carnivore at the top of the chain – a **top carnivore**.

Trophic level 3

Plants capture the sun's energy and convert it to glucose, herbivores eat plants and carnivores eat herbivores. The different feeding levels are named (Greek for food is *trophe*):

Trophic level 1 – producer
Trophic level 2 – herbivore (primary consumer)
Trophic level 3 – carnivore (secondary consumer)
Trophic level 4 – carnivore (tertiary consumer)

Trophic level 2

It is possible to classify the way organisms obtain energy into two categories:

- **producers or autotrophs** (green plants) which manufacture their own food from simple inorganic substances
- **consumers or heterotrophs** which feed on autotrophs or other heterotrophs to obtain energy (herbivores, carnivores, omnivores, detritivores (or detrivores) and decomposers).

Trophic level 1

Fig. 3.18

But within the consumers there is a hierarchy of feeding.

Some consumers are **decomposers** which obtain their energy from dead organisms as their food source. Bacteria and fungi are decomposers and obtain the nutrients by secreting enzymes that break down the organic matter.

Detritivores are consumers that derive their food from detritus or decomposing organic material. This may be a dead organism or faeces or part of an organism, e.g. shed skin from a snake, a crab carapace. They often increase the surface area of organic material by shredding it into smaller pieces then decomposers can attack it. Detritivores include snails, slugs, blowfly maggots, vultures.

Fig. 3.19 Decomposer fungi in a woodland

The first trophic level, the autotrophs, supports the energy requirements of all the other trophic levels above.

Food webs

Ecosystems have a hierarchy of feeding relationships (trophic levels) that determine the pathway of energy flow in the ecosystem. The energy flow in the ecosystem can be illustrated as a food chain but ecosytems are rarely this simple as there are many more organisms involved.

It is possible to construct food chains for an entire ecosystem, but this starts to create a problem.

The food chains below are from a European oak woodland. In fact they are based on real food chains at Wytham Wood in Oxford, UK, where some pioneer ecologists worked in the 1920s.

herbs	→	insects	→	spiders	→	parasites
herbs	→	insects	→	voles	→	owls
hazel	→	winter moth	→	voles	→	owls
hazel	→	winter moth	→	titmice	→	weasels

In the four different food chains, only ten species are listed and some of them are in more than one food chain. If we continued to list all the species in the wood and their interactions in every food chain, the list would run for many pages.

Food chains only illustrate a direct feeding relationship between one organism and another in a single hierarchy. The reality though is very different. The diet of almost all consumers is not limited to a single food species. So a single species can appear in more than one food chain.

A further limitation of representing feeding relationships by food chains arises when a species feeds at more than one trophic level. Voles are omnivores and, as well as eating insects, they also eat plants. Humans eat plants and animals which may be herbivores or carnivores. We would then have to list all the food chains that contain voles or humans twice, showing them in both the second and third trophic levels.

The reality is that there is a complex network of interrelated food chains which create a **food web**.

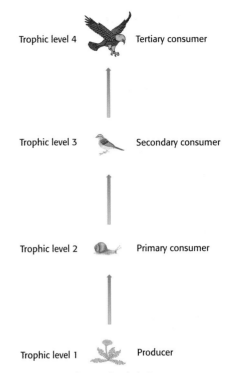

Fig. 3.20 A longer food chain

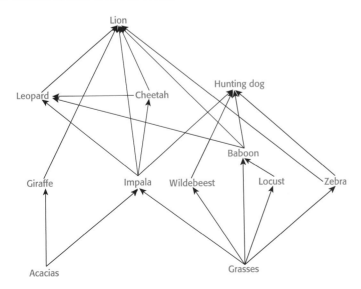

Fig. 3.21 Food web on the African savanna

The earliest food webs were published in the 1920s by Elton (on Bear Island, Norway) and Hardy (on plankton and herring in the North Sea). Figure 3.22 is Elton's food web.

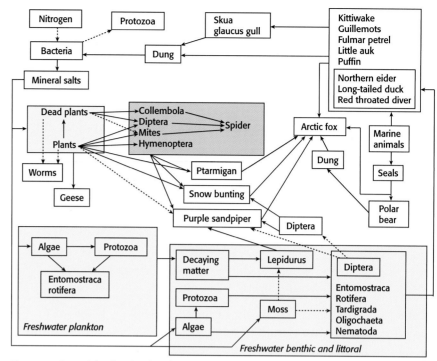

Fig. 3.22 One of the first food webs observed by Elton on Bear Island, Norway

To do

1 Make a table of the differences between a food chain and a food web.
2 Read the information on a food web in the tundra and answer the questions that follow.

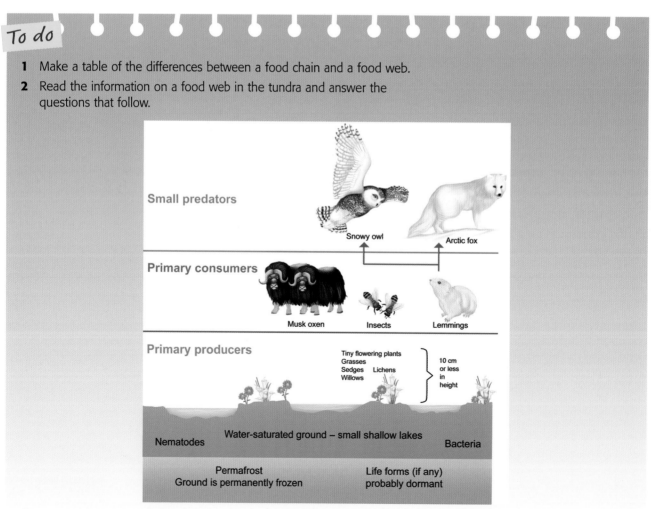

Fig. 3.23 A food web in the tundra

There are several species of bear in the tundra. Polar bears live further north, but are also found in the tundra searching for food. The brown Kodiak is the largest bear in the tundra and it is not as fierce as its reputation suggests. Kodiaks seldom eat meat. Wolves are the top predators of the tundra. They travel in small families and prey on caribou and other large herbivores that are too slow to stay with their groups. Some wolves change to a bright white colour in the winter. Otters live near rivers and lakes so they can feed on fish. Shrews are the smallest carnivores of the tundra. Even bats are found in the tundra during the summer. They feed on the swarms of insects that fill the air.

The primary production is not sufficient to support animal life if only small areas of tundra are considered. The large herbivores and carnivores are dependent on the productivity of vast areas of tundra and have adopted a migratory way of life. Small herbivores feed and live in the vegetation mat eating roots, rhizomes and bulbs. The populations of small herbivores like lemmings show interesting fluctuations that also affect the carnivores dependent on them, such as the arctic fox and snowy owl.

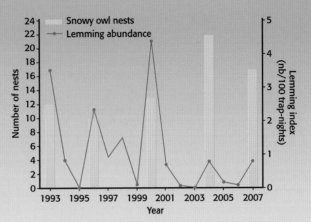

Fig. 3.24 Snowy owl nest numbers and lemming abundance on Bylot Island, northern Canada, 1993–2007

a Draw a food web for the tundra including **only** the animals mentioned.

b Describe and explain the relationship between snowy owl nest numbers and lemming abundance in Fig. 3.24.

 To do

Study the food web in Fig 3.25 and answer the following questions.

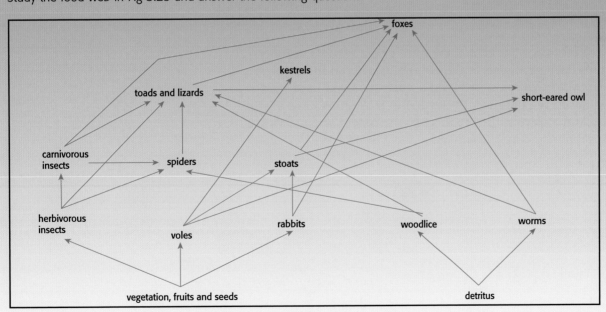

Fig. 3.25 A simplified food web from the acid heathland at Studland, Dorset, UK

1   What is the longest food chain in this food web?

2   Name two species that are found at two trophic levels.

3   If all the kestrels die, what may happen to (a) voles and (b) short-eared owls?

4   If there is a great increase in the rabbit population, what may happen to (a) rabbit predators and (b) the vegetation?

5   If a pesticide is added to kill spiders, what may happen to the foxes?

Ecological pyramids

These are graphical models of the quantitative differences between the amounts of living material stored at each trophic level of a food chain. They allow us to easily examine energy transfers and losses. They give an idea of what feeds on what and what organisms exist at the different trophic levels. They also help to demonstrate that ecosystems are unified systems; that they are in balance.

There are three types of ecological pyramid which measure number, biomass and productivity differences. Pyramids of numbers and biomass are snapshots at one time and place.

Pyramid of numbers

A **pyramid of numbers** shows the number of organisms at each trophic level in a food chain.

Fig. 3.26 Pyramids of numbers

The length of each bar gives a measure of the relative numbers. Pyramids have producers at the bottom, usually in the greatest number. Most pyramids are broad at their base and have many individuals in the producer (P) level. But some may have a large single plant, a tree, as the producer so the base is one individual which supports many consumers.

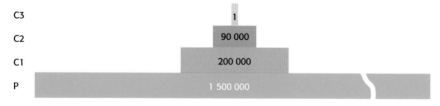

Fig. 3.27 Pyramid of numbers for a grazing ecosystem

In Fig. 3.27 why is the bar for the producers split?

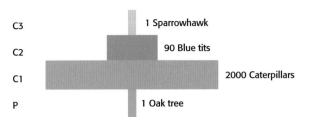

Fig. 3.28 Pyramid of numbers for an oak wood

Advantage of pyramid of numbers: This is a simple easy method of giving an overview and is good for comparing changes in population numbers over different times or seasons.

Disadvantages of pyramid of numbers: All organisms are included regardless of their size, therefore a pyramid based on an oak tree would be inverted (have a narrow base and get larger as it goes up the trophic levels). Also they do not allow for juveniles or immature forms. Numbers can be too great to represent accurately.

There is the question of where to put animals that feed at more than one trophic level (omnivores), for all types of pyramid.

Pyramid of biomass

A **pyramid of biomass** contains the biomass (mass of each individual × number of individuals) at each trophic level. Biomass is the quantity of (dry) organic material in an organism, a population, a particular trophic level or an ecosystem.

The units of a pyramid of biomass are units of mass per unit area, often grams per square metre ($g\,m^{-2}$) or volume of water ($g\,m^{-3}$) or sometimes as energy content (joules, J). A pyramid of biomass is more likely to be a pyramid shape but there are some exceptions, particularly in oceanic ecosystems where the producers are phytoplankton (unicellular green algae). Phytoplankton reproduce fast but are present in small amounts at any one time. As a pyramid represents biomass at one time only, e.g in winter, the phytoplankton bar may be far shorter than that of the zooplankton which are the primary consumers.

Fig. 3.29 Pyramids of biomass (units $g\,m^{-2}$)

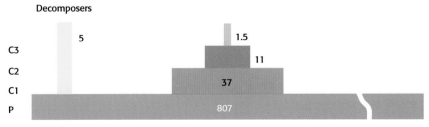

Fig. 3.30 Pyramid of biomass for a lake ($g\,m^{-2}$)

Advantage of pyramid of biomass: Overcomes the problems of pyramids of number.

Disadvantages of pyramid of biomass: Only uses samples from populations, so it is impossible to measure biomass exactly. Organisms must be killed to measure dry mass. Also the time of the year that biomass is measured affects the result. In the case of algae, their biomass changes by large amounts during the year therefore the shape of the pyramid would depend on the season. The giant redwood trees of California have accumulated their biomass over many years yet algae in a lake at the equivalent trophic level may only have needed a few days to accumulate the same biomass. This pyramid will not show these differences. Also, pyramids of total biomass accumulated per year by organisms at a trophic level would usually be pyramidal in shape. But two organisms with the same mass do not have to have the same energy content. A dormouse contains around 37 kJ g^{-1} of potential chemical energy in the form of stored fat, while a carnivore of equivalent mass would contain around 17 kJ g^{-1} potential energy, mainly in the form of carbohydrates and proteins. Also some organisms contain a high proportion of non-digestible parts such as the exoskeletons of marine crustaceans.

Pyramid of productivity

A pyramid of productivity contains the flow of energy through each trophic level. It shows the energy being generated and available as food to the next trophic level during a fixed period of time. So, unlike pyramids of numbers and biomass, which are snapshots at one time, these pyramids show the flow of energy over time. They are always pyramid-shaped in healthy ecosystems and measured in units of energy per unit area per period of time, often joules per square metre per year ($Jm^{-2}yr^{-1}$). Productivity values are rates of flow, whereas biomass values are stores existing at one particular time.

Supermarket analogy: The turnover of two supermarkets cannot be compared by just looking at the goods displayed on the shelves; the rate at which goods are being stocked and sold needs to be known. Both shops may have well-stocked shelves but the rate of removal of goods from a city centre shop may be considerably more than in a village shop. In the same way, pyramids of biomass simply represent the stock on the shelves, whereas pyramids of productivity show the rate at which that stock is being removed by customers and restocked by shop assistants.

The bars are drawn in proportion to the total energy utilized at each trophic level. As only about 10% of the energy in one level is passed on to the next, in pyramids of productivity, each bar will be about 10% of the lower one. Sometimes the term "pyramid of energy" is used which can be either the standing stock (biomass) or productivity. We shall avoid it here as it is confusing.

Advantages of pyramid of productivity: Shows the actual energy transferred and allows for rate of production. Allows comparison of ecosystems based on relative energy flows. Pyramids are not inverted. Energy from solar radiation can be added.

Disadvantages of pyramid of productivity: It is very difficult and complex to collect energy data as the rate of biomass production over time is required. There is still the problem (as in the other pyramids) of assigning a species to a particular trophic level when they may be omnivorous.

To do

On graph paper, draw and label pyramids from the data in the table. Comment on these.

	Number pyramid	Biomass pyramid / kJ m^{-2}	Productivity pyramid / 1000 kJ m^{-2} yr^{-1}
Primary producers	100 000	2500	500
Primary consumers	10 000	200	50
Secondary consumers	2 000	15	5
Tertiary consumers	500	1	–

Consequences of pyramids and ecosystem function
1 The concentration of toxic substances in food chains
2 The vulnerability of top carnivores
3 The limited length of food chains

Bioaccumulation and biomagnification
If a chemical in the environment (e.g. a pesticide or a heavy metal) breaks down slowly or does not break down at all, plants may take it up and animals may take it in as they eat or breathe. If they do not excrete or egest it, it accumulates in their bodies over time. Its concentration builds up. Eventually, the concentration may be high enough to cause disease or death. This is **bioaccumulation**.

If a herbivore eats a plant that has the chemical in its tissues, the amount of the chemical that is taken in by the herbivore is greater than that in the plant that is eaten – because the herbivore grazes many plants over time. If a carnivore eats the herbivores, it too will take in more of the chemical than each herbivore contained as it eats several herbivores over time. In this way the chemical's concentration is magnified from trophic level to trophic level. While the concentration of the chemical may not affect organisms lower in the food chain, the top trophic levels may take in so much of the chemical that it causes disease or their death as there is less biomass for each successive trophic level. This is **biomagnification**.

A serious problem with pesticides is how long they persist in the environment once they are sprayed. Some decompose into harmless chemicals as soon as they touch the soil. Glyphosate (first sold by Monsanto as Roundup, but is no longer under patent) is one of these: once it touches the soil, it is inactivated. Others are persistent and do not break down in this way. They enter the food web and move through it from trophic level to trophic level as they do not break down even inside the bodies of organisms. They are non-biodegradable. Many early insecticides such as DDT, dieldrin and aldrin fall into this group and they are stored in the fat of animals. Seals and penguins in Antarctica and polar bears in the Arctic have been found with pesticides in their tissues. The nearest land is 1000 km away. How may the pesticides have reached them?

To do

In this food web, the smaller fish (minnows) eat plankton (microscopic plants and animals) in the water. The minnow is eaten by the larger fish called pickerel.

These are eaten by heron, ospreys and cormorants and heron eat the minnows as well. The numbers give the percentage concentration of DDT.

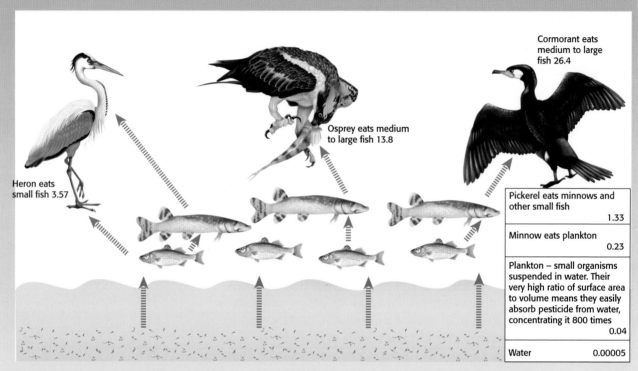

Cormorant eats medium to large fish 26.4

Osprey eats medium to large fish 13.8

Heron eats small fish 3.57

Pickerel eats minnows and other small fish	1.33
Minnow eats plankton	0.23
Plankton – small organisms suspended in water. Their very high ratio of surface area to volume means they easily absorb pesticide from water, concentrating it 800 times	0.04
Water	0.00005

Fig. 3.31 Food web in a freshwater ecosystem

1 How many trophic levels are in this food web?
2 How many times more concentrated is the DDT in the body of the cormorant than in the water? Explain how this happens.
3 In which species does bioaccumulation occur?
4 In which species does biomagnification occur?

Test yourself

1 An ecosystem consists of one oak tree on which 10 000 herbivores are feeding. These herbivores are prey to 500 spiders and carnivorous insects. Three birds of the same species are eating these spiders and carnivorous insects. The oak tree has a mass of 4000 kg, the herbivores insects have an average mass of 0.05 g, the spiders and carnivorous insects have an average mass of 0.2 g and the three birds have an average mass of 10 g.
   a Construct, to scale, a pyramid of numbers.
   b Construct, to scale, a pyramid of biomass.
   c Explain the differences between these two pyramids.

2 Explain whether the energy "loss" between two subsequent trophic levels is in contradiction with the first law of thermodynamics (see page 75).

3 Assuming an ecological efficiency of 10%, 5% and 20% respectively (see Fig. 3.32), what will be the energy available at the tertiary consumer level (4th trophic level), given a net primary productivity of 90 000 kJ m⁻² yr⁻¹? What percentage is this figure of the original energy value at the primary producer level?

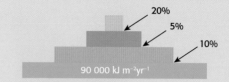

Fig. 3.32 Pyramid of biomass, not to scale

Case study 1

The story of Minamata Bay

Minamata is a small factory town in Japan, dominated by one factory, the Chisso Factory. Chisso make petrochemical-based substances from fertilizer to plastics. Between 1932 and 1968 Chisso released an estimated 27 tons of mercury into Minamata Bay. Beginning in the 1950s, several thousand people living locally started to suffer from mercury poisoning.

What had happened? Some bacteria can change mercury to a modified form called methylmercury. Methylmercury is easily absorbed into the bodies of small organisms such as shrimp. When the shrimp are eaten by fish, the methylmercury enters the fish. The methylmercury does not break down easily and can stay in the fish bodies for a long time. As the fish eat more and more shrimp, the amount of methylmercury in them increases. The same increase in concentration happens when people then eat the fish. Fish are a major part of the diet of people around Minamata Bay.

There is a slow build-up along the food chain: very many bacteria absorb very small amounts of mercury; many shrimp eat a lot of bacteria building up the mercury concentration; lots of fish eat lots of shrimp again building up the concentration; and finally a small number of humans at the top of the food chain eventually eat a lot of fish and absorb high levels of methylmercury.

Why top carnivores are in trouble

It is often the highest trophic level in a food chain (the top carnivore) that is the most susceptible to alterations in the environment. In the UK, the peregrine falcon (a bird of prey) population crashed in the late 1950s due to agricultural chemicals such as DDT accumulating and then magnifying in the food chain. This appeared to caused eggshell thinning and reduced breeding success. These chemicals were banned and from the mid 1960s, the peregrine population began to slowly recover despite persecution and the threat from egg collectors.

The top of the food chain is always vulnerable to the effects of changes further down the chain. Top carnivores often have a limited diet so a change in their food prey has a knock-on effect. Their population numbers are low because of the fall in amount of energy available along a food chain, therefore their ability to withstand negative influences is more limited than species lower in the food chain with larger populations.

Polar bears and "the new DDT"

Fig. 3.33 Polar bears

"The new DDT" could be polybrominated diphenyl ether (PBDE). It is manufactured in the United States and was widely used in the 1990s as a flame retardant to coat electrical appliances, sofas, carpets and car seats. The problem is that this chemical was designed to last the lifetime of the product, but in fact it lasts much longer. When sofas, carpets and car seats were thrown away,

PBDE entered the rivers, the oceans and the atmosphere. The Arctic, where all the world's polar bears live, is one of the great sinks of the planet. Chemical pollutants such as PBDE are carried towards the Arctic Ocean by the great rivers of Russia and Canada. PBDE already in the sea is taken north by ocean currents and carried by the wind. As it moves through the food chain from plankton to predator, PBDE bioaccumulates and is biomagnified so that long-lived top carnivores such as the polar bear accumulate the most concentrated amounts of PBDE. High amounts of PBDE have now been found in the body tissue of polar bears and killer whales. The long-term environmental effect of PBDE is unknown, but it will probably damage the immune systems, brain functions and bone strength. It also messes up the polar bear's sex hormones. One female bear on Spitzbergen had both male and female organs, a condition called imposex which is often linked to chemical pollution.

The length of food chains

As a rule of thumb, only 10% of the energy in one trophic level is transferred to the next – the **trophic efficiency** is 10%. A major part of the energy is used in respiration to keep the organism alive and is finally lost as heat to the environment. This is a result of the second law of thermodynamics (page 75) which states that energy is degraded to lower quality and finally to heat. Energy is lost because herbivores destroy more plant material than they actually eat – by trampling on it or rejecting it because it is too tough, old or spiky. Some material is not eaten at all and some dies and decomposes before it can be eaten. The 90% loss of energy in going from one trophic level to the next means there is very little energy available after about four trophic levels in terrestial ecosystems and five trophic levels in aquatic ecosystems.

Top carnivores are vulnerable because of the loss of energy from each trophic level. There is only so much energy available and that is why large top predators (often big, fierce animals) are rare. It is hard for them to accumulate enough energy to grow to a large size and to maintain their bodies.

To do

Model of the structure of an ecosystem

A model is a simplified diagram that shows the structure and workings (functions) of a system.

Copy Fig. 3.34, label the boxes and add arrows with labels to show the processes involved in this model of an ecosystem.

Fig. 3.34 Model for students to label

To do

Constructing a food web from information

Make a list of all the organisms described below and construct a food web diagram to show all feeding relationships.

The Aigamo paddy farming system is a self-sustaining agro-ecosystem based on rice, ducks (aigamo) and fish. The ducks eat up insect pests and golden snails, which attack rice plants. They also eat the seeds and seedlings of weeds, using their feet to dig up the weed seedlings, thereby oxygenating the water and encouraging the roots of the rice plants to grow more strongly. The "pests" and "weeds" are food sources for rearing the ducks. The ducks are left in the fields 24 hours a day and are completely free range until the rice plants form ears of grain in the field. At that point, the ducks have to be rounded up (otherwise they would eat the rice grains) and are fed exclusively on waste grain. There they mature, lay eggs, and fatten up for the market.

The ducks are not the only inhabitants of this system. The aquatic fern *Azolla*, or duckweed, which harbours a mutualistic blue-green bacterium that can fix atmospheric nitrogen, is also grown on the surface of the water. The *Azolla* is an efficient nitrogen fixer, and is readily eaten by the ducks, as well as attracting insects to be eaten by the ducks. The plant is very prolific, doubling itself every three days, so it can be harvested for cattle-feed as well. In addition, the plants spread out to cover the surface of the water, providing hiding places for another inhabitant, the roach (a fish), and protecting them from the ducks. The roach feed on duck faeces, on *Daphnia* (a crustacean) and various worms, which in turn feed on the plankton. Both fish and ducks provide manure to fertilize the rice plants throughout the growing season, and the rice plants in turn provide shelter for the ducks.

The Aigamo paddy field, then, is a complex, well-balanced, self-maintaining, self-propagating ecosystem. The only external input is the small amount of waste grain fed to the ducks, and the output is a nutritious harvest of organic rice, duck and roach. It is amazingly productive. A 2-hectare farm of which 1.5 ha are paddy fields can yield annually seven tonnes of rice, 300 ducks, 4,000 ducklings and enough vegetables to supply 100 people.

To think about

The **Millennium Ecosystem Assessment (MEA)**, funded by the UN and started in 2001, is a research programme that focuses on how ecosystems have changed over the last decades and predicts changes that will happen. In 2005, it released the results of its first four-year study of the Earth's natural resources. It was not happy reading. The report said that natural resources (food, fresh water, fisheries, timber, air) are being used faster than they can be replaced (unsustainably) and in ways that degrade them.

Key facts reported are:

- 60% of world ecosystems have been degraded.
- About 25% of the Earth's land surface is now cultivated.
- We use between 40 and 50% of all available fresh water running off the land, and water withdrawals have doubled over the past 40 years.
- Over 25% of all fish stocks are overharvested.
- Since 1980, about 35% of mangroves have been destroyed.
- About 20% of coral reefs have been lost in 20 years and another 20% degraded.
- Nutrient pollution has led to eutrophication of waters and to dead coastal zones.

- Species extinction rates are now 100–1 000 times above the background rate.

We have had more effect globally on the ecosystems of Earth in the last 50 years than ever before.

Some recommendations were to:

- remove subsidies to agriculture, fisheries and energy sources that harm the environment
- encourage landowners to manage property in ways that enhance the supply of ecosystem services, such as carbon storage, and the generation of fresh water
- protect more areas from development, especially in the oceans.

Questions

1 Are you optimistic or pessimistic about the results of the impact of humans on the Earth?

2 If you had the power, what actions would you force governments to take now to safeguard the environment but also protect humans from suffering?

Ecosystem function and ecological energetics

Now review the sections on photosynthesis, respiration and productivity at the start of this chapter. Also review pyramids and food webs.

Ecological energetics is the study of energy flow and storage in food chains and webs. From this study, energy flow diagrams can be drawn and these are very useful to show energy entering and leaving each trophic level, and where the energy goes and is lost to the system. By ecosystem function we mean how ecosystems work. In this section we will examine in detail how energy flows through ecosystems.

Energy flows

Almost all ecosystems are driven by energy from the sun – solar radiation. One important thing to remember is that energy continually flows through systems. We all give out energy all the time as respiration releases heat. It is only stored temporarily as chemical energy. Materials (matter), whether nutrients, oxygen, carbon dioxide or water, are cycled and recycled within ecosystems. They can be stored in the long and shorter term.

So remember: **energy flows, materials cycle**.

The systems of the biosphere are dependent on the amount of energy reaching the ground, not the amount reaching the outer atmosphere. This amount varies according to the time of day, the season, the amount of cloud cover and other factors. Most of this energy is not used to power living systems; it is reflected from soil, water or vegetation or absorbed and reradiated as heat.

This radiation energy from the sun follows the first and second laws of thermodynamics. The first law states that energy is neither created nor destroyed but may be changed from one form to another: light energy can be transformed into chemical energy, sugars, by green plants, and can then be changed to energy of motion or heat (respiration). This is also known as the law of conservation of energy. The second law states that the efficiency of energy conversion to useful work is never perfect: when energy changes from one form, some of the energy is not available to do useful work in the system. It leaves the system mainly as useless heat. This is true for all energy changes, even those involving living organisms. There is more about these laws in Chapter 4.

Only a very small part of the total sunlight reaching the Earth's surface is ever transformed into energy used by living organisms. A study of an Illinois cornfield suggested only 1.6% of sunlight energy was actually used by the corn. Living organisms can use energy in several forms. Organisms that gain energy from inorganic sources and fix it as energy-rich organic molecules are autotrophs, mostly green plants. Some bacteria obtain energy directly from inorganic chemicals: chemosynthetic autotrophs. Organisms that utilize energy-rich organic molecules, edible food, for their energy supply are heterotrophs. The heterotrophs can be split into two kinds, consumers that obtain their food from living organisms and decomposers that obtain their food from dead organisms or from organic material dispersed in the environment.

The fate of solar radiation hitting the Earth

Energy from the sun (solar radiation) is made up of visible wavelengths (light) and those wavelengths that humans cannot see (ultraviolet and infrared).

Our sun is about 4.5 billion years old and halfway through its lifespan. It has burned up about half of its hydrogen in nuclear fusion to make helium and release energy. This energy is in packets called photons and it takes eight minutes for a photon leaving the sun to reach the Earth. The energy leaving the sun is about 63 million joules per second per square metre ($J\ s^{-1}\ m^{-2}$). The solar energy reaching the top of the atmosphere of Earth is $1400\ J\ s^{-1}\ m^{-2}$ (or 1400 watts per second). This is the Earth's solar constant. Some 60% of this is intercepted by the atmospheric gases and dust particles. Nearly all the ultraviolet light is absorbed by ozone. Most of the infrared light (heat) is absorbed by carbon dioxide, clouds and water vapour in the atmosphere. The systems of the biosphere are dependent on the amount of energy reaching the ground, not the amount reaching the outer atmosphere. This amount varies according to the time of day, the season, the amount of cloud cover and other factors. Most of this energy is not used to power living systems, it is reflected from soil, water or vegetation or absorbed and reradiated as heat.

Of the energy reaching the Earth's surface, about 35% is reflected back into space by ice, snow, water and land. Some energy is absorbed and heats up the land and seas. Of all the energy coming in, only about 1–4% of it is available to plants on the surface of the Earth.

The only way in which that life can turn this energy into food is through photosynthesis by green plants. For a crop plant, such as wheat, which is an efficient converter, the figures are as follows. The plant can only absorb about 40% of the energy that hits a leaf. About 5% is reflected, 50% is not available and 5% passes straight through the leaf. Plants only use the red and blue wavelengths of light in photosynthesis and reflect the other colours (which is why plants look green). So of the 40%, just over 9% can be used. This is the GPP of the plant. Just under half of this is required in respiration to stay alive so 5.5% of the energy hitting a leaf becomes NPP.

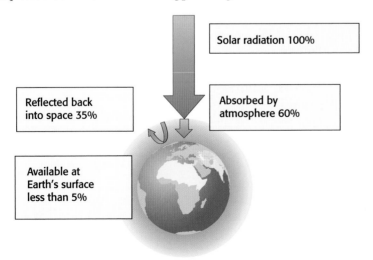

Fig. 3.35 The fate of solar radiation hitting the Earth

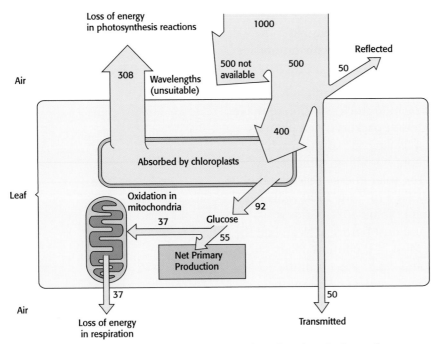

Fig. 3.36 Photosynthetic efficiency of a crop plant. This is based on the input of 1000 units of solar radiation

Of all the solar radiation falling on the Earth, plants only capture 0.06% (GPP) and use some of that to stay alive. What is left over (NPP) is the amount of food available to all the animals including humans. The efficiency of conversion of energy to food is low at about 2–3% in terrestrial systems but even lower at about 1% in aquatic systems as water absorbs more of the light before it reaches the plants.

The energy available theoretically for consumption by higher trophic levels declines rapidly as you go up the food chain. The maximum available energy is assimilated (absorbed by the organism), but much of this is used in respiration. Many organisms die or lose tissue as they age (e.g. male deer shed their antlers each year), and a lot of energy is lost through excretion. So far less than is assimilated is available to the next trophic level.

This decline in biomass and energy for the next trophic level has been expressed as an ecological tithe, a tithe being a tax of one-tenth. This equates to the general rule that only 10% of the energy consumed at one trophic level is available to the next trophic level. (Incidentally, this was first proposed by a man named Lindemann who died aged 26 in 1942. He worked at Lake Mendota, Minnesota, USA and his seminal paper was published posthumously.) The 10% rule applies to whole trophic levels and not between species. It also only applies to the higher consumer levels and not to the energy flow between producer and primary consumer.

The loss of productivity along a food chain illustrates a major feature of community structure. The biomass at higher trophic levels is usually significantly less than at lower levels; this is commonly reflected in numbers of organisms. Within any ecosystem it would be reasonable to find a small number of carnivores consuming a much greater number of herbivores, which in turn consume greater numbers of plants.

All ecosystems also contain the energy flows within decomposer food chains. Decomposition mainly occurs within the soil or in aquatic sediments. While primary and secondary production take place in the upper layers of an ecosystem, decomposition mainly occurs lower down: within the soil or aquatic sediments. An example is the leaf litter of a woodland. Dead leaves are at first broken down by earthworms, mites and fungi. They use the energy released for their life processes but also increase the amount of small detritus on which more fungi and bacteria can act. During the decomposition process, essential nutrients are released back to the soil to be available again for primary productivity. As the decomposers work, heat energy is lost from the ecosystem through their respiration. Without the decomposer food chain, dead organic matter would build up, locking up nutrients within organic molecules and so preventing their eventual return to the organic nutrient pool.

Energy flow diagrams

Energy flow diagrams allow easy comparison of various ecosystems. These show the energy entering and leaving each trophic level. Energy flow diagrams also show loss of energy through respiration and transfer of material and energy to the decomposer food chain.

While energy flows through an ecosystem in one direction, starting as solar radiation and finally leaving as heat released through the respiration of decomposers, the chemical nutrients in the biosphere cycle: nutrients are absorbed by organisms from the soil and atmosphere and circulate through the trophic levels and are finally released back to the ecosystem, usually via the detritus food chain. These are the biogeochemical cycles.

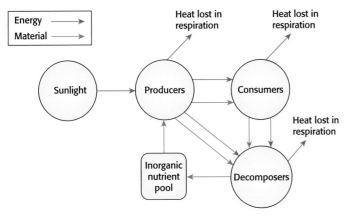

Fig. 3.37 Generalized energy flow diagram through an ecosystem

There are around 40 elements that cycle through ecosystems, though some exist only in trace amounts. All the biogeochemical cycles have both organic (when the element is in a living organism) and inorganic (when the element is in a simpler form outside living organisms) phases. Both are vital: the efficiency of movement through the organic phase determines how much is available to living organisms. Yet the major reservoir for all the main elements tends to be outside the food chain as inorganic molecules in rock and soils. Flow in this inorganic phase tends to be much slower than the movement of these nutrients through organisms, the organic phase. The major biogeochemical cycles are those of water, carbon, nitrogen, sulfur and phosphorus, all of which follow partially similar routes.

There are many different ways to draw energy flow diagrams and you need to be able to interpret these.

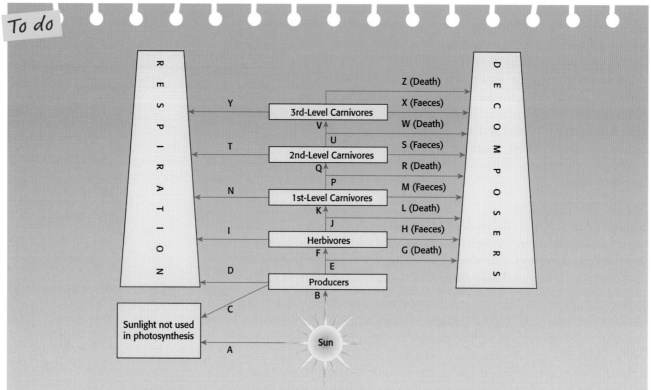

Fig. 3.38 Generalized energy flow diagram through a food web

Answer these questions using the letters (A–Z) in the diagram.

1   What is the total amount of energy from the sun?

2   What energy is reflected or lost as heat from the producers?

3   What is the GPP?

4   What is the NPP?

5   Explain what happens to NPP.

6   What is the GSP of the herbivores?

7   What is the NSP of the herbivores?

8   What energy is available to the 1st level carnivores?

9   Why do the boxes for respiration and decomposers get smaller as you go up the trophic levels?

10   If the NPP of the producers is 1000 and assuming that 10% of energy in one level is passed on to the next, what is the NPP of the 3rd level carnivores?

Case study 2

Grazing on the Serengeti

In the 1970s research was first carried out into the grazing habits on the Serengeti[1] savanna in Africa to look into the efficiency of energy transfer. Zebra, wildebeest and Thomson's gazelle are migratory and move on to the Serengeti grasslands in a specific sequence, based on the growth of the grass. The first species to arrive in the growing season are zebra. They enter a long grass community and remove mainly long grass stems. Wildebeest follow and eat grass leaves, but the large herds also trample the grass down. Thomson's gazelle finally arrive and eat a large amount of herbaceous vegetation.

In a comparison of food quality, researchers found that grass stems contain little protein but a large amount of lignin, structural material that is hard to digest. Grass leaves contain more protein and less lignin than stems. The herbaceous vegetation has a lot of protein and little lignin, making it easy to digest. This means that even though they are the first species to arrive on the Serengeti, zebra have the poorest diet. Also zebra are the largest of the three herbivores at about 220 kg.

Both wildebeest and Thomson's gazelle are ruminants, like cows and sheep, having a specialized stomach containing bacteria and protozoa that break down cellulose, so therefore they absorb nutrients more efficiently.

So how do zebra survive on the poorest diet? They eat a greater volume of food relative to their size compared to both wildebeest and gazelle. This allows them to make up for the energy deficiency of the food that they eat.

[1] Bell, R. H. V. (1971). A grazing ecosystem in the Serengeti. *Scientific American* **225**: 86–94.

Test yourself

1 "Energy flows but nutrients cycle." Draw systems to illustrate this statement.

2 State the equations for photosynthesis and respiration in words.

3 What are autotrophs and heterotrophs?

4 What four things can happen to light energy striking leaves?

5 State the first law of thermodynamics. How does it apply to ecosystems?

6 Explain with a diagram and an equation the relation between gross and net primary productivity.

7 Explain, with an equation, what all consumers do with the food that they ingest.

8 Copy and fill in the missing words in the passage on the right:

Energy enters ecosystems as _____ energy. Only 1–4% of this energy is then trapped by _____ and converted into, and stored as _____ energy by the process of _____ . In this process, _____ energy is transformed into _____ energy. Plants use only a little of the incoming solar radiation and most of the insolation is either reflected from the surfaces of leaves, absorbed by leaf cells to warm up leaves or is transmitted through them. The total amount of energy fixed by plants is called _____

_____ _____ . Plants will use some of this energy for various life processes, e.g. _____ and respiration. Respiration is the process of breaking down food to release _____ . Some of the energy is lost to the environment in the form of _____ energy.

The classical energy flow example

Silver Springs in central Florida is famous among ecologists as the place where Howard T. Odum researched energy flow in the ecosystem in the 1950s. Odum (1924–2002) was a pioneer ecologist working on ecological energetics. This was the first time an energy budget measurement was attempted when Odum measured primary productivity and losses by respiration. (Later, near the end of a long and illustrious career, he and David Scienceman developed the concept of **emergy** (embodied energy) which is a measure of the quality and type of energy and matter that go into making an organism.)

Figure 3.39 shows the energy flows and biomass stores measured by Odum at Silver Springs. Many different versions of this diagram have been produced. This simple community consists of algae and duckweed (producers); tadpoles, shrimps and insect larvae (herbivores); water beetles and frogs (first carnivores); small fish (top consumers); and bacteria, bivalves and snails (decomposers and detritivores). Dead leaves also fall into the water and spring water flows out, exporting some detritus.

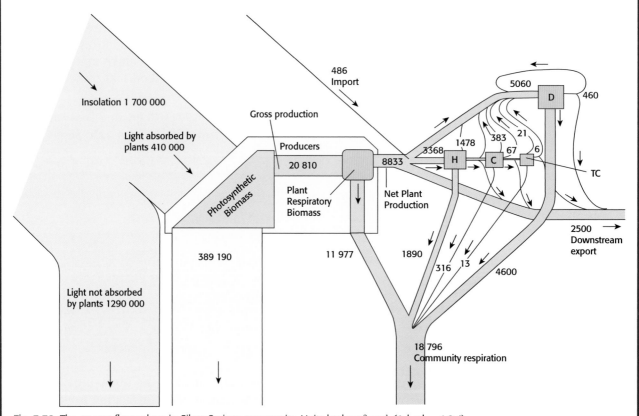

Fig. 3.39 The energy flow values in Silver Springs community. Units kcal m^{-2} yr^{-1} (1 kcal = 4.2 J)

Questions

1 Why does the width of the energy flow bands become progressively narrower as energy flows through the ecosystem?

2 Suggest an explanation for the limit on the number of trophic levels to four or five at most in a community.

3 How is the energy transferred between each trophic level?

4 Insolation (light) striking leaves is 1 700 000 units but only 410 000 are absorbed. What happens to the unabsorbed light energy?

5 A further 389 190 units escapes from producers as heat. Why is this?

6 Account (mathematically) for the difference between gross and net primary productivity.

7 Draw a productivity pyramid from the data given.

8 Would it be possible to draw a biomass pyramid from the data given?

9 Does the model support the first law of thermodynamics? Show your calculations.

Population interactions

Review of population definitions

A **population** is a group of organisms of the same species living in the same area at the same time and capable of interbreeding. Population **density** is the average number of individuals in a stated area, e.g. gazelles km^{-2}, bacteria cm^{-3}.

Four factors affect population size: natality (birth rate), mortality (death rate), immigration and migration.

Competition

All the organisms in any ecosystem have some effect on every other organism in that ecosystem. Also any resource in any ecosystem exists only in a limited supply. When these two conditions apply jointly, competition takes place.

Competition between members of the same species is **intraspecific competition**. When the number of individuals in a population is small, there is little real competition between individuals for resources. Provided the number is not too small for individuals to find mates, population growth will be high.

In a seagull colony on an oceanic outcrop, as the population grows, so the pressure for good nesting sites increases. This can affect the number of eggs that each female can successfully hatch, and so affects the birth rate of the population as a whole. As the population grows, so does the competition between individuals for the same resources until eventually the **carrying capacity** (the maximum number of a species that an environment can sustainably support) of the ecosystem is reached. In this situation, often the stronger individuals claim the larger share of the resources.

Some species, e.g. deer, deal with intraspecific competition by being territorial. An individual or pair hold an area and fend off rivals. Individuals that are the most successful reproductively will hold the biggest territory and hence have access to more resources.

Intraspecific competition tends to stabilize population numbers. It produces a sigmoid or logistic growth curve which is S-shaped (see pages 161–2).

Competition does not only occur between individuals of the same species. Individuals of different species could be competing for the same resource. This is **interspecific competition**.

Interspecific competition may result in a balance, in which both species share the resource. The other outcome is that one species may totally outcompete the other; this is the principle of **competitive exclusion**. An example of both of these outcomes can be seen in a garden that has become overrun by weeds. A number of weed species coexist together, but often the original domestic plants have been totally excluded.

In a temperate deciduous woodland light is a limiting resource. Plant species that cannot get enough light will die out in a woodland. This is especially true of small flowering plants on the woodland floor that are shaded out not only by trees but also by shrubs and bushes.

Beech trees have very closely overlapping leaves, resulting in an almost bare woodland floor.

But even in beech woods flowers manage to grow in the spring. Carpets of snowdrops, primroses and bluebells are an integral part of all Northern European deciduous woodlands in the spring. The key to the success of these species is that they grow, flower and reproduce before the shrub and tree species burst into leaf. They avoid competing directly with species that would outcompete them for light by completing the stages of their yearly cycle that require the most energy and therefore the greatest photosynthesis when competition is less.

Competition reduces the carrying capacity for each of the competing species, as both species use the same resource(s).

Predation

Predation happens when one animal, the predator, eats another animal, the prey. Examples are plenty, like lions eating zebras and wolves eating moose. The predator kills the prey. Be aware that not only do animals eat other animals; some plants (insectivorous plants) consume insects and other small animals. Look at the example of the Canadian lynx and the snowshoe hare on pages 81–2 as an example of negative feedback control.

Sometimes, however, a wider definition of predation is used: predation is the consumption of one organism by another. This broad definition includes predation as well as herbivory and parasitism.

Herbivory

Herbivory is defined as an animal (herbivore) eating a green plant. Some plants have defence mechanisms against this, e.g. thorns or spines (some cacti), a stinging mechanism (stinging nettles) or toxic chemicals (poison ivy). Herbivores may be large – elephants, cattle, or small – larvae of leaf miner insects that eat the inside of leaves, or in between – rabbits.

Parasitism

Parasitism is a relationship between two species in which one species (the parasite) lives in or on another (the host), gaining its food from it. Normally parasites do not kill the host, unlike in predation. However, high parasite population densities can lead to the host's death. Examples of parasites are vampire bats and intestinal worms.

Mutualism

Mutualism is a relation between two or more species in which both or all benefit and none suffer. It is a form of symbiosis (living together). The other types of symbiosis are parasitism, and commensalism (when one partner is helped and the other is not significantly harmed, e.g. an epiphyte such as an orchid or fern growing half-way up a tree trunk).

Lichens are an example of mutualism. A lichen is a close association of a fungus underneath and a green alga (or cyanobacterium) on top. The fungus benefits by obtaining sugars from the photosynthetic alga. The alga or bacterium benefits from minerals and water that the fungus absorbs and passes on to it.

Fig. 3.40 A lichen

Another example is the relationship between leguminous plants (beans, clover, vetch, peas) and nitrogen-fixing bacteria, *Rhizobium*. The bacteria live inside root nodules in the legumes. They absorb nitrogen from the air and make it available to the plant in the form of ammonium compounds. The plants in turn supply the bacteria with sugar from photosynthesis. This mutualistic relationship enables legumes to live on very poor soils. As a consequence, leguminous plants are among the earliest pioneer species during succession on poor soil. Clover is also often used to increase the nutrient content of agricultural soil.

Mycorrhizal fungi and tree roots are another example. The fungi form a sheath around the feeding roots of many trees. They provide the tree with phosphates that they take up from the soil. The tree provides the fungi with glucose that it produces from photosynthesis. Both grow better than they do without the other one.

Sea anemones and clown fish are also mutualistic. The clown fish provide food for the sea anemone in the form of their faeces. The anenome's stinging tentacles protect the clown fish from predators, but do not affect the clown fish.

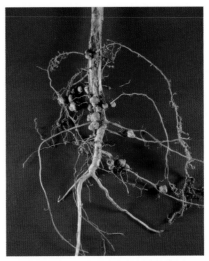

Fig. 3.41 Nitrogen-fixing root nodules on a legume root

Test yourself

1 Copy the table and fill in the last column.

Type of interaction		Species A	Species B	Example
Competition	Neither species benefits, both species suffer	−	−	
Predation	One species kills the other species for food	+	−	
Parasitism	The parasite benefits at the cost of the host	+	−	
Mutualism	Both species benefit	+	+	

2 What four things do all organisms need to survive? (*Hint*: think back to your first biology lesson.)

3 What is the difference between interspecific and intraspecific competition?

4 What effect does intraspecific competition have on the individuals of a species?

5 Why is species diversity believed to be beneficial for a community?

Human activities and ecosystems

A process, effect or activity derived from humans is known as **anthropogenic**. The enhanced greenhouse effect is anthropogenic. Do not confuse this with anthropomorphic which means giving human characteristics to other animals, plants or inanimate objects, e.g. a pet or a doll.

The concept of energy subsidy

Generally, when humans have an influence on an ecosystem, be it farming or living within it, we tend to simplify it and make it less diverse. Usually, this is on purpose. We cut down a forest to grow crops and often this is just one species, e.g. wheat. So the complex food web that may have been there in a deciduous temperate forest becomes:

wheat → human

or

improved pasture grasses → cattle → human

Much of what we do in agriculture is also aimed at reducing the number of species – killing pests and getting rid of weeds as these either eat or compete with the crops we want. Our aim is to maximize the NPP of the organisms we grow to maximize our profit. What happens is that we have to become ever more sophisticated in our farming practices – agribusiness – so we use artificial means to maintain the system. The Green Revolution which brought improved varieties of rice and other crops also brought the need to buy fertilizers for them or pesticides to kill the pests to which they were susceptible.

All farming practices require an **energy subsidy** which is the additional energy that we have to put into the system above that which comes from the sun's energy. It may be the human labour, animal labour or machines using fuel to power the tractors and ploughs, pump water for cattle, make fertilizers and other chemicals, and transport the crop. In spite of all this, some agricultural systems are very productive with high NPP, e.g. sugar cane.

As humans lived in larger groups and population density increased, they needed more food so farming methods became more sophisticated and used more energy. The advantage of an energy subsidy is that we can feed more people because food production seems more efficient but the energy has to come from somewhere (first law of thermodynamics). As communities become more complex, the energy subsidy increases. Hunter-gatherers have to add little energy to the system apart from their own work. Subsistence farming may involve draught animals, wind power or water power to irrigate or grind corn. All these are subsidized by human effort. Commercial farming now involves major use of fossil fuels to power machines, make chemicals to put on the crops or produce feedstuffs for animals. It is estimated that we use 50 times as much energy in MEDCs (more economically developed countries) as a hunter-gatherer society does, and this is rising all the time.

Energy-to-yield ratio

In economic terms, we can look at a farming system as inputs and outputs or costs and profits. So we can look at the ratio of energy in to energy out in the form of food. It seems that as agriculture has got more sophisticated, this ratio goes down. A simpler slash-and-burn type of agriculture (when land is cleared in the rainforest and then a variety of crops grown by a subsistence farmer) may have an energy:yield ratio of 1:30 or 1:40 (30–40 units of food energy for each one unit of input energy as work). With increasing input of energy, this could reduce to 10:1 for battery chicken or egg production so far more energy is put into the system than is taken out. But the important thing is that the energy is in the form of high-energy foods, concentrated energy as protein and meat, not lower-energy cereals. We are producing high-energy foodstuffs.

The issue to remember is that energy has to keep flowing through ecosystems whether natural or influenced by humans. If it does not, the system alters rapidly. Blocking sunlight from reaching a plant stops photosynthesis and the plant dies. Stopping the energy subsidy to agriculture will result in chaos. In a natural ecosystem, the large number and variety of food chains and energy paths mean the

To do

1 List the three trends that can be seen in relation to the impact of human activities over time on ecosystems.

2 Describe the two main reasons for the increase in the impact of human activities on the environment over time.

system is complex and less likely to fail completely. If one species goes, others can take its role. The system is resilient. If there is only one species in an ecological niche, e.g. wheat, its failure can have a bigger impact.

Test yourself

To test yourself, try these questions.

1 The data refer to carbon (in biomass) flows in a freshwater system at 40°N latitude:

	g C m^{-2} yr^{-1}
Gross productivity of phytoplankton	132
Respiratory loss by phytoplankton	35
Phytoplankton eaten by zooplankton	31
Faecal loss by zooplankton	6
Respiratory loss by zooplankton	12

From the data, write down word equations and calculate:

a net productivity of phytoplankton
b gross productivity of zooplankton
c net productivity of zooplankton
d % assimilation of zooplankton
e % productivity of zooplankton

2 Look at ecosystem I in Fig. 3.42.
a For ecosystem I, copy and draw a rectangle on the diagram to show the ecosystem boundary.
b Explain why the storage boxes reduce in size as you go up the food chain.
c Name three decomposers and explain how they lose heat.

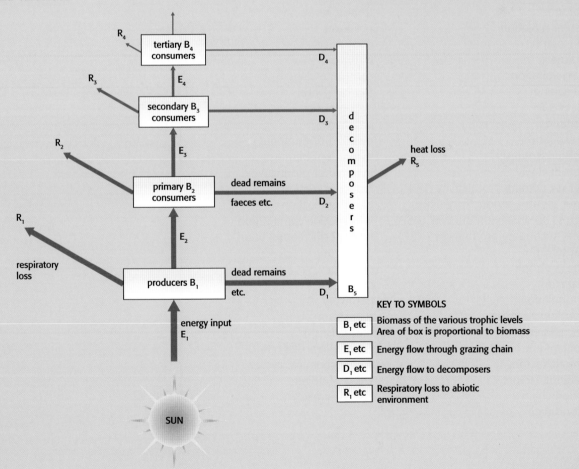

Fig. 3.42 Energy flow diagram of an ecosystem I

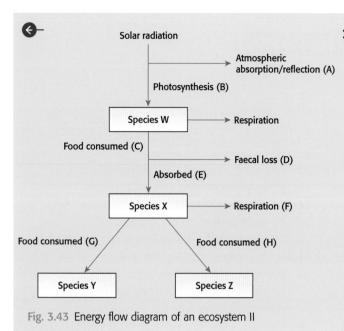

Fig. 3.43 Energy flow diagram of an ecosystem II

3 Look at ecosystem II in Fig. 3.43, identify from the diagram the letter(s) referring to the following energy flow processes and explain what happens to this energy at each stage as it passes through the ecosystem.

a loss of radiation through reflection and absorption

b conversion of light to chemical energy in biomass

c loss of chemical energy from one trophic level to another

d efficiencies of transfer

e overall conversion of light to heat energy by an ecosystem

f re-radiation of heat energy to atmosphere

Key words

biome	trophic level
solar radiation	producer
insolation	consumer
climate	herbivore
precipitation	carnivore
leaching	top carnivore
latitude	decomposer
altitude	detritivore
respiration	pyramid of numbers
photosynthesis	pyramid of biomass
productivity	pyramid of productivity
gross primary productivity (GPP)	bioaccumulation
net primary productivity (NPP)	biomagnification
gross secondary productivity (GSP)	intraspecific competition
net secondary productivity (NSP)	interspecific competition
autotroph	predation
heterotroph	predator
species	prey
population	herbivory
ecological niche	parasitism
ecosystem	mutualism
biosphere	energy flow diagram
biotic factors	biogeochemical cycles
abiotic factors	energy budget
food chain	energy subsidy
food web	

Key points

- A system is an assemblage of parts, working together, forming a functioning whole.
- Systems can be small or large.
- Open, closed and isolated systems exist in theory though most living systems are open systems.
- The first and second laws of thermodynamics and the concepts of positive and negative feedback mechanisms apply to both living and non-living systems.
- Most living systems are in a steady state equilibrium, not a static one though they may be stable or unstable.
- Material and energy undergo transfers and transformations in flowing from one storage to the next.
- Models have their limitations but can be useful in helping us to understand systems.

"Nature does nothing uselessly."

Aristotle (384–322 BC)

The human place in the biosphere

The biosphere is a fragile skin on the planet Earth. It includes the air (atmosphere), rocks (lithosphere) and water (hydrosphere) within which life (the ecosphere) occurs. Humans and all other organisms live within this thin layer yet we know little about how it is regulated or self-regulates, or about the effects the human species is having upon it.

Biosphere = atmosphere + lithosphere + hydrosphere + ecosphere

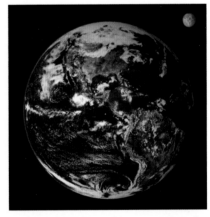

Fig. 4.1 The Earth and its moon viewed from space

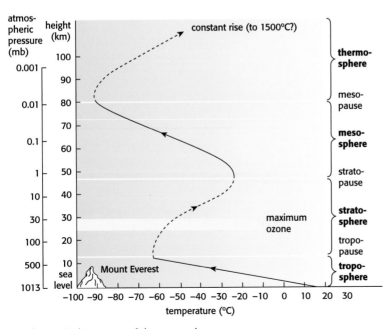

Fig. 4.2 The vertical structure of the atmosphere

The **atmosphere** is a mixture of nitrogen (78%), oxygen (21%), and traces (remaining 1%) of carbon dioxide, argon, water vapour and other components. Although the atmosphere is approximately 1100 km in depth, the stratosphere (10 to 50 km above the Earth's surface) and the troposphere (up to 10 km from the Earth's surface) are where most reactions affecting life occur. Heat and pollutants are carried across the Earth by air currents in the atmosphere.

The **hydrosphere** is water on Earth in all its states (solid, liquid and gas) and the elements dissolved it in (sodium, magnesium, calcium, chloride and sulfate). Water covers 71% of the Earth's surface but only 2.6% of all water on Earth is fresh water. Some 97% of the water is in the oceans, 2% is ice (north and south poles) and 1% in rivers, lakes, groundwater and water vapour. Heat is dispersed from the equator to the poles by water currents, and pollutants also diffuse around Earth in water.

The **lithosphere** (rocks) is the thin crust between the mantle and the atmosphere. Although the lithosphere is around 100 km thick, only the top 1 km of it interacts with the biosphere. Its main constituents are oxygen (47%), silicon (28%), aluminium (8%) and iron (5%). The lithosphere is the main source of natural pollutants. Some are naturally released through sources like volcanic eruptions, while others, like fossil fuels, are the result of artificial extraction and combustion.

The **ecosphere** is made up of all living organisms, including animals and plants. They are temporary accumulators of pollutants (e.g. lead) as well as being sources of pollutants (e.g. natural forest burning) in a very complex set of relationships with the atmosphere, hydrosphere and lithosphere.

Why systems?

This course is called *Environmental Systems and Societies* and not Environmental Science or Studies. Have you considered why this is? There is a difference in emphasis. In the systems approach, the environment is seen as a set of complex systems: sets of components that function together and form integrated units. You study plants, animals, soils, rocks or the atmosphere not separately (as is sometimes the case in other sciences such as biology, geology or geography), but together as components of the complex environments of which they form parts. You also study them in relation to other elements of the system of which they are a part. The course takes an integrated view, and this emphasis on relationships and linkages distinguishes the systems approach.

A system is defined as "an assemblage of parts and their relationship forming a functioning entirety or whole."

A system can be living or non-living and on any scale. A cell is a system as are you, a bicycle, a car, your home, a pond, an ocean, a computer, a farm and an iPod.

We mostly consider ecosystems in this book and they can be on many scales from a drop of pond water to an ocean, a tree to a forest, a coral reef to an island continent. A biome can be seen as an ecosystem, though it helps if an ecosystem has clear boundaries. The biosphere is an ecosystem as well.

A system may be an abstract concept as well as something tangible. It is a way of looking at the world. Usually, we can draw a system as a diagram.

Social systems and economic systems include the legal, financial and bureaucratic frameworks that make them work and regulate them. The environmental value system that you hold consists of your opinions on the environment and how you evaluate it.

A system may remain stable for a long time or may change quickly. Systems occur within their own environment which may be made up of other systems or ecosystems, and they usually exchange inputs and outputs – energy and matter in living systems, information in non-living ones – with their environment. Systems are all more than the sum of their parts, e.g. a computer is more than the materials used to make it.

All systems have:	Represented by
STORAGES or stores of matter or energy	A box
FLOWS into, through and out of the system	Arrows
INPUTS	Arrows in
OUTPUTS	Arrows out
BOUNDARIES	Lines
PROCESSES which transfer or transform energy or matter from storage to storage	e.g respiration, precipitation, diffusion

Fig. 4.3 Systems terminology

Types of system

Systems can be thought of as one of three types: open, closed and isolated.

An **open system** exchanges matter and energy with its surroundings (see Fig. 4.4).

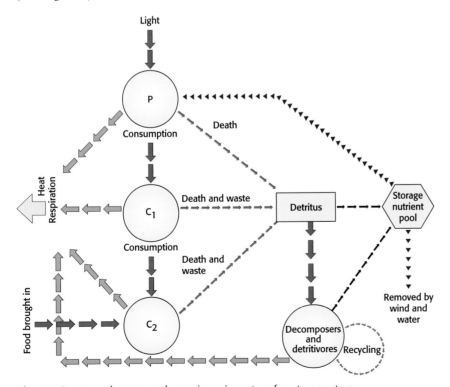

Fig. 4.4 Energy and matter exchange in an immature forest ecosystem

Most systems are open systems. All ecosystems are open systems exchanging matter and energy with their environment.

In forest ecosystems, plants fix energy from light entering the system during photosynthesis. Nitrogen from the air is fixed by soil bacteria. Herbivores that live within the forest may graze in adjacent ecosystems such as a grassland, but when they return they enrich the soil with faeces. After a forest fire topsoil may be removed by wind and rain. Mineral nutrients are leached out of the soil and transported in groundwater to streams and rivers. Water is lost through evaporation and transpiration from plants. Heat is exchanged with the surrounding environment across the boundaries of the forest.

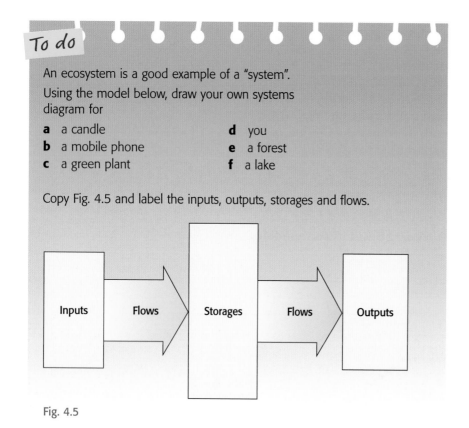

To do

An ecosystem is a good example of a "system".
Using the model below, draw your own systems diagram for

a a candle   **d** you
b a mobile phone   **e** a forest
c a green plant   **f** a lake

Copy Fig. 4.5 and label the inputs, outputs, storages and flows.

| Inputs | Flows | Storages | Flows | Outputs |

Fig. 4.5

Open system models can even be applied to the remotest oceanic island – energy and matter are exchanged with the atmosphere, surrounding oceans and even migratory birds.

A **closed system** exchanges energy but not matter with its environment.

Closed systems are extremely rare in nature. No natural closed systems exist on Earth but the planet itself can be thought of as an "almost" closed system.

Light energy in large amounts enters the Earth's system and some is eventually returned to space as longwave radiation (heat). (Because a small amount of matter is exchanged between the Earth and space, it is not truly a closed system. What types of matter can you think of that enter the Earth's atmosphere and what types that leave it?)

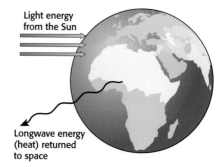

Light energy from the Sun

Longwave energy (heat) returned to space

Fig. 4.6 A closed system – the Earth

Most examples of closed systems are artificial, and are constructed for experimental purposes. An aquarium or terrarium may be sealed so that only energy in the form of light and heat but not matter can be exchanged. Examples include bottle gardens or sealed terraria but they usually do not survive for long as the system becomes unbalanced, e.g. not enough food for the animals, not enough oxygen or carbon dioxide, and organisms die. An example of a closed system that went wrong is Biosphere 2 (see box).

An **isolated system** exchanges neither matter nor energy with its environment. Isolated systems do not exist naturally though it is possible to think of the entire universe as an isolated system.

System	Energy exchanged	Matter exchanged
Open	Yes	Yes
Closed	Yes	No
Isolated	No	No

Biosphere 2

Biosphere 2, a prototype space city, was a human attempt to create a habitable closed system on Earth. Built in Arizona at the end of the 1980s, Biosphere 2 was a 1.27-hectare greenhouse intended to explore the use of closed biospheres in space colonization. Two major "missions" were conducted but both ran into problems. The biosphere never managed to produce enough food to adequately sustain the participants and at times oxygen levels became dangerously low and needed augmenting – the inhabitants opened the windows, so making it an open system.

Fig. 4.7 Biosphere 2

Inside were various ecosystems: a rainforest, coral reef, mangroves, savanna, desert, an agricultural area and living quarters. Electricty was generated from natural gas and the whole building was sealed off from the outside world.

For two years, eight people lived in Biosphere 2 in a first trial. But oxygen levels dropped from 21% to 14% and of the 25 small animal species put in, 19 became extinct, while ants, cockroaches and katydids thrived. Bananas grew well but there was not enough food to keep the eight people from being hungry. Oxygen levels gradually fell and it is thought that soil microbes respired much of this. Carbon dioxide levels fluctuated widely. A second trial started in 1994 but closed after a month after two of the team had vandalized the project, opening up doors to the outside. Cooling the massive greenhouses was a problem; it took three units of energy from air conditioners to cool the air for the input of every one unit of solar energy. So there were social, biological and technological problems with the project as the team split into factions and questions were asked as to whether this was a scientific, business or artistic venture.

The result showed how difficult it is to make a sustainable closed system when the complexities of the component ecosystems are not fully understood.

Questions

1 Why do you think this was called Biosphere 2?
2 Biosphere 2 has been described as a "closed system". What does this means?
3 Biosphere 2 was designed to include some of the major ecosystems of the Earth. List the ecosystems and divide them into terrestrial (land based) and marine (sea-water based).

To think about

All matter is made up of atoms. You are taught this in some of your first science lessons. Living things are made up of atoms, grouped into molecules and macromolecules, organelles, cells, tissues, organs and systems.

Read these two excerpts and think about what makes you you.

THE DAILY TELEGRAPH 20 November 2007

Science: Quantum theory and relativity explained

Quantum theory has made the modern world possible, giving us lasers and computers and iPod nanos, not to mention explaining how the sun shines and why the ground is solid.

Take the fact that you are constantly inhaling fragments of Marilyn Monroe. It is stretching it a bit to say that this is a direct consequence of quantum theory.

Nevertheless, it is connected to the properties of atoms, the Lego bricks from which we are all assembled, and quantum theory is essentially a description of this microscopic world.

The important thing to realise is that atoms are small. It would take about 10 million of them laid end to end to span the full stop at the end of this sentence. It means that every time you breathe out, uncountable trillions of the little blighters spread out into the air.

Eventually the wind will spread them evenly throughout the Earth's atmosphere. When this happens, every lungful of the atmosphere will contain one or two atoms you breathed out.

So, each time someone inhales, they will breathe in an atom breathed out by you—or Marilyn Monroe, or Alexander the Great, or the last *Tyrannosaurus rex* that stalked the Earth.

Fig. 4.8 Is Marilyn part of you?

Bill Bryson's *A Short History of Nearly Everything*

"Why atoms take this trouble is a bit of a puzzle. Being you is not a gratifying experience at the atomic level. For all their devoted attention, your atoms do not actually care about you—indeed, they do not even know that you are there. They don't even know that they are there. They are mindless particles, after all, and not even themselves alive. (It is a slightly arresting notion that if you were to pick yourself apart with tweezers, one atom at a time, you would produce a mound of fine atomic dust, none of which had ever been alive but all of which had once been you.) Yet somehow for the period of your existence they will answer to a single overarching impulse: to keep you you.

The bad news is that atoms are fickle, and their time of devotion is fleeting indeed. Even a long human life adds up to only about 650,000 hours, and when that modest milestone flashes past, for reasons unknown, your atoms will shut you down, silently disassemble, and go off to be other things. And that's it for you. Still, you may rejoice that it happens at all. Generally speaking in the universe, it doesn't…so far as we can tell."

Bill Bryson continues to say that "life is simple in terms of chemicals – oxygen, hydrogen, carbon, nitrogen make up most of all living things and a few other elements too – sulfur, calcium and some others. But in combination and for a short time, they can make you and that is the miracle of life."

Energy in systems

Energy in all systems is subject to the laws of thermodynamics.

According to the **first law of thermodynamics**, energy is neither created nor destroyed. What this really means is that the total energy in any isolated system, such as the entire universe, is constant. All that can happen is that the form the energy takes changes. This first law is often called the law of conservation of energy.

In the food chain in Fig. 4.9 the energy enters the system as light energy; during photosynthesis it is converted to stored chemical energy (glucose). It is the stored chemical energy that is passed along as food. No new energy is created; it is just passed along the food chain and transformed from one form to another (light to chemical energy) by photosynthesis and then transferred as chemical energy from the tree to the caterpillar.

If we look at the sunlight falling on Earth, not all of it is used for photosynthesis.

About 30% is reflected back into space, around 50% is converted to heat, and most of the rest powers the hydrological cycle: rain, evaporation, wind, etc. Less than 1% of incoming light is used for photosynthesis.

The **second law of thermodynamics** states that the **entropy** of an isolated system not in equilibrium will tend to increase over time. What this really means is that the energy conversions are never 100% efficient. When energy is transformed into work, some energy is always dissipated (lost to the environment) as waste heat. Entropy refers to the spreading out or dispersal of energy. As energy is dispersed to the environment, there will always be a reduction in the amount of energy passed on to the next trophic level.

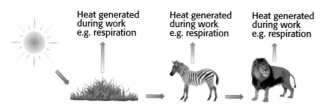

Fig. 4.11 Loss of energy to the environment in a food chain

If you examine the food chain above in terms of the second law, when the lion chases the zebra, the zebra attempts to escape, changing the stored chemical energy in its cells into useful work. But during its attempted escape some of the stored energy is converted to heat and lost from the food chain.

This process can be summarized by a simple diagram showing the energy input and outputs.

The second law can also be thought of as a simple word equation:

energy = work + heat (and other wasted energy)

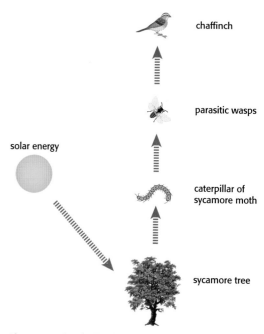

Fig. 4.9 A simple food chain

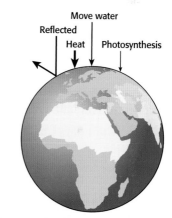

Fig. 4.10 The fate of the sun's energy hitting the Earth

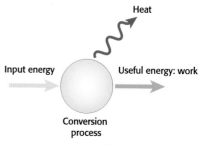

Fig. 4.12 The second law of thermodynamics

Using the above example the energy spreads out. The useful energy consumed by one level is less than the total energy at the level below. Energy transfer is never 100% efficient. Depending on the type of plant, the efficiency at converting solar energy to stored sugars is around 1–2%. Herbivores on average only asssimilate (turn into animal matter) about 10% of the total plant energy they consume. The rest is lost in metabolic processes. A carnivore's efficiency is also only around 10% (see Chapter 3).

So the carnivore's total efficiency in the chain is $0.02 \times 0.1 \times 0.1 = 0.0002\%$. This means the carnivore only uses 0.02% of incoming solar energy that went into the grass. The rest of the energy is dispersed into the surrounding environment.

Life is a battle against entropy and, without the constant replenishment of energy, it cannot exist. Consider this pictorial view (Fig. 4.13) of rowing upstream. Stop for a moment and you are swept back downstream by the current of entropy. In a simple example, a room tidied up is put in order relative to an untidy room. Thus, the tidy room has order, while the untidy one has disorder. In the language of physics, we would say that the tidy room has low entropy (high order = low disorder), while the untidy room has high entropy (high disorder). What is the natural route? For the tidy room to get untidy, if no one is taking care of it, or for an untidy room to get tidy? Evidently, the latter occurs only in fairy tales! This situation obeys the second law of thermodynamics, since the tidy room of low entropy becomes untidy, a situation of high entropy. In the process, entropy increases spontaneously.

Solar energy powers photosynthesis; chemical energy, through respiration, powers all activities of life; electrical energy runs all home appliances; the potential energy of a waterfall turns a turbine to produce electricity, etc. These are all high-quality forms of energy, because they power useful processes. They are all ordered forms of energy. Solar energy reaches us via photons in solar rays; chemical energy is stored in the bonds of macromolecules, like sugars; the potential energy of falling water is due to the specific position of water, namely that it is high and falls. These ordered forms have low disorder, so low entropy.

On the contrary, heat may not power any process; it is a low-quality form of energy. Heat is simply dispersed in space, being capable only of warming it up. Heat dissipates to the environment without any order; it is disordered. In other words, heat is a form of energy characterized by high entropy.

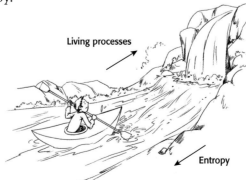

Fig. 4.13 A representation of life against entropy

Implications of the second law for environmental systems

1 The way we experience the second law in our everyday lives, namely the fact that entropy or disorder tends to increase spontaneously, is by realizing that all things, including living things, tend to change from order to disorder. In other words, the inescapable fate of all living creatures is death. Life is an anti-entropic system. Organisms manage to "survive" against the odds, that is, against the second law of thermodynamics, which dictates that they should die and their systems become more chaotic.

2 The way living creatures manage to maintain their order, that is, manage to stay alive, defying entropy, is by a continuous input of energy. As in the example of the room, the only way to keep the room tidy is to continuously clean it, that is, to expend energy. Similarly, organisms need to continuously obtain chemical energy from organic compounds via respiration to maintain their order. This is why energy is required even at rest, and not only when an organism is active. If any organism stops respiring, it dies.

3 Actually, living activities conform very well to the second law, because they contribute to the increase of entropy in the universe, an isolated system. In any living process that maintains the low entropy of organisms, the entropy of the universe is increasing, exactly like the law predicts. How does this happen? In any process, some of the useful energy turns into heat, namely a low-entropy (high-quality) form degrades into high-entropy (low-quality) heat. Thus, while the entropy of the living system is maintained at a low level, the entropy of the environment is increasing. As energy moves through an ecosystem, solar energy is transformed into chemical energy, and chemical energy into mechanical energy. In both cases, a large amount of initial high-quality, low-entropy solar or chemical energy is "lost" as low-quality, high-entropy heat, increasing the entropy of the environment, in which heat dissipates.

4 As a consequence, no process can be 100% efficient. Efficiency is defined as the useful energy, the work or output produced by a process divided by the amount of energy consumed (the input to the process):

efficiency = work or energy produced / energy consumed

efficiency = useful output / input

Multiply by 100%, if you want to express efficiency as a percentage.

5 A last philosophical implication is that, according to physics, the fate of all the energy that exists today in the universe is to degrade into high-entropy heat. When all energy has turned into heat, the whole universe will have a balanced temperature, and no process will be possible any longer, since heat may not turn into something of higher entropy. This is referred to as the thermal death of the universe. Do you have an opinion on what may happen after that?

Equilibrium

Equilibrium is the tendency of the system to return to an original state following disturbance; at equilibrium, a state of balance exists among the components of that system.

We can think of systems as being in dynamic (steady-state) or static equilibria as well as in stable or unstable equilibria. We discuss each of these here. Note that the term steady-state equilibrium is used instead of dynamic equilibrium in this book.

Open systems tend to exist in a state of balance or equilibrium. Equilibrium avoids sudden changes in a system, though this does not mean that all systems are non-changing. If change exists it tends to exist between limits.

A steady-state equilibrium is a characteristic of open systems where there are continuous inputs and outputs of energy and matter, but the system as a whole remains in a more-or-less constant state (e.g. a climax ecosystem).

In a steady-state equilibrium there are no long-term changes but there may be small fluctuations in the short term, e.g. in response to weather changes, and the system will return to its previous equilibrium condition following the removal of the disturbance. Some systems may undergo long-term changes to their equilibrium while retaining integrity to the system. Successions (see Chapter 14) are good examples of this.

Examples of a steady-state equilibrium

1 A water tank. If it fills at the same rate that it empties, there is no net change but the water flows in and out. It is in a steady state.

2 In economics, a market may be stable but there are flows of capital in and out of the market.

3 In ecology, a population of ants or any organism may stay the same size but individual organisms are born and die. If these birth and death rates are equal, there is no net change in population size.

4 A mature, climax ecosystem, like a forest, is in steady-state equilibrium as there are no long-term changes. It usually looks much the same for long periods of time, although all the trees and other organisms are growing, dying and being replaced by younger ones. However, there are flows in and out of the system – light inputs from the sun, energy outputs as heat lost through respiration; matter inputs in rainwater and gases, outputs in salts lost in leaching and rain washing away the soil. However, over years, the inputs and outputs balance.

5 Another example of a steady-state equilibrium is people maintaining a constant body weight, thus "burning" all the calories (energy) we get from our food. In cases of increasing or decreasing body weight there is no steady state.

6 The maintenance of a constant body temperature is another example. We sweat to cool ourselves and shiver to warm up but our body core temperature is about 37°C.

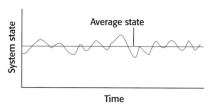

Fig. 4.14 Steady-state equilibrium

Maintenance of a steady-state equilibrium is achieved through negative feedback mechanisms, as we shall see later.

Static equilibrium

Another kind of equilibrium is called a **static equilibrium**, in which there is no change over time, e.g. a pile of books which does not move unless toppled over. When a static equilibrium is disturbed it will adopt a new equilibrium as a result of the disturbance. A pile of scree material (a mass of weathered rock fragments) piled up against a cliff could be said to exist in static equilibrium. The forces within the system are in balance, and the components (the rock fragments, the cliff and the valley floor) remain unchanged in their relationship to one another for long periods of time.

Fig. 4.15 Static equilibrium

Most non-living systems like a pile of rocks or a building are in a state of static equilibrium. This means that they do not change their position or state, i.e. they look the same for long periods of time and the rocks or bricks stay in the same place.

This cannot occur in living systems as life involves exchange of energy and matter with the environment.

Systems can also be **stable** or **unstable**.

Stable and unstable equilibria

In a stable equilibrium the system tends to return to the same equilibrium after a disturbance.

In an unstable equilibrium the system returns to a new equilibrium after disturbance.

Some people think that this is happening to our climate and that the new state will be hotter.

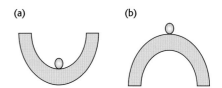

Fig. 4.16 Diagrams of (a) stable and (b) unstable equilibrium

Feedback

Systems are continually affected by information from outside and inside the system. Feedback loops can be positive or negative. Feedback mechanisms either change a system to a new state or return it to its original state. Simple examples of this are:

1 If you start to feel cold you can either put on more clothes or turn the heating up. The sense of cold is the information, putting on clothes is the reaction.
2 If you feel hungry, you have a choice of reactions as a result of processing this "information": eat food, or do not eat and feel more hungry.
3 If a teacher provides you with knowledge, knowledge is the information and "knowledge acquisition" is the reaction. The interaction may happen both ways, if you are encouraged by the teacher and respond positively.

If the students respond positively by learning and showing interest in the course, then the teacher realizes that the methodology is successful, and continues in the same way or improves it in response to constructive comments. In other words, the teaching process is reinforced and strengthened. This is called **positive feedback**.

On the other hand, students might respond negatively showing indifference, distraction or even dissent. Such a "result" should clearly indicate to the teacher that the methodology is not appropriate, at least for the specific group of students, and the teacher should change the style of teaching. This is **negative feedback**.

Natural systems act in exactly the same way. The information starts a reaction which in turn may input more information which may start another reaction. This is called a feedback loop.

Negative feedback tends to damp down, neutralize or counteract any deviation from an equilibrium, and it stabilizes systems or results in steady-state (dynamic) equilibrium. It results in self-regulation of a system.

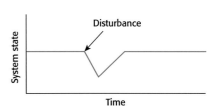

Fig. 4.17 Stable equilibrium

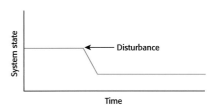

Fig. 4.18 Unstable equilibrium

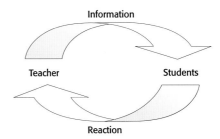

Fig. 4.19 A teacher-students interaction loop; feedback in the teaching process

Examples of negative feedback

1 Your body temperature starts to rise above 37°C because you are walking in the tropical sun and the air temperature is 45°C. The sensors in your skin detect that your surface temperature is rising so you start to sweat and go red as blood flow in the capillaries under your skin increases. Your body attempts to lose heat.

2 A thermostat in a central heating system is a device that can sense the temperature. It switches a heating system on when the temperature decreases to a predetermined level, and off when it rises to another warmer temperature. So a room, a building, or a piece of industrial plant can be maintained within narrow limits of temperature.

3 Predator-prey interactions. The Lotka-Volterra model (proposed in 1925 and 1926) is also known as the predator-prey model and shows the effect of changing numbers of prey on predator numbers. When prey populations (e.g. mice) increase, there is more food for the predator (e.g. owl) so they eat more and breed more, resulting in more predators which eat more prey so the prey numbers decrease. If there are fewer prey, there is less food and the predator numbers decrease. The change in predator numbers lags behind the change in prey numbers. The snowshoe hare and Canadian lynx is a well-documented example of this (see box, next page).

4 Some organisms have internal feedback systems, physiological changes occurring that prevent breeding when population densities are high, promoting breeding when they are low. It is negative feedback loops such as these that maintain "the balance of nature".

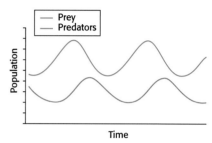

Fig. 4.20 Cycles of predator and prey in the Lotka-Volterra model

Positive feedback results in a further increase or decrease in the output, that is feedback enhances the change in the system and it is destabilized and pushed to a new state of equilibrium. The process may speed up, taking ever-increasing amounts of input until the system collapses. Alternatively, the process may be stopped abruptly by an external force or factor. Positive feedback results in a "vicious circle".

Examples of positive feedback

1 You are lost on a high snowy mountain. When your body senses that it is cooling below 37°C, various mechanisms such as shivering help to raise your body core temperature again. But if these are insufficient to restore normal body temperature, your metabolic processes start to slow down, because the enzymes that control them do not work so well at lower temperatures. As a result you become lethargic and sleepy and move around less and less, allowing your body to cool even further. Unless you are rescued at this point, your body will reach a new equilibrium: you will die of hypothermia.

2 In some developing countries poverty causes illness and contributes to poor standards of education. In the absence of knowledge of family planning methods and hygiene, this contributes to population growth and illness, adding further to the causes of poverty: "a vicious circle of poverty".

3 Global warming – higher temperatures may cause more evaporation so lead to more water vapour in the atmosphere. Water vapour is a greenhouse gas so will trap more heat so the atmosphere will warm more.

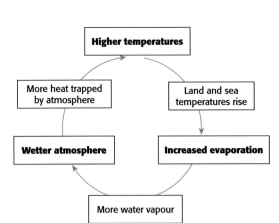

Fig. 4.21 Positive feedback in global warming

Whether a system is viewed as being in static or steady-state equilibrium may be a matter of the timescale. An ecosystem undergoing succession (see Chapter 14) is in a state of flux – it changes constantly.

In the case of a community in succession, the system undergoes long-term changes, so cannot have an equilibrium. However, the system retains its long-term integrity, since it is functioning properly, in a balanced, natural way. A better way to describe this situation is that the system shows **stability** and all systems in nature show stability by default.

Both natural and human systems are regulated by feedback mechanisms. Generally, we wish to preserve the environment in its present state, so negative feedback is usually helpful and positive feedback is usually undesirable. However there are situations where change is needed and positive feedback is advantageous, e.g. if students enjoy their Environmental Systems and Societies lessons, they want to learn more, so attend classes regularly and complete assignments. Consequently they move to a new equilibrium of being better educated about the environment.

We shall come back to feedback loops in various chapters of this book, particularly climate change and sustainable development.

Predator-prey interactions and negative feedback

The Hudson Bay Trading Company in Northern Canada kept very careful records of pelts (skins) brought in and sold by hunters over almost a century. This is a classic set of data and shows this relationship because the hare is the only prey of the lynx and the lynx its only predator. Usually things are more complicated.

Figure 4.23 (adapted from Odum, *Fundamentals of Ecology*, Saunders, 1953) shows a plot of that data.

Fig. 4.22 Canadian lynx chasing snowshoe hare

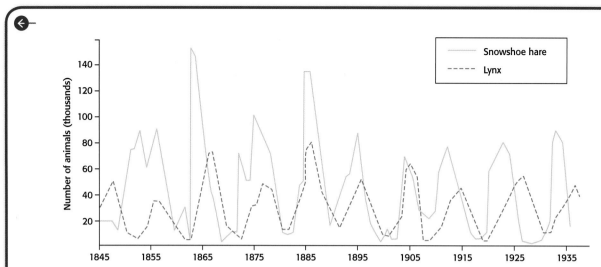

Fig. 4.23 Snowshoe hare and Canadian lynx population numbers, 1845–1940

We have to assume that the numbers of animals trapped were small compared with the total populations, and that the numbers trapped were roughly proportional to total population numbers. Also assumed is that the prey always has enough food so does not starve. Given that, the cycles are remarkably constant with the lynx populations always smaller than and lagging behind the hare populations.

Questions

1   On average, what was the cycle length of the lynx population?
2   On average, what was the cycle length of the hare population?
3   Why do lynx numbers lag behind hare numbers?

Things are never as straightforward in ecology as we expect though. In regions where lynx died out, hare populations still continued to fluctuate. Why do you think this was?

Review

Here are a number of examples of how both positive and negative feedback mechanisms might operate in the physical environment. No one can be sure which of these effects is likely to be most influential, and consequently we cannot know whether or not the Earth will manage to regulate its temperature, despite human interference with many natural processes.

Label each example as either positive or negative feedback.

Draw diagrams of one example of positive feedback and one example of negative feedback using the examples given, to show how feedback affects a system. Include feedback loops on your diagrams.

1   As carbon dioxide levels in the atmosphere rise, the temperature of the Earth rises. As the Earth warms the rate of photosynthesis in plants increases, more carbon dioxide is therefore removed from the atmosphere by plants, reducing the greenhouse effect and reducing global temperatures.

2   As the Earth warms, ice cover melts, exposing soil or water. Albedo decreases (albedo is the fraction of light that is reflected by a body or surface). More energy is absorbed by the Earth's surface. Global temperature rises. More ice melts.

3   As the Earth warms, upper layers of permafrost melt, producing waterlogged soil above frozen ground. Methane gas is released into the environment. The greenhouse effect is enhanced. The Earth warms, melting more permafrost.

4   As the Earth warms, increased evaporation produces more clouds. Clouds increase albedo, reflecting more light away from the Earth. The temperature falls. Rates of evaporation fall.

5   As the Earth warms, organic matter in soil is decomposed faster, more carbon dioxide is released, the enhanced greenhouse effect occurs, the Earth warms further and rates of decomposition increase.

6 As the Earth warms, evaporation increases. Snowfall at high latitudes increases. Icecaps enlarge. More energy is reflected by increased albedo of ice cover. The Earth cools. Rates of evaporation fall.

7 As the Earth warms, polar icecaps melt, releasing large numbers of icebergs into oceans. Warm ocean currents such as the Gulf Stream are disrupted by additional freshwater input into oceans. Reduced transfer of energy to the poles reduces temperature at high latitudes. Ice sheets reform and icebergs retreat. Warm currents are re-established.

Transfers and transformations

Both matter (or material) and energy move or flow through ecosystems. A transfer happens when the flow does not involve a change of form or state, e.g. water moving from a river to the sea, chemical energy in the form of sugars moving from a herbivore to a carnivore. A transformation happens when a flow involves a change of form or state, e.g. liquid to gas, light to chemical energy. Both types of flow require energy; transfers, being simpler, require less energy and are therefore more efficient than transformations.

Transfers can involve:

- the movement of material through living organisms (carnivores eating other animals)
- the movement of material in a non-living process (water being carried by a stream)
- the movement of energy (ocean currents transfering heat).

Transformations can involve:

- matter to matter (soluble glucose converted to insoluble starch in plants)
- energy to energy (light converted to heat by radiating surfaces)
- matter to energy (burning fossil fuels)
- energy to matter (photosynthesis).

Flows and storages

Both energy and matter flow (as inputs and outputs) through ecosystems but, at times, they are also stored (as storages or stock) within the ecosystem.

Energy flows from one compartment to another, e.g. in a food chain. But when one organism eats another organism, the energy that moves between them is in the form of stored chemical energy: the body of the prey organism.

Energy flows through an ecosystem in the form of carbon–carbon bonds within organic compounds. These bonds are broken during respiration when carbon joins with oxygen to produce carbon dioxide. Respiration releases energy that is either used by

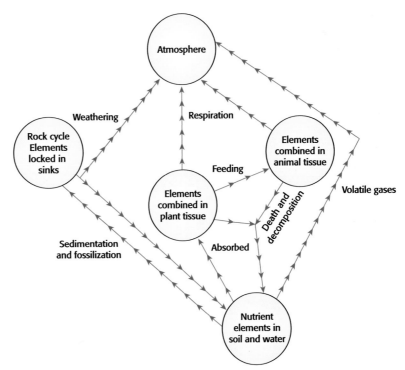

Fig. 4.24 Biogeochemical cycle illustrating the general flows in an ecosystem

organisms (in life processes) or is lost as heat. The origin of all the energy in an ecosystem is the sun and the fate of the energy is eventually to be released as heat.

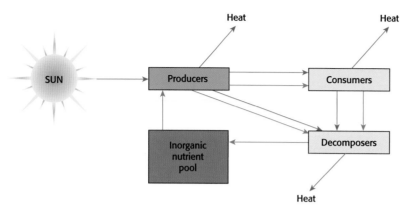

Fig. 4.25 Energy flow and flow of matter through an ecosystem

In Fig. 4.25 the flow of energy is shown by the red arrows and the flow of matter by the blue arrows.

Unlike energy, *matter* cycles round the system as minerals (blue arrows). Plants absorb mineral nutrients from the soil. These nutrients are combined into cells. Consumers eat plants and other consumers, ingesting the minerals they contain and recombining them in cells. Eventually decomposers break down dead organic matter (DOM) and return the minerals to the soil. These minerals may be taken out of the soil quickly by plants or can eventually, through geological processes, become locked within rocks until erosion eventually returns them to new soil.

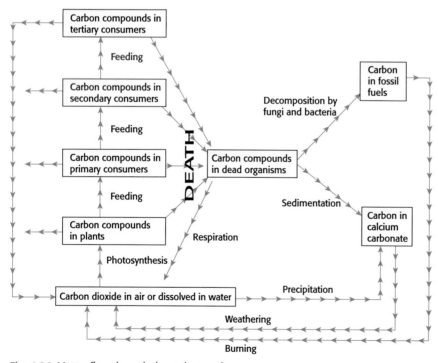

Fig. 4.26 Matter flow through the carbon cycle

The geochemical cycles (e.g. Fig. 4.24) are a way of showing the flows and storage of energy and matter. The carbon cycle (Fig. 4.26) and nitrogen cycle (Fig. 4.27) only show the flow and storage of matter.

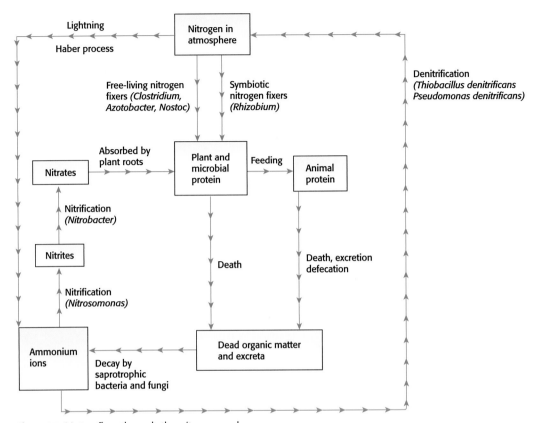

Fig. 4.27 Matter flow through the nitrogen cycle

Complexity and stability

Most ecoystems are very complex. There are many feedback links, flows and storages. It is likely that a high level of complexity makes for a more stable system which can withstand stress and change better than a simple one can, as another pathway can take over if one is removed. Imagine a road system where one road is blocked by a broken-down truck; vehicles can find an alternative route on other roads. If a community has a number of predators and one is wiped out by disease, the others will increase as there is more prey for them to eat and prey numbers will not increase. Tundra ecosystems are fairly simple and populations in them may fluctuate widely, e.g. lemming population numbers. Monocultures (farming systems in which there is only one major crop) are vulnerable to the sudden spread of a pest or disease through a large area with devastating effect. The spread of potato blight through Ireland in 1845–8 provides an example; potato was the major crop grown over large areas of the island, and the biological, economic and political consequences were severe.

Models of systems

Simplified models of a system can help predict changes in the system by modelling reality, as systems work in predictable ways, following rules. We just do not always know what these rules are. A model can take many forms. It could be:

- a physical model, e.g. a wind tunnel or river, a globe or model of the solar system, an aquarium or terrarium
- a software model, e.g. of climate change or evolution (Lovelock's Daisy world)
- mathematical equations
- data flow diagrams.

Climate models

Modelling climate change is a complex business requiring huge computing resources. Simple models of the climate system have been developed to predict changes with a range of emissions of greenhouse gases. The models solve complex equations but have to use approximations. They have improved over 30 years. The early ones included rain but not clouds. Now they have interactive clouds, rain, oceans, land and aerosols. The latest climate models predict similar possible global average temperature changes to those predicted by models five or 10 years ago, with increases ranging from 1.6 to 4.3°C. (See Chapter 7 for more about these.)

Models have their limitations as well as strengths. While they may omit some of the complexities of the real system (through lack of knowledge or for simplicity), they allow us to look ahead and predict the effects of a change to an input to the system.

Review

1 Define a system.
2 Fill in the gaps. The terms "open", "closed" and "isolated" are used to describe particular kinds of systems. Match the above names to the following definitions:

a *A _____ system exchanges matter and energy with its surroundings (e.g. an ecosystem).*

b *A _____ system exchanges energy but not matter.* The "Biosphere 2" experiment was an attempt to model this. These systems do not occur naturally on Earth, although the biosphere (or Gaia) itself can be considered a _____ system.

c *A _____ system exchanges neither matter nor energy.* (No such systems exist, with the possible exception of the entire universe.)

All ecosystems are _____ systems, because of the input of _____ energy and the exchange of _____ with other ecosystems.

3 Consider the following simple systems and complete the table:

List	Burning candle	Boiling kettle	A plant	Animal population
Inputs				
Outputs				
Energy and material transfers				
Energy and material transformations				

4 Look at the example of feedback control in Fig. 4.28.
  a How is the growth of the animal population regulated in the diagram?
  b Explain why it is an example of negative feedback control.

5 Explain with a named example how positive feedback may contribute to global warming.

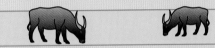

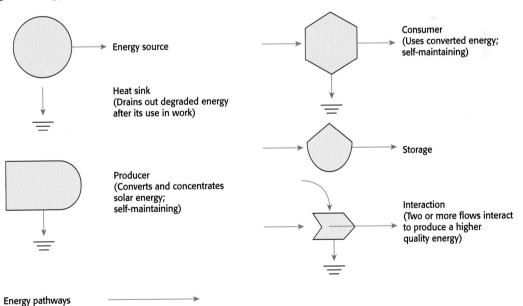

Fig. 4.28 Negative feedback among grazing animals

Energy diagrams

Odum, in researching Silver Springs, developed a system for representing ecosystems that gave more information on modelling the energy flow in the system.

Energy source

Heat sink
(Drains out degraded energy after its use in work)

Producer
(Converts and concentrates solar energy; self-maintaining)

Consumer
(Uses converted energy; self-maintaining)

Storage

Interaction
(Two or more flows interact to produce a higher quality energy)

Energy pathways

Fig. 4.30 Some of Odum's symbols

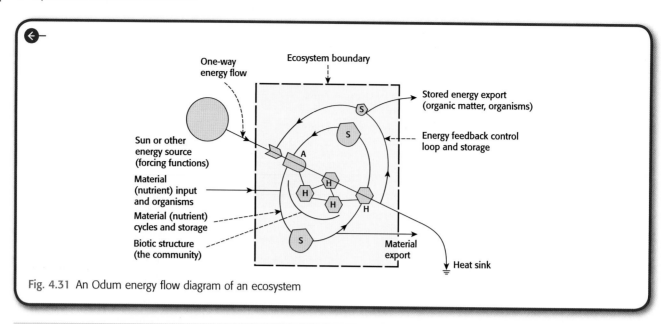

Fig. 4.31 An Odum energy flow diagram of an ecosystem

Test yourself

1 "Energy flows but nutrients cycle." Draw a systems diagram to illustrate this phrase.

2 Describe the transfers and transformations involved when light energy strikes a leaf.

3 State the first law of thermodynamics. How does it apply to ecosystems?

Gaia

The "Great Aerial Ocean" was Alfred Russel Wallace's description of the atmosphere.

In 1979, James Lovelock published his "Gaia hypothesis" in *Gaia: A new look at life on Earth*. In it he argued that the Earth is a planet-sized organism and the atmosphere is its organ that regulates it and connects all its parts. (Gaia is an ancient Greek Earth goddess.) Lovelock argued that the biosphere keeps the composition of the atmosphere within certain boundaries by negative feedback mechanisms.

He based his argument on these facts:

1 The temperature at the Earth's surface is constant even though the sun is giving out 30% more energy than when the Earth was formed.

2 The composition of the atmosphere is constant with 79% nitrogen, 21% oxygen and 0.03% carbon dioxide. Oxygen is a reactive gas, but its proportion does not change.

3 The oceans' salinity is constant at about 3.4% but rivers washing salts into the seas might be expected to increase this.

Lovelock was much criticized over this hypothesis. Lynn Margulis who worked with him also supported his views though used less emotive language about the Earth as an organism. Lovelock has defended his hypothesis for 30 years and many people now accept some of his views. He developed "Daisyworld" as a mathematical simulation to show that feedback mechanisms can evolve from the activities of self-interested organisms, black and white daisies in this case.

In Lovelock's 2006 book, *The Revenge of Gaia*, he makes a strong case for the Earth being an "older lady", more than half-way through her existence as a planet and so not able to bounce back from changes as well as she used to. He suggests that we may be entering a phase of positive feedback when the previously stable equilibrium will become unstable and we will shift to a new, hotter, equilibrium state. Controversially, he suggests that the human population will survive but with a 90% reduction in numbers.

The Gaia Model

Guest essay by James Lovelock
(published 16 January 2006)

The Earth is about to catch a morbid fever that may last as long as 100 000 years.

Each nation must find the best use of its resources to sustain civilization for as long as they can.

Imagine a young policewoman delighted in the fulfilment of her vocation; then imagine her having to tell a family whose child had strayed that he had been found dead, murdered in a nearby wood. Or think of a young physician newly appointed who has to tell you that the biopsy revealed invasion by an aggressive metastasizing tumour. Doctors and the police know that many accept the simple awful truth with dignity but others try in vain to deny it.

Whatever the response, the bringers of such bad news rarely become hardened to their task and some dread it. We have relieved judges of the awesome responsibility of passing the death sentence, but at least they had some comfort from its frequent moral justification. Physicians and the police have no escape from their duty.

This article is the most difficult I have written and for the same reasons. My Gaia theory sees the Earth behaving as if it were alive, and clearly anything alive can enjoy good health, or suffer disease. Gaia has made me a planetary physician and I take my profession seriously, and now I, too, have to bring bad news.

The climate centres around the world, which are the equivalent of the pathology lab of a hospital, have reported the Earth's physical condition, and the climate specialists see it as seriously ill, and soon to pass into a morbid fever that may last as long as 100 000 years. I have to tell you, as members of the Earth's family and an intimate part of it, that you and especially civilization are in grave danger.

Our planet has kept itself healthy and fit for life, just like an animal does, for most of the more than three billion years of its existence. It was ill luck that we started polluting at a time when the sun is too hot for comfort. We have given Gaia a fever and soon her condition will worsen to a state like a coma. She has been there before and recovered, but it took more than 100 000 years. We are responsible and will suffer the consequences: as the century progresses, the temperature will rise 8 degrees centigrade in temperate regions and 5 degrees in the tropics.

Much of the tropical land mass will become scrub and desert, and will no longer serve for regulation; this adds to the 40% of the Earth's surface we have depleted to feed ourselves.

Curiously, aerosol pollution of the northern hemisphere reduces global warming by reflecting sunlight back to space. This "global dimming" is transient and could disappear in a few days like the smoke that it is, leaving us fully exposed to the heat of the global greenhouse. We are in a fool's climate, accidentally kept cool by smoke, and before this century is over billions of us will die and the few breeding pairs of people that survive will be in the Arctic where the climate remains tolerable.

By failing to see that the Earth regulates its climate and composition, we have blundered into trying to do it ourselves, acting as if we were in charge. By doing this, we condemn ourselves to the worst form of slavery. If we choose to be the stewards of the Earth, then we are responsible for keeping the atmosphere, the ocean and the land surface right for life. A task we would soon find impossible – and something before we treated Gaia so badly, she had freely done for us.

To understand how impossible it is, think about how you would regulate your own temperature or the composition of your blood. Those with failing kidneys know the never-ending daily difficulty of adjusting water, salt and protein intake. The technological fix of dialysis helps, but is no replacement for living healthy kidneys.

My new book *The Revenge of Gaia* expands these thoughts, but you still may ask why science took so long to recognize the true nature of the Earth. I think it is because Darwin's vision was so good and clear that it has taken until now to digest it. In his time, little was known about the chemistry of the atmosphere and oceans, and there would have been little reason for him to wonder if organisms changed their environment as well as adapting to it.

Had it been known then that life and the environment are closely coupled, Darwin would have seen that evolution involved not just the organisms, but the whole planetary surface. We might then have looked upon the Earth as if it were alive, and known that we cannot pollute the air or use the Earth's skin — its forest and ocean ecosystems — as a mere source of products to feed ourselves and furnish our homes. We would have felt instinctively that those ecosystems must be left untouched because they were part of the living Earth.

So what should we do? First, we have to keep in mind the awesome pace of change and realize how little time is left to act; and then each community and nation must find the best use of the resources they have to sustain civilization for as long as they can. Civilization is energy-intensive and we cannot turn it off without crashing, so we need the security of a powered descent. On these British Isles, we are used to thinking of all humanity and not just ourselves; environmental change is global, but we have to deal with the consequences here in the UK.

Unfortunately our nation is now so urbanized as to be like a large city and we have only a small acreage of agriculture and forestry. We are dependent on the trading world for sustenance; climate change will deny us regular supplies of food and fuel from overseas.

We could grow enough to feed ourselves on the diet of the Second World War, but the notion that there is land to spare to grow biofuels, or be the site of wind farms, is ludicrous. We will do our best to survive, but sadly I cannot see the United States or the emerging economies of China and India cutting back in time, and they are the main source of emissions. The worst will happen and survivors will have to adapt to a hell of a climate.

Perhaps the saddest thing is that Gaia will lose as much or more than we do. Not only will wildlife and whole ecosystems go extinct, but in human civilization the planet has a precious resource. We are not merely a disease; we are, through our intelligence and communication, the nervous system of the planet. Through us, Gaia has seen herself from space, and begins to know her place in the universe.

We should be the heart and mind of the Earth, not its malady. So let us be brave and cease thinking of human needs and rights alone, and see that we have harmed the living Earth and need to make our peace with Gaia. We must do it while we are still strong enough to negotiate, and not a broken rabble led by brutal warlords. Most of all, we should remember that we are a part of it, and it is indeed our home.

Key words

system	equilibrium
open system	negative feedback
closed system	positive feedback
isolated system	transfers
first law of thermodynamics	transformations
second law of	flows
thermodynamics	storages
entropy	models

5 Biodiversity

Key points

- We do not know how many species are alive on Earth but we do know that many are becoming extinct or are endangered.
- Some species are more prone to extinction than others.
- Threatened species are on the IUCN Red List.
- The mechanism for speciation is isolation of populations and natural selection.
- Movement of the Earth's plates over geological time has increased biodiversity.
- Biodiversity is lost through natural hazards, habitat degradation, agriculture, introduction of non-native species, pollution, hunting and harvesting.
- Tropical rainforests are the biome with the most biodiversity and they are under threat.

"It should not be believed that all beings exist for the sake of the existence of man. On the contrary, all the other beings too have been intended for their own sakes and not for the sake of something else."

Maimonides, *The Guide for the Perplexed* 1:72, c. 1190

Total world biodiversity

How many species are there on Earth today? That should be a straightforward question as you might expect that we have explored just about every region of the Earth and logged and catalogued what is there. But, in fact, we have very little idea of how many groups of organisms there are and certainly no clear idea of how many are becoming extinct.

Between about 1.4 and 1.8 million species of organisms have been described and named, so are "known to science". How we reach an estimate of the total number of species depends on the characteristics of each species. It is easier to see large animals than smaller ones. Big, furry animals grab our attention. It is relatively easy to identify large animals or animals that do not run away. Most mammals and birds are known. It is also relatively easy to study plants as they cannot move. But many groups of smaller organisms such as insects, nematodes, fungi and bacteria have not been found, identified and named. Some experts think that there are 50 times more species on Earth than have been named, making up to 100 million species. Others think there are about 8 to 10 million species on Earth. That is quite a range of estimates. Beetles (Coleoptera) are the group with the most identified and named species. Beetles are insects and comprise about 25% of all named species; the estimated total insect species number is between 5 and 8 million. One way of finding and identifying insects is by "fogging" the canopies of rainforest trees with short-lived insecticides. Then the organisms fall out of the trees and are collected and counted and the numbers per unit area are estimated.

Groups	Species	Total estimated species	Percentage identified and named
Animals			
Vertebrates	46 500	50 000	93
Molluscs	70 000	200 000	35
Arthropods: Insects	840 000	8000 000	11
Arachnids	75 000	750 000	10
Crustaceans	30 000	250 000	12
Protozoa	40 000	200 000	20
Algae	15 000	500 000	3
Fungi	70 000	1 000 000	70
Plants	256 000	300 000	85

Table 5.1 Numbers of identified species and total estimated species for various groups

But the number of species alive on Earth has not been constant over geological time. Some become extinct while others evolve into new species. Extinction, when a species ceases to exist after the last individual in that species dies, is a natural process. Eventually all species become extinct. The average lifespan for a species varies – most mammals have a species lifespan of 1 million years; some arthropods 10 million years or more. The rate at which extinctions occur is not constant and depends on the background extinction rate as well as mass extinctions, when a sudden loss of species occurs in a relatively short time.

Biodiversity headlines

Fig. 5.1 A golden-mantled tree kangaroo

"Lost World" of New Species Found in Indonesia (2005)

The golden-mantled tree kangaroo is one of dozens of species discovered in late 2005 by a team of scientists on the island of New Guinea.

Scientists find 24 new species in Suriname (2007) including a fluorescent purple toad and 12 kinds of dung beetles

New jellyfish species found in the Monterey Submarine Canyon at depths of 645 meters (2003)

Two new species of wobbegongs, otherwise known as carpet sharks, have been found in Western Australian waters (2008)

Question

Does it surprise you that we know so little about life on Earth?

Background and mass extinctions

The background extinction rate is the natural extinction rate of all species. We think it is about one species per million species per year, so between 10 and 100 species per year. However, we do not know how many species there are alive today, so this rate is estimated from the fossil record. Some species will become extinct before we even know that they exist. We think there are about 5000 mammal species alive today. Their background extinction rate would be one per 200 years but the past 400 years have seen 89 mammalian extinctions, much more than the background rate. Another 169 mammal species are listed as critically endangered or the "living dead". These species have such small populations that there is little hope they will survive. They may have lost a species that they depend on, e.g. a pollinator insect for a flowering plant, a prey species for a predator. E. O. Wilson, a well-known biologist from Harvard, thinks that the current rate of extinction is 1000 times the background rate and that it is caused by human activities. He suggests that 30–50% of species could be extinct within 100 years. The rate is estimated to be about three species per hour. But the rate is not equally spread over the Earth; it is far greater in some areas called hotspots, where species are more vulnerable to extinction – tropical rainforests are one hotspot and coral reefs are another. Some species are more likely to become extinct than others. Those we humans like to hunt or eat or wear or those that are dangerous to us or our crops may be more vulnerable to extinction than others.

If species are becoming extinct at a rate far greater than the background rate, what else is causing this? Mass extinctions have occurred over geological time and there have been five major ones. We know this from the fossil record when fossils suddenly disappear from the rock strata and there is an abrupt increase in rates of extinction. We think this is due to a rapid change of climate, perhaps a natural disaster (volcanic eruption, meteorite impact) which results in many species dying out as they cannot survive the change in conditions. Most biologists now think we are in the sixth mass extinction, called the **Holocene extinction event**. The Holocene epoch is the part of the Quaternary geological period we are in now. This extinction event started at the end of the last ice age about 10 000 years ago when large mammals such as the woolly mammoth and sabre-toothed tiger became extinct, probably through hunting. But the extinction rate has accelerated in the last 100 years. While it may be due to climate change, the big difference is that it has been caused by one species, *Homo sapiens*.

A little bit about the geological timescale

The Earth formed about 4.6 billion (4.6×10^9) years ago. The first life forms are thought to have been simple cells like bacteria, living about 4 billion years ago. Some 65 million years ago, the dinosaurs became extinct. The human species has been recognizably human for the last 200 000 years. To put this timescale into perspective, if you consider geological time as a 24-hour day, humans appeared a few seconds before midnight.

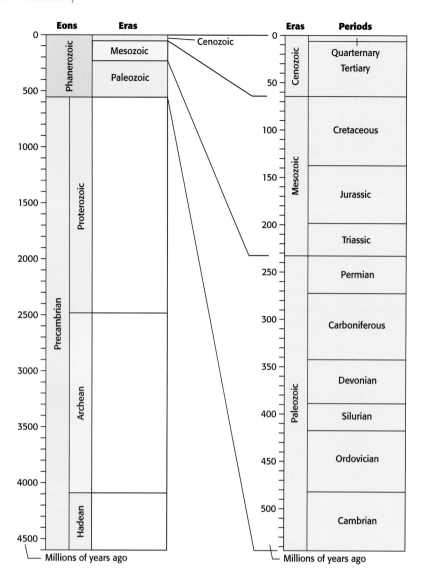

Fig. 5.3 The geological timescale

The last five mass extinctions were spread over 500 million years. You may have heard of some of them, particularly the one when the dinosaurs became extinct. This was the Cretaceous-Tertiary or K-T extinction, about 65 million years ago. In this mass extinction, most of the large animals on land and sea as well as small oceanic plankton died out, but most small animals and plants survived. The causes have been argued about for years but the general view now is that it was either a result of climate change over a long period or caused by a volcanic eruption or meteor impact putting huge amounts of dust into the atmosphere. The evidence for the impact is the Chicxulub crater in the Yucatan peninsula, Mexico. The crater is 180 km in diameter and the igneous rock underneath contains high levels of iridium which is not normally found on Earth. The volcanoes of the Deccan plateau in what is now India erupted for a million years at the time of the K-T boundary. Both these events would have caused dust clouds to block much of the incoming solar radiation so plants would have been unable to photosynthesize and so would have died, and food webs would have collapsed.

All the mass extinctions are listed in Table 5.2.

Mass extinction	MYA (million years ago)	Geological period	Estimate of losses
6th	now	Holocene	unknown
5th	65	Cretaceous-Tertiary	17% of all families and all large animals including dinosaurs
4th	199–214	End Triassic	23% of all families, some vertebrates
3rd	251	Permian-Triassic	95% of all species, 54% of all families
2nd	364	Devonian	19% of all families
1st	440	Ordovician-Silurian	25% of all families

Table 5.2 The six mass extinctions

After each of the previous five mass extinctions, there was a burst of adaptive radiation (when one species rapidly gave rise to many species) to fill the ecological niches left vacant. This did not happen overnight but over tens of millions more years.

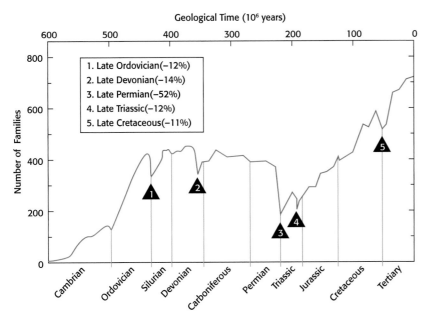

Fig. 5.4 The five major extinction episodes of life on Earth shown by family diversity of marine vertebrates and invertebrates. (after E. O. Wilson, 1988)

The sixth mass extinction

We think that the current, sixth mass extinction, which is caused by the actions of humans, may be far greater than any in the past. And the rate of this extinction is far greater too.

We have wiped out many large mammal and flightless bird species – woolly mammoths and ground sloths, moas in New Zealand, dodos in Mauritius. One estimate is that 25% of all plant and animal species will be extinct between 1985 and 2015.

The UN estimated 12 October 1999 to be the day when there were 6 billion people alive on Earth. The estimate for 12 October 2008 was 6.7 billion. Humans alter the landscape on an unprecedented scale. Some organisms do well in the environments that we create (urban rats, domesticated animals, some introduced species) but most do not. We call the successful ones "weedy" species, both animals and plants. Many weedy species will probably survive, and thrive, in the current mass extinction. But others, many never identified, are likely to die out. The question we should ask is whether we are a weedy species or not. It has taken 5 to 10 million years for the planet's biodiversity to recover after past mass extinctions. That would be more than 200 000 generations of humankind and there have been about 7500 generations since the emergence of *Homo sapiens*. Within this, human civilization started only 500 generations ago.

The previous mass extinctions were due to physical (abiotic) causes over long timespans, often millions of years. The current mass extinction is caused by humans so has a biotic cause and is accelerating. Humans are the direct cause of ecosystem stress because we:

● transform the environment – with cities, roads, industry, agriculture
● overexploit other species – in fishing, hunting and harvesting
● introduce alien species (species moved by humans to areas where they are not native) – which may not have natural predators
● pollute the environment – which may kill species directly or indirectly.

WWF, the Worldwide Fund for Nature (www.panda.org) produces a periodic report called the **Living Planet Report** on the state of the world's ecosystems. The 2006 report painted a grim picture of loss and degradation.

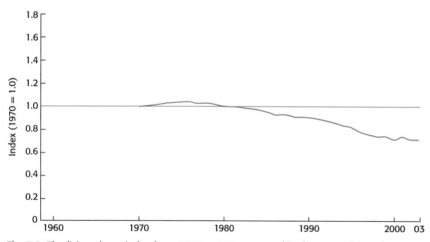

Fig. 5.5 The living planet index from 1970 to 2003 reported in the *WWF Living Planet Report* 2006

The Living Planet Index measures trends in the Earth's biological diversity. It follows populations of 1313 vertebrate species (fish, amphibians, reptiles, birds, mammals) from all around the world and so gives a numerical index of changes in biodiversity. It fell about 30% between 1970 and 2003.

One estimate of about 30 000 species a year becoming extinct may even be an underestimate.

There are two phases to the sixth extinction. The first happened when modern humans spread over the Earth about 100 000 years ago. The second started when humans became farmers about 10 000 years ago when there were about one to ten million humans living on Earth. In the first phase, *Homo sapiens* outcompeted Neanderthals in Europe and probably led to their extinction. In North America, the arrival of humans about 12 500 years ago resulted in the hunting to extinction of woolly mammoths and mastodons (elephant-like prehistoric animals). In Australia, the megafauna (largest animals) disappeared after the arrival of humans 40 000 years ago. In Madagascar, where humans only colonized 2000 years ago, the same disappearance of larger animals occurred soon after their arrival. Only in Africa did the large mammals survive. What do you think the reasons for that may be?

In the second phase, the invention of agriculture meant that humans did not compete with other species in the same way for food but could manipulate their environment and other species for their own use. By importing goods to where they were living, they could also exceed the carrying capacity (how many individuals of a population the environment can support) of the local environment and so live in larger communities. This meant that humans could concentrate on using a few species – crop plants and domestic animals – and consider the others as weeds or pests.

Hotspots

Biodiversity is not equally distributed on Earth. Some regions have more biodiversity than others – more species and more of each species than in other areas. These are hotspots where there are also unusually high numbers of **endemic species** (those only found in that place). About 30 hotspot areas have been recognized. They include about ten in tropical rainforests but also include areas in most other biomes. What they all have in common is that they are threatened areas where 70% of the habitat has already been lost and which contain more than 1500 species of plants which are endemic. These hotspots cover only 2.3% of the land surface. They tend to be nearer the tropics and are often tropical forests because there are more limiting factors as latitude increases. Hotspots tend to have large densities of human habitation nearby.

To do

Hotspot questions

Hotspot	Plant species	Endemic plant species	Endemics as a percentage of world total
Atlantic Forest, Brazil	20 000	8 000	2.7
California Floristic Province	3 488	2 124	0.7
Cape Floristic Region, South Africa	9 000	6 210	2.1
Caribbean Islands	13 000	6 550	2.2
Caucasus	6 400	1 600	0.5
Cerrado	10 000	4 400	1.5
Chilean Winter Rainfall - Valdivian Forests	3 892	1 957	0.7
Coastal Forests of Eastern Africa	4 000	1 750	0.6
East Melanesian Islands	8 000	3 000	1.0
Eastern Afromontane	7 598	2 356	0.8

Guinean Forests of West Africa	9 000	1 800	0.6
Himalaya	10 000	3 160	1.1
Horn of Africa	5 000	2 750	0.9
Indo - Burma	13 500	7 000	2.3
Irano - Anatolian	6 000	2 500	0.8
Japan	5 600	1 950	0.7
Madagascar and the Indian Ocean Islands	13 000	11 600	3.9
Madrean Pine-Oak Woodlands	5 300	3 975	1.3
Maputaland - Pondoland - Albany	8 100	1 900	0.6
Mediterranean Basin	22 500	11 700	3.9
Mesoamerica	17 000	2 941	1.0
Mountains of Central Asia	5 500	1 500	0.5
Mountains of Southwest China	12 000	3 500	1.2
New Caledonia	3 270	2 432	0.8
New Zealand	2 300	1 865	0.6
Philippines	9 253	6 091	2.0
Polynesia - Micronesia	5 330	3 074	1.0
Southwest Australia	5 571	2 948	1.0
Succulent Karoo	6 356	2 439	0.8
Sundaland	25 000	15 000	5.0
Tropical Andes	30 000	15 000	5.0
Tumbes - Choc	11 000	2 750	0.9
Wallacea	10 000	1 500	0.5
Western Ghats and Sri Lanka	5 916	3 049	1.0

Table 5.3 Biodiversity hotspots showing the total and endemic plant species

1 How many of these hotspots can you locate on a world map?
2 Why is the number of plant species important for biodiversity?
3 What are the criteria for defining hotspots?
4 What is an endemic species?
5 For what reasons do you think these hotspots are threatened areas?

Keystone species

All species are not equal in that some have a bigger effect on their environment than others, regardless of their abundance or biomass. They act as the keystone in an arch, holding the arch together. This means that their disappearance from an ecosystem can have an impact far greater than and not proportional to their numbers or biomass. The loss of a small population of a keystone species could destroy the ecosystem or imbalance it greatly, far more than the loss of other species. The difficulty for researchers is in identifying the keystone species. They are often predators or engineer species in the ecosystem (transforming the ecosystem by their actions). A small predator can keep a herbivore population in check, without which the herbivores would increase and eat all the producers, so causing the loss of all food in the ecosystem. An example of this is the sea otter eating sea urchins in kelp forests. If there are no sea otters, the urchins need only eat the holdfast (anchor) of the kelp and it floats away. Beavers are engineers, making dams which turn a stream into a swampy area. Swamp-loving species then move in. Without the beaver and its dams, the habitat changes. Elephants in the African savanna are also engineers, removing trees, after which grasses can grow.

Types of diversity

So far we have talked about biodiversity as the numbers of species of different animals and plants in different places. But it is more than this. **Species diversity** is both the range and number of organisms found in a single place or habitat – how many species and how many of each.

For example, a forest has 15 different species with 100 individuals of one species and one individual of each of the other 14 species. Another forest also has 15 species, but this one has seven individuals of each of the 15 species. The first forest has more individuals than the second one, *but* it is less diverse – has lower biodiversity. In the second forest, the total number of individuals is spread more evenly between each species.

In terms of biodiversity, the spread of individuals between species is more important than the total number of individuals in a habitat.

Biodiversity may be considered at three levels which are **genetic diversity, species diversity** and **habitat diversity**.

Genetic diversity is the range of genetic material present in a species or a population, the gene pool. Domestication and plant breeding lead to a loss of genetic variety, hence the importance of "gene banks". Genetic diversity reflects the genetic variety that exists within a species, which determines the amount of variation between different individuals. A small population normally has a lower genetic diversity than a larger one. However, humans can alter this by artificially breeding or genetically engineering populations with reduced variation in their genotypes, or even producing identical genotypes – clones. This can be an advantage if it produces a high-yielding crop or animal but a disadvantage if disease strikes and the whole population is susceptible.

Species are made up of both individuals and populations. Naturally, each individual in a species has a slightly different set of genes from any other individual in that species. And if a species is made up of two or more different populations in different places, then each population will have a different total genetic make-up. Therefore to conserve the maximum amount of genetic diversity, different populations of a species need to be conserved.

Not all species have the same amount of genetic diversity. Almost the entire world population of grey seals exist on the Farne islands off the northeast coast of England with a few small scattered populations in other places. This is an example of an organism with a small amount of genetic diversity.

Organisms with a large genetic diversity include the European red fox, which is found right across Europe, and also humans, as we exist planet-wide in many populations.

Species diversity is the number of different species within a given area or habitat. It usually includes how many species there are and how many individuals of each species – their abundance. Species diversity varies from habitat to habitat. Some habitats such as coral reefs and rainforests have high species diversity. Urban habitats and polar regions have much lower species diversity by comparison.

Habitat diversity is the number of different habitats per unit area that a particular ecosystem or biome contains. Tropical rainforests are high in habitat diversity as there are many ecological niches. Tundra has a lower level of habitat diversity.

Diversity index

A quantitative method of estimating biodiversity is the Simpson's diversity index which is a measure of species diversity. It measures both number of species present (species richness) and the abundance of each (species evenness). If there are two ecosystems, one with 300 individuals of species A, 29 of species B and 1 of species C, it would be less "even" than the ecosystem with 110 individuals of each of species A, B and C. So it would be less diverse.

In fact there are three related Simpson's diversity indices. You should know the one called **Simpson's reciprocal index** in which 1 is the lowest diversity and a higher value means greater diversity. The maximum value is equal to the number of species in the sample. (See Chapter 16.)

$$D = \frac{N(N-1)}{\Sigma\, n(n-1)}$$

where N = the total number of organisms of all species and

n = the total number of organisms of a particular species.

How new species form

Charles Darwin proposed the theory of evolution which is outlined in *The Origin of Species*, published in 1859. Evidence, in the form of the fossil record, discovery of the structure of DNA and mechanisms of mutations, supports the theory. It is summarized below.

Species are formed by gradual change over a long time. This is called **speciation**. When populations of the same species become separated, they cannot interbreed and may start to diverge if the environments they inhabit change. Separation may have geographical or reproductive causes. Humans can speed up speciation by artificial selection of animals and plants and by genetic engineering but the natural process of speciation is a slow one.

Each individual is different (apart from identical twins) due to their particular set of inherited genes and to mutations. Each will be slightly differently adapted (or fitted) to its environment. Resources are limited for any population and there will be competition for these resources, e.g. for food, light, space, water. For example, a giraffe with a slightly longer neck than the other giraffes may be able to reach tree leaves that are out of reach of the others and so get more food. This gives that giraffe an advantage. These small differences mean that some individuals will be more successful and likely to survive and to breed than others and so to pass their genes on to the next generation. Over time the population gradually changes. This is **natural selection** where those more adapted to their environment have an advantage and those less adapted do not survive, flourish and reproduce. So the fittest survive; "survival of the fittest" is a term you may have heard.

Over many generations, if a population is separated from others, the differences may increase to such an extent that, should the populations be reunited, they will be unable to interbreed. Then a new species has formed. This isolation may occur on an island, on a mountain or in a body of water. Isolation can also occur in populations that can mix freely if their mating seasons are not synchronized or their flowers mature at different times.

Physical barriers

Species may develop into two or more new species if their population is split by some kind of physical barrier, e.g. a mountain range or ocean. The physical barrier will split the gene pool: the genes of the populations on both sides of the barrier will not be mixing any more and the two populations can develop in different directions. Examples of speciation due to physical barriers:

1 Large flightless birds (e.g. emu, cassowary, ostrich, rhea) only occur on those continents that were once part of Gondwana (Africa, Australia including New Zealand, South America). However, because Gondwana split up a very long time ago, the large flightless birds are not closely related.

2 Australasia (and Antarctica) split off from Gondwana a long time ago. At that time, both marsupials (see page 105) and early placental mammals lived in the same area. In South America and Africa, the placental mammals prevailed and outcompeted the marsupials. In Australia the opposite happened and now marsupials are only found in Australia and nearby Papua New Guinea. Placental mammals are rare in Australia. These mammals came by sea (seals) or by air (bats) or were introduced by humans (dogs, cats, rabbits and rats).

3 The cichlid fish in the lakes of East Africa are one of the largest families of vertebrates. In Lake Victoria there are 170 species of cichlids (99% endemic); in Lake Tanganyika 126 species (100% endemic); in Lake Malawi 200 species (99% endemic). These populations have probably been isolated from each other for millions of years and have different selection pressures within them due to their slightly different environments. So the fish have adapted to the different environments. As long as the population is large enough, isolated populations may continue to diverge from other populations, but some isolated populations become too small and die out.

Land bridges allow species to invade new areas. North and South America have been separated for a long time and have therefore rather different species. However, the relatively recent land bridge (Central America) formed by continental drift between these continents allowed species to move to the North or South. Bears for example moved from North to South America. Land bridges may also be the result of lowered seawater levels. Examples include the (now disappeared) land bridges between England and Europe, and between North America and Asia (Bering Straits).

Continental drift (see below) has also resulted in new and diverse habitats. As the original supercontinent (called Pangaea) broke up, the resulting continents have drifted to different climate zones. The changing climatic conditions forced species to adapt and resulted in an increased biodiversity. A dramatic example of this is the changes on Antarctica. Once this continent had a tropical climate and was covered by forest. When Antarctica moved southwards, the forest gradually disappeared and the snow and ice-covered landscape was formed and new, cold-adapted species arrived or evolved. The only remains of the former forest are some tree fossils.

How plate activity influences biodiversity

The planet Earth consists of different layers – the core, mantle and crust. The outer core is hot, molten rock and convection currents within it and the molten inner mantle cause the more solid outer mantle and crust (together called the lithosphere) to move or float. The lithosphere is divided into seven large tectonic plates and many smaller ones and these drift around, moving at about 50 to 100 mm per year. This is called **continental drift** and the study of the movement of the plates is **plate tectonics**. Where the plates meet, they may either slide past each other (e.g. San Andreas fault line, California), diverge (e.g. mid-Atlantic ridge where the plates are moving slowly apart) or converge. Convergent plates may either collide and both be forced upwards as mountains (e.g. formation of the Himalayas and Alps) or collide and one (the heavier oceanic plate) sinks underneath the lighter continental plate. This results in a subduction zone where deep ocean trenches and volcanic island chains are formed (e.g. where the Nazca plate under the Pacific meets the west coast of South America and the Andes form).

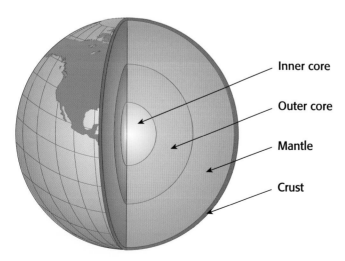

Inner core
Outer core
Mantle
Crust

Fig. 5.6 Simplified diagram of the Earth

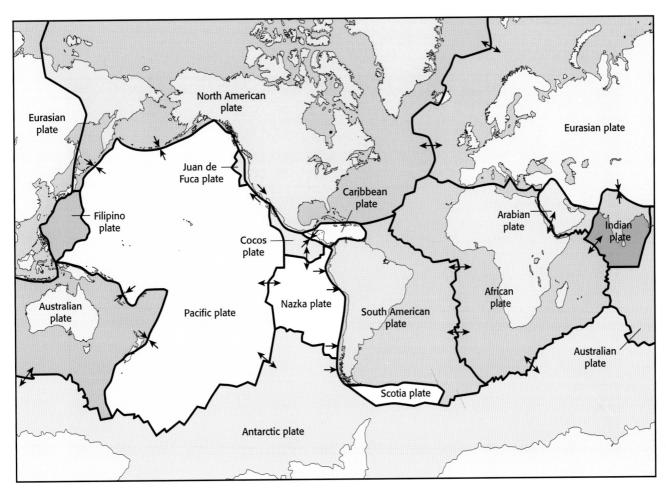

Fig. 5.7 The plates of the Earth

To think about

Jigsaws and continents

Looking at a world map or a globe, you may notice that the west coast of the African continent and the east coast of the South American continent appear to fit together like pieces of a jigsaw. Many others noticed this in the past since the first world maps were drawn. Abraham Ortelius, from Antwerp, noticed (see Fig. 5.8). He was a cartographer and drew one of the first atlases in 1570. Francis Bacon (1620), a British philosopher and scientist, did too as did Benjamin Franklin (USA) and Antonio Snider-Pellegrini (France) who in 1858 drew a map of how they may have fitted together. In 1912, Alfred Wegener also pointed out that fossils on the west coast of Africa were similar to those on the east coast of South America, and coal fields in Europe matched those in North America. He proposed that this was because the continents had once been joined and had floated apart. This was thought outrageous and few

supported his hypothesis; many people heaped criticism on it. In 1930, Wegener died during an expedition to the Arctic and still no one believed his hypothesis. It took until the late 1960s for his idea to be accepted when research was done on the deep ocean floor vents at the edges of the tectonic plates.

For at least a millennium, sailors have made long ocean voyages using the sun and stars to guide them in rudimentary navigation. There is debate about the dates of the earliest maps of the world but Zhu Ziben in the Yuan Dynasty in China may have produced the map in Fig. 5.9 as long ago as 1320. This not only shows the shape of the Earth but also the continents including Australasia, Antarctica and the Americas.

If the shapes of the continents has been known by some for so long, why do you think Wegener was laughed at when he proposed that the continents move? ➔

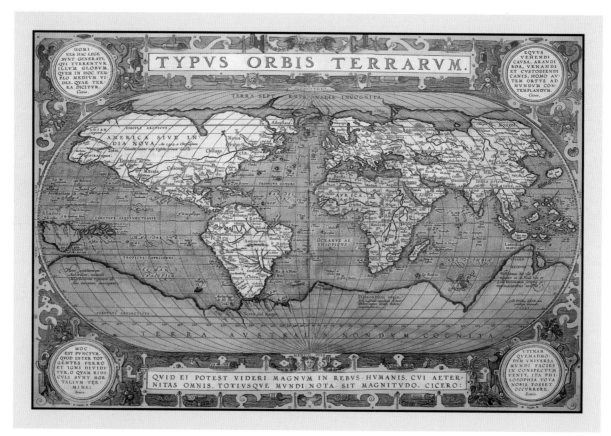

Fig. 5.8 Ortelius' world map, 1570

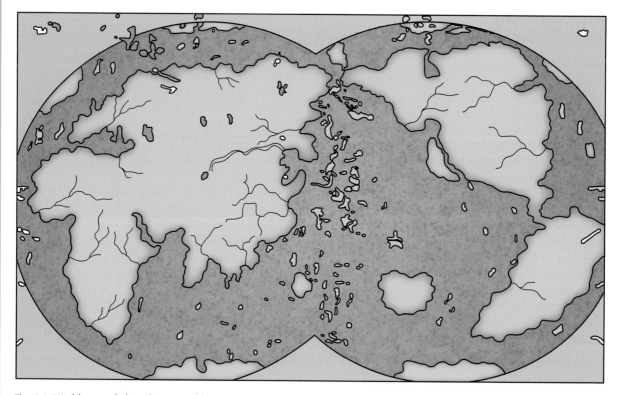

Fig. 5.9 World map of Zhu Ziben, possibly 1320

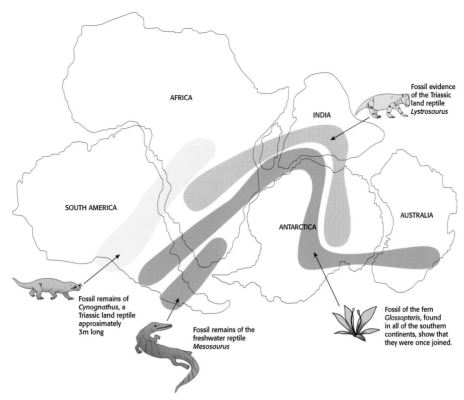

Fig. 5.10 How groups of organisms may have spread between continents

The similarities between groups of animals on various continents has not gone unnoticed. Here are a few. You can probably think of others.

Llamas and camels are both domesticated animals fulfilling similar roles, llamas in South America, camels in Africa (one hump – dromedary) and central Asia (two humps – Bactrian camel). They are distant cousins. Their common ancestor was a rabbit-sized mammal. Both were domesticated about 5000 years ago and are used by humans as pack animals and for meat, milk and an indication of wealth. African and Indian elephants are another example, and there are plant species that also show much similarity between continents.

Kangaroos fulfil a similar ecological role in Australia as cattle do in the rest of the world. They are both large herbivores, eating grass and converting it to meat. Kangaroos are marsupial mammals (as are wombats and koalas). All mammals suckle their young with milk that the mother produces. The placental mammals have a placenta to provide food for the embryo in the uterus, and are the most successful in evolutionary terms. The marsupials have a pouch in which the young embryo – joey – continues to develop. The monotremes (e.g. duck-billed platypus) lay eggs instead of giving birth to live young and are the third group of mammals. Marsupials and monotremes are mostly confined to Australasia and it seems that the placental mammals outcompeted the other two groups in most other continents.

Factors that help to maintain biodiversity

Complexity of the ecosystem

The more complex a food web, the more resilient it is to the loss of one species or reduction in its population size. If one type of prey or food source or predator is lost, the others will fill the gaps left. This resilience of more complex communities and ecosystems is a good thing for biodiversity overall. But it may not maintain species diversity if one species is lost while the community continues.

Stage of succession

When plants and animals colonize a bare piece of land, there are few species at first. Species diversity increases with time until a climax community (see Chapter 14) is reached when the species composition is stable. It may fall slightly once the climax community is reached but, normally, species diversity increases as succession proceeds. So communities in young ecosystems that are undergoing succession may be more vulnerable than those in older ones which are more resilient and stable.

Limiting factors

If it is difficult for the organisms in an ecosystem to get enough raw materials for growth, e.g. water is limiting in a desert, then any change that makes it even harder may result in species disappearing. If the abiotic factors required for life are available in abundance (water, light, heat, nutrients), then if one is reduced, species may withstand the reduction.

Inertia

Inertia is the property of an ecosystem to resist change when subjected to a disruptive force. Along with resilience (ability to recover from change or disruption) and stability (maintaining equilibrium), it is key to helping environmental managers know which site will either resist change or recover most quickly.

Factors that lead to loss of biodiversity

Natural hazards include events such as volcanic eruptions, earthquakes, landslides, natural fires, avalanches, tsunami (Fig. 5.11), drought. These are outside the control of humans. We can mitigate their effects but are unlikely to stop them happening. They can cause natural disasters such as the eruption of Mount St Helens (Fig. 5.12) in Washington State, USA in 1980, the Yangtze River floods in China that left 14 million people homeless, the southern Asia floods of 2007 and Hurricane Katrina in New Orleans, USA in 2005.

Environmental disasters are usually thought of as caused by human activity and would include loss of tropical rainforest on a massive scale and oil spills.

The major cause of loss of biodiversity is **loss of habitat**. There is loss of diversity on a small and large scale due to human activities. In many parts of the world, humans have destroyed or changed most of the natural, original habitat. Habitat destruction and degradation occur when we develop or build on a piece of land. In the Philippines, Vietnam, Sri Lanka and Bangladesh, where human population levels are high, we have lost most of the wildlife habitat and most of the primary rainforest. In the Mediterranean region, only 10% of original forest cover remains. In Madagascar, by 2020, it is reckoned that no tropical rainforest will be left except for the small area under protection. Madagascar is the only place where

Fig. 5.11 Tsunami wave – a natural hazard

lemurs occur and there are many endemic species. Loss of their habitat is highly likely to make them extinct.

Fragmentation of habitat is the process whereby a large area is divided up into a patchwork of fragments, separated from each other by roads, towns, factories, fences, power lines, pipelines, fields. The fragments are isolated in a modified or degraded landscape and they act as islands within an inhospitable sea of modified ecosystems. There are edge effects to the islands with higher edge-to-area ratios as the fragments get smaller and there are greater fluctuations of light, temperature, humidity at the edge than the middle. Invasion of the habitat by pest species or humans increases and there is the possibility that domestic and wild species come into contact and spread diseases between populations.

Pollution caused by human activities can also degrade or destroy habitats. Local pollution, e.g. spraying of pesticides, may drift into wild areas; oil spills may kill many seabirds and smaller marine species. Environmental pollution of emissions from factories and transport can lead to acid deposition or photochemical smog. Run-off of fertilizers into waterways can cause eutrophication; toxic chemicals can accumulate in food chains. Climate change alters weather patterns and shifts biomes away from the equator. All degrade habitats and may make them unsuitable to support the range of species that a pristine ecosystem can support.

Overexploitation has escalated as human populations have expanded and technology has allowed us to get better and better at catching and hunting and harvesting. Chainsaws have replaced handsaws in the forests, factory ships with efficient sonar and radar find and process fish stocks, and bottom trawling scoops up all species of fish whether we then eat them or not. We are just too good at getting our food. The Grand Banks off Newfoundland were one of the richest fishing grounds in the world but are now fished out. If we exceed the maximum sustainable yield of any species (the maximum which can be harvested each year and replaced by natural population growth), then the population is not sustainable. The difficulty is knowing what this figure is and how much is left. But growing rural poverty mixed with improved methods of hunting and harvesting mean that more and more humans living at a subsistence level overexploit the environment. With lower population densities, traditional customs and practices prevented overexploitation, but faced with the choice between starvation and eating bush meat (wild animal meat) or felling the last tree, there is no real choice.

Introducing non-native (exotic) species can drastically upset a natural ecosystem. Humans have done this through colonization of different countries, bringing their own crops or livestock. Sometimes it works – potatoes from the Americans to Europe, rubber trees from the Amazon to South-East Asia. Sometimes it is a disaster. Sometimes the species is introduced by accident or escapes from gardens or zoos. Rhododendrons were introduced to Europe from Nepal by plant collectors as they have large flowers. But they have escaped into the wild and taken over many areas as they

Fig. 5.12 Mount St Helens eruption in 1980

outcompete the native plants and are toxic. Dutch elm disease came from imported American logs to Europe and decimated elm populations. Australia has been particularly unfortunate with rabbits, cane toads, red foxes, camels, blackberry, prickly pear, crown of thorns starfish, to name just a few. The very different flora and fauna of Australia was well adapted to the environment but unable to compete with the aggressive invasive species.

Spread of disease may decrease biodiversity. The last population of black-footed ferrets in the wild in North America was wiped out by canine distemper in 1987 and the Serengeti lion population is reduced due to distemper. Diseases of domesticated animals can spread to wild species and vice versa, particularly if population densities are high. In zoos, disease is a constant threat where species are kept close together. Mutation of diseases so they can infect across the species barrier is an ever-present consideration. Bird flu passing from birds to humans and BSE passing from cattle to humans are two current issues for the human species.

Modern agricultural practices also reduce diversity with monocultures, genetic engineering and pesticides. Fewer and fewer species and varieties of species are grown commercially and more pest species are removed.

To do

Rabbits in Australia

Fig. 5.13 Rabbits at a watering hole in Australia

Rabbits normally get their water from the food that they eat but they had overgrazed so much here that they had to drink from the watering hole.

Rabbits are not indigenous to Australia. In 1859, rabbits were introduced into Australia for sport and for meat by shipping a few rabbits from Europe. With no predator there, they multiplied exponentially and after ten years, there were estimated to be two million. The rabbits ate all the grass and forage so there was none for sheep and the farming economy collapsed. Rabbits caused erosion as the topsoil blew away but also probably caused the extinction of many Australian marsupials which could not compete with them.

It was so bad that in 1901-7, Western Australia built a rabbit-proof fence from north to south over 3200 km long to keep the rabbits out. Other control methods are shooting, trapping and poisoning.

In 1950, the **myxomatosis virus** was brought from Brazil to control the rabbits and released. The epidemic that followed killed off up to 500 million of the 600 million rabbits. But the rabbits were not eliminated as a very few were resistant to the virus. They bred and now rabbits are again a problem with the population recovering to 200–300 million within 40 years.

In 1996, the government released rabbit hemorrhagic disease or RHD as a second disease to attempt to control the rabbits. But neither of these biological control measures has eliminated the rabbits.

1 What factors contributed to the success of the rabbits in Australia?

2 Find out what a biological control measure is and give two examples.

3 Research the introduction of cane toads and prickly pear (*Opuntia*) into Australia.

4 Rabbits, cane toads and prickly pear are all pests in Australia. What characteristics do they have in common to make them successful pests?

Vulnerability of tropical rainforests

Over half of all species of plants and animals alive on the Earth today live in tropical rainforests, yet these forests cover only 6% of the land area of the planet. They also produce about 40% of the oxygen that animals use. They have been called the lungs of the Earth. One hectare of rainforest may have 300 species living in it. Overall tropical rainforests have high species diversity and habitat diversity yet some areas have even more than others. These are called hotspots. Malaysia has one of the most species-rich rainforests. These are 8000 species of flowering plants and many Dipterocarp trees (e.g. mahogany) which have great commercial value. Denmark, of similar size to Malaysia, has 45 species of mammals compared with Malaysia's 203.

Fig. 5.14 Tropical rainforest in Sabah, Malaysia showing canopy and emergent tree

Tropical rainforests have been lost at a massive rate within the last 50 years. They covered up to 14% of the Earth's land surface in 1950 but have been cleared by humans at unprecedented rates. Some estimate that they will be gone completely in another 50 years due to human activity. Perhaps 1.5 hectares of rainforest is cleared per second. There is about 50% of the Earth's timber in tropical rainforests and timber is the next biggest resource after oil in the world today. But ranching and logging commercially are not the biggest threats to the remaining rainforest, it is individual people. At least two billion humans live in the wet tropics and many of these rely on the rainforests for subsistence agriculture. A low density of human population is sustainable as they clear a small area of forest, grow crops for two or three years, then move on to the next area as the soil is exhausted. This is called shifting cultivation. This works as long as there is enough time for the forest to regenerate before the same area is cleared again. With increasing population pressures, the forest does not fully regrow before it is cleared and there is a gradual degradation of nutrients and of biodiversity.

Of course, clearance of the forest does not mean that nothing then grows on that land. There may be animals grazing grasslands, oil palm plantations, subsistence agriculture as well as urban development. Forest cleared and then not managed does regrow in time. But it is estimated that it takes 1000 years for the biodiversity of the primary forest (before logging or clearing) to be recovered and the secondary forest that does grow up is impoverished in many ways.

The rainforests are so diverse because of the many ecological niches they provide. High levels of heat, light and water mean that photosynthesis is rapid and not limited by lack of raw materials. The four layers of the rainforest (see page 38) allow for many different habitats and niches and these are filled by diverse species. All year round growth means that food can be found at any time of the year. Some rainforests are old both in living and geological terms. The lack of disturbance may mean that the system has had time to become more complex.

The fast rate of respiration and decomposition means that the forests appear to be very fertile with high levels of biomass in standing crop (trees and other plants). But most nutrients are held in the plants, not the soil or leaf litter. Once the plants are cleared or burned, fertility reduces rapidly because the heavy rainfall washes the nutrients and soil away and the vegetation is not there to lock up the nutrients nor protect the soil. So more forest is cleared to use the short-term fertility for crop growth.

What makes a species prone to extinction?

Some species are more vulnerable to extinction than others. Even in the same ecosystem, some species survive habitat loss or degradation, others do not. So some species are more likely to be in danger of extinction.

Narrow geographical range

If a species only lives in one place and that place is damaged or destroyed, the habitat has gone. It may be possible to keep breeding populations of the species in zoos or reserves but that is not the real solution. There are many examples of species lost or threatened by this – the golden lion tamarin, birds on oceanic islands, fish in lakes that dry up.

Fig. 5.15 Island species are prone to extinction

Small population size or declining numbers

A small population has a smaller genetic diversity and is less resilient to change, i.e. cannot adapt as well. As individual numbers fall, there is more inbreeding until populations may reduce further and they become extinct. Large predators and extreme specialists are commonly in this category, e.g. snow leopard, tiger, lonesome George (page 93).

Low population densities and large territories

If an individual of a species requires a large territory or range over which to hunt and only meets others of the species for breeding, then habitat fragmentation can restrict its territory. If there is not a large enough area left for each individual or if they are unable to find each other because there is a city, road, factory or farm splitting up the territory, they are less likely to survive as a species.

Few populations of the species

Similarly, the more populations, the more likely a species is to survive. One or more populations may be wiped out but the others continue.

A large body

As we mentioned earlier (page 54), the 10% rule (where only about 10% of the energy is passed on to the next trophic level and 90% is lost to the environment) means that large top predators are rare. Whether aquatic or terrestrial, they tend to have large ranges, low population densities and need a lot of food. They also compete with humans for food (e.g. wolves and tigers), may be a danger to humans and are hunted for sport.

Low reproductive potential

Reproducing slowly and infrequently means the population takes a long time to recover. Whales fall into this category. Many of the larger species of seabirds, e.g. gannets, albatrosses or some species of penguins, only produce one egg per pair per year, and do not breed until several years old.

Fig. 5.16 Flightless fairy penguin

Seasonal migrants

Species that migrate can face problems. Not only do they often have long migration routes (swallows between southern Africa and Europe, songbirds between Canada and Central and South America) and a hazardous journey, they need both habitats. If one is destroyed, they get there to find no food or habitat. Barriers on their journey can prevent them completing it (salmon trying to swim upriver to spawn, large African herbivores following the rains).

Poor dispersers

Species that cannot move easily to new habitats are also vulnerable. Plants can only rely on seed dispersal or vegetative growth to move. That takes a long time and climate change resulting in biome shift may mean the plant dies out before it can move. The flightless birds of New Zealand are mostly extinct as they cannot escape hunters nor fly to another island.

Specialized feeders or niche requirements

The giant panda (about 2000 left) mostly eats bamboo shoots in forests of central China. The koala only eats eucalyptus leaves and lives in the coastal regions of southern and western Australia. Bog plants such as the sundew or pitcher plant can only survive in damp places.

Hunted for food or sport

Overhunting or overharvesting can eradicate a species quickly, especially if that species lives in large groups – herds of bison in North America, shoals of fish, flocks of passenger pigeons. Then, with human technology, it is easier to exploit the species as, once located, many can be caught.

Fig. 5.17 Giant panda

Fig. 5.18 Hunting "big cats"

Fig. 5.19 Minke whale harvest

Populations suffering from any of the above environmental stresses can be prone to extinction. Often though, more than one pressure operates on an organism, or the nearer to extinction an organism becomes, the greater the number of increasing pressures.

Island organisms can be particularly vulnerable to extinction. Dependent on the size of the island, populations tend to be small. Islands have a high degree of endemic species. Genetic diversity tends to be low in small unique island populations. Also island populations tend to be vulnerable to introductions of non-native predators. The island species have no fear of these so they do not flee and are killed. The dodo is a classic example of an island species that became extinct.

Sadly, many of these characteristics of extinction-prone species can often be found in the same species. Large top carnivorous mammals (tigers, leopards, bears, wolves) have many of these characteristics. Species that live in large groups and are an attractive food source (bison, cod) have many too.

The **minimum viable population size** that is needed for a species to survive in the wild is a figure that scientists and conservationists consider. But there is no magic number. It depend on so many factors such as genetic diversity in the individuals left, rate of reproduction, mortality rate, growth rate, threats to habitat and so on. For large carnivores, 500 individuals is sometimes thought of as the absolute minimum below which there is little hope for the species. There may be 400 tigers in Sumatra and about 3500 in the world today; it is estimated that there were 40 000 in India alone at the start of the 20th century). Some 700 blue whales are thought to live in the Antarctic Ocean; it is estimated that there were 250 000 about 60 years ago.

The case studies that follow are about species that are **extinct**, **endangered** or have **recovered** their numbers. A species is extinct if there is no reasonable doubt that the last individual has died. A species is endangered when it faces a very high risk of extinction in the wild. A species has recovered when the decline in numbers of an endangered or threatened species is stopped and the threats to its survival are removed or reduced so that it should survive in the wild. These and other categories of species which are under threat are listed in the IUCN (the International Union for Conservation of Nature) Red List which lists all threatened species (see Chapter 6).

Case study 1

Recovered species

Australian saltwater crocodile

Eighteen species of crocodiles out of 23 worldwide were once endangered but have since recovered sufficiently to be removed from the endangered list. Many species are thriving.

Description

The Australian saltwater crocodile can grow up to 5 m long. It is a bulky reptile with a broad snout. It lays up to 80 eggs each year which take up to 3 months to hatch. The crocodiles take 15 years to mature. The Australian saltwater

Fig. 5.20 Australian saltwater crocodile

crocodile was listed as a protected species in Australia in 1971, and is protected under CITES which banned trade in endangered animals.

Ecological role

The habitat of the saltwater crocodiles is estuaries, swamps and rivers. Nests are built on river banks in a heap of leaves. The eggs are food for goannas, pythons, dingoes and other small animals. Older crocodiles eat young crocodiles, mud crabs, sea snakes, turtle eggs and catfish. Baby crocodiles eat tadpoles, crabs and fish. The Australian saltwater crocodile is a top predator.

Pressures

The saltwater crocodile was overexploited for skin (leather), meat and body parts through illegal hunting, poaching and smuggling. It was hunted for sport. Crocodiles were also deliberately killed because of attacks on humans.

Method of restoring populations

To restore the crocodile populations there was a sustainable use policy with limited culling of wild populations, ranching (collecting eggs and hatchlings from the wild and raising them in captivity) and closed-cycle farming (maintaining breeding adults in captivity and harvesting offspring at 4 years of age). The exploitation of farmed animals reduces the hunting of wild crocodiles. Visitors tour areas to see wild crocodiles, so they are now a valued species. This policy was supported by the Species Survival Commission (SSC) of IUCN but was viewed by others as treating crocodiles inhumanely.

Case study 2

Extinct species

Thylacine (Tasmanian tiger) *Thylacinus cynocephalus*

Fig. 5.21 The last thylacine

Description

The thylacine or Tasmanian tiger was a marsupial, similar in appearance to a wolf, but with a rigid tail. During the last few hundred years it was found only on the Australian island of Tasmania but prior to that it had existed on the mainland. The thylacine was a strong fighter with jaws that could open more widely than any other known mammal. Thylacines had a life expectancy of 12 to 14 years and gave birth to two to four young per year.

Ecological role

The thylacines' habitat was open forest and grassland but they became restricted to dense rainforest as the population declined. Thylacines lived in rocky outcrops and large, hollow logs. They were nocturnal. Their typical prey were small mammals and birds, but they also ate kangaroos, wallabies, wombats and echidnas (spiny anteaters) and sheep.

Pressures

Thylacines competed with dingoes on the mainland of Australia and became extinct there hundreds of years ago. In Tasmania they were hunted by farmers whose stock of sheep was the thylacines' prey. Private and government bounties were paid for scalps, leading to a peak kill in 1900. Bounties continued until 1910 (government) and 1914 (private). There was an intense killing spree by hunting, poisoning and trapping. Shooting parties were organized for tourists' entertainment. The severely depleted population was affected by disease and competition from settlers' dogs.

The last known wild thylacine was believed to have been killed in 1930 and the last captive animal died at the Hobart Zoo on 7 September 1936. The thylacine was legally protected earlier the same year. It was classified by IUCN as endangered in 1972 (as some people thought it was still surviving in the wild in remote areas), and is now listed as extinct.

Consequences of disappearance

The thylacine was a carnivorous marsupial from a unique marsupial family. In Tasmania where there were no dingoes, the thylacine was a significant predator. Introduced dogs have taken over the ecological role of the thylacine.

Case study 3

Endangered species

Rafflesia

Fig 5.22 Rafflesia flower

Description

A tropical parasitic plant in the forests of South-East Asia, parasitic on a vine.

Ecological role

These plants are single sexed (either male or female) and pollination must be carried out when the plants are in bloom (flowering). Therefore a male and female in the same area must both be ready for pollination at same time. The seeds are dispersed by small squirrels and other rodents and they must reach a "host" vine.

Pressures

The Rafflesia plants are vulnerable because they need very specific conditions to survive and carry out their life cycle. They are vulnerable due to deforestation and logging which destroy their habitat. Humans damage them and fewer plants means less chance of breeding.

Methods of restoring populations

In Sabah, Sumatra and Sarawak there are Rafflesia sanctuaries. There are many "locating, protecting and monitoring" programmes being set up, and educating the public through "save Rafflesia" campaigns.

Recovered species – or not?

Golden lion tamarin (GLT)
Leontopithecus rosalia

Fig. 5.23 Golden lion tamarin

Description

Small monkey. Endemic to Atlantic coastal rainforests of Brazil. Among rarest animals in the world. About 1000 in the wild and 500 in captivity. Some say there are only 400 in the wild. Life expectancy about 8 years.

Ecological role

Omnivores. Prey to large cats, birds of prey. Live territorially in family groups in the wild in tropical rainforest in the canopy.

Pressures

Only 2% of their native habitat is left. Poachers can earn US$20000 per skin. Predation is great in the wild and their food source is not dependable, as well as habitat destruction.

Method of restoring populations

A captive breeding programme (breeding in zoos) for the last 40 years or more. Over 150 institutions are involved in this and exchange individuals to increase genetic diversity. Some are reintroduced to the wild but with only a 30% success rate as the habitat is threatened and predators (including humans) take many. It seems unlikely that this species would be alive today if not for captive breeding. The long-term future in the wild is uncertain.

Extinct species

Dodo *Raphus cucullatus*

Fig. 5.24 The dodo

Description

Large flightless bird endemic to the island of Mauritius.

Ecological role

No major predators on Mauritius so dodo had no need of flight. Dodo was a ground-nesting bird.

Pressures

In 1505 Portuguese sailors discovered Mauritius and used it as a restocking point on their voyages to get spices from Indonesia. They ate the dodo as a source of fresh meat. Later, the island was used as a penal colony (gaol) and rats, pigs and monkeys were introduced. These ate dodo eggs and humans killed the birds for sport and food. Crab-eating macaque monkeys introduced by sailors also seem to have had a large impact, stealing dodo eggs. Conversion of forest to plantations also destroyed their habitat. Known to be extinct by 1681.

Consequences of disappearance

Island fauna impoverished by its loss. Became an icon due to its apparent stupidity (just standing there and being clubbed to death – as it had no fear of humans, never having had need to fear predators) and its untimely extinction. A few skeletons remain and a stuffed dodo in an Oxford museum which was thrown out for burning around 1755. A foot and mummified head were salvaged and are now all that remains of the species. "Dead as a dodo" is now a common saying.

Test yourself

1 What is biodiversity?
2 What is an endangered species? List several examples.
3 What are the *two* main reasons for extinction of species?
4 List four examples of *human activity* that threaten species.
5 Consider the case studies of organisms in the previous pages and the list of characteristics that make species more prone to extinction. Make a table to compare the characteristics shared by these species. What do they have in common?

Key words

biodiversity
species diversity
habitat diversity
diversity index
genetic diversity
extinction rates
background extinction rate
mass extinction
hotspots
endemic species
exotic species
plate tectonics
continental drift
keystone species

natural selection
extinct
endangered
vulnerable
lithosphere
core
mantle
crust
subduction
mid-Atlantic ridge
evolution
physical barriers
land bridges

6 Conservation of biodiversity

Key points

- Species and habitats should be preserved for many reasons, based on economic, commercial, ethical and aesthetic grounds.
- Both governmental and non-governmental organizations work to preserve biodiversity and protect ecosystems. They act in different ways but have similar goals.
- Criteria are used to design protected areas based on the principles of island biogeography.
- Species-based conservation has both strengths and weaknesses.

"I wonder what the Maori who killed the last moa said. Perhaps the Polynesian equivalent of 'Your ecological models are untested, so conservation measures would be premature'? No, he probably just said, 'Jobs, not birds', as he delivered the fatal blow."

Jared Diamond, writing about the extermination of the Moa in New Zealand

Setting the scene

The relatively new discipline of conservation biology has brought together experts on various subjects in the last 20 years. All are concerned about the loss of biodiversity and want to act together to save species and communities from extinction. Their rationale is that:

- diversity of organisms and ecological complexity are good things
- untimely extinction of species is a bad thing
- evolutionary adaptation is good
- biological diversity has intrinsic value and is worth conserving.

Some acronyms will help you make sense of this chapter:

UN	United Nations
IUCN	International Union for Conservation of Nature
UNEP	United Nations Environment Programme
CITES	Convention on International Trade in Endangered Species
WCS	World Conservation Strategy
UNESCO	UN Educational, Scientific & Cultural Organization
WWF	World Wide Fund for Nature
WRI	World Resources Institute
NGO	non-governmental organization
GO	governmental organization
MDG	millennium development goals

Fig. 6.1 Deforestation by burning in southern Mexico

Fig. 6.2 Yellowfin tuna for sale in a market in the Maldives

Why conserve biodiversity?

Before you read the section on why conserve biodiversity, think about this yourself for a few moments and make a list of reasons why we should conserve and reasons why we should not. These questions may help you make your lists:

- Do humans need other species? In what ways?

- Do other species exist for human use? Do they only exist for human use?
- Do other species have a right to exist? Does a great ape have more rights than a mosquito?
- Are your reasons based on rational thought or emotion or both? Does this affect how valid they are?

Here are some of the many reasons to conserve biodiversity.

Economic values of biodiversity

One way of classifying the values of biodiversity is into direct and indirect categories. Direct values are easier to calculate as they are goods harvested which are eaten or sold. In economic terms these are private goods. Indirect values are harder to calculate as they provide benefit and the goods are not harvested or destroyed. They may be services or processes and do not appear in a country's economic figures yet are vital for the economy, e.g. climate regulation or water quality. These are public goods in economic terms.

Direct values

Food sources: Humans eat other species both animals and plants. Even though we tend to eat only small number of food crops, there are many varieties of these. We need to preserve old varieties in case we need them in the future. Pests and diseases can wipe out non-resistant strains. Breeders are only one step ahead of the diseases and require wild strains from which they may find resistant genes. Wheat, rice and maize provide one half of the world's food. In the 1960s, wheat stripe rust disease wiped out a third of the yield in the US. It was the introduction of resistant genes from a wild strain of wheat in Turkey that saved the crops. Maize is particularly vulnerable to disease, as it is virtually the same genetically worldwide. A perennial maize was found in a few hectares of threatened farmland in Mexico. The plants also contain genes that confer resistance to four of the seven major maize diseases and could give the potential of making maize a perennial crop that would not need sowing.

Natural products: Many of the medicines, fertilizers and pesticides we use are derived from plants and animals. Guano, which is seabird droppings, is a fertilizer high in phosphate. Oil palms give us oil for anything from margarine to toiletries. Rubber (latex) is from rubber trees, linen from flax, rope from hemp, cotton from cotton plants, and silk from silkworms. Honey, beeswax, rattan, natural perfumes, and timber are all from plants or animals.

Indirect values

Ecosystem productivity: Soil aeration depends on worms. Fertilization and pollination of some food crops depend on insects. Ecosystem productivity gives our environment stability and recycles materials. Plants capture carbon and store it in their tissues. They release oxygen which all organisms need for respiration. Agriculture captures 40% of the productivity of the terrestrial ecosystem. Soil and water resources are protected by vegetation. Climate is regulated by the rainforests and vegetation cover. Waste is recycled by decomposers. Preservation of as many species and as much natural or semi-natural habitat as possible, may render the environment more stable, and less likely to be affected by spread of disease (plant, animal or human) or some other environmental catastrophe.

Scientific and educational value: Humans have studied and classified the natural world and continue to do so. Each year many new species are still being discovered. As we learn more about the interactions of species with each other, we pass this knowledge on in educational programmes. The encyclopaedia of life (see Chapter 20) intends to document all the 1.8 million species that are named on the Earth.

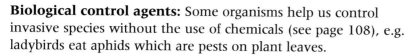

Biological control agents: Some organisms help us control invasive species without the use of chemicals (see page 108), e.g. ladybirds eat aphids which are pests on plant leaves.

Genes: Wild animals and plants are sources of genes for hybridization and genetic engineering.

Environmental monitors: Plants and animals can be environmental monitors as an early warning system. Miners used to take canaries into the mines with them. If the birds died, they knew there was a toxic gas around and they should get out. Indicator species, e.g. lichens, can show air quality.

Recreational: Many people take vacations in areas of outstanding natural beauty and national parks, go skiing, scuba diving and hiking.

Human health: The first antibiotics (such as penicillin) were obtained from fungi. A rare species of yew (*Taxus*) from the Pacific Northwest of the USA has recently been found to produce a chemical that may prove of value in the treatment of certain forms of cancer. Drugs derived from the rosy periwinkle, from the Madagascan forest, are used to treat children with leukemia.

Human rights: If biodiversity is protected, indigenous people can continue to live in their native lands. If we preserve the rainforests, indigenous tribes can continue to live in them.

Ethical/intrinsic value: Each species has a right to exist – a bioright (see page 21) unrelated to human needs. Biodiversity should be preserved for its own sake and humans have a responsibility to act as stewards of the Earth.

Human recreation and ecotourism enjoyment: Biodiversity is often the subject of esthetic interest. People rely on wild places and living things in them for spiritual fulfillment.

Biorights self-perpetuation: Biologically diverse ecosystems help to preserve their component species, reducing the need for future conservation efforts on single species.

Conservation and preservation of biodiversity

Many people use these two terms interchangeably but they have different meanings.

Conservation biology is the sustainable use and management of natural resources. Conservation biologists do not necessarily want to exclude humans from reserves or from interacting with other organisms. They will even consider harvesting or hunting of species as long as it is sustainable. They recognize that it is very difficult to exclude humans from habitats and development is needed to help people out of poverty. But they want the development not to be at the expense of the environment. So a conservation biologist would look for ways to create income for local people from ecotourism or manage a reserve to allow education of the public by opening it up to access.

Preservation biology attempts to exclude human activity in areas where humans have not yet encroached. This is a non-anthropocentric viewpoint which puts value on nature for its own intrinsic worth, not as a resource that humans can exploit. Deep green ecologists argue that, whatever the cost, species should be preserved regardless of their value or usefulness to humans. So the smallpox virus should not be destroyed in preservation biology even though it causes disease in humans. Preservation is often a more difficult option than conservation.

Look back in Chapter 1 to find a pioneer conservationist and a pioneer preservationist.

How conservation organizations work

Efforts to conserve and preserve species and habitats rely on citizens, conservation organizations and governments to act in various ways. Because conservation is mostly seen as for the "public good", politicians act to change public policy to align with conservation aims. There will be tension between what is seen as good for the economy and human needs, and good for the environment, but the two can coincide and this is being recognized more and more. A recent report on climate change (the Stern Review in 2006 for the UK government) proposed that 1% of the gross domestic product (GDP) per annum should be invested in climate change mitigation activities in order to save a drop of 20% in GDP later on.

It is easy to think that you, just one individual, can have little or no impact on what is a global issue. But everyone can have an effect locally in a specific place as an individual and perhaps a larger effect as a member of a group. The group may act locally or globally or both and major changes have happened in thinking about and acting for the environment in the last few decades. "Think globally, act locally" is a slogan you may have heard. It was first used in the early 1970s and is as relevant today as it was then. Your environmental conscience starts at home, whether in using less plastic packaging and fossil fuel or acting to help a local nature reserve. You may not be personally able to save a whale but you can act to save some form of biodiversity.

Sustainable development – meeting the needs of the present without negatively impacting the needs of future generations and biodiversity – can and has been applied to conservation projects. Providing revenue to local people within a national park is an example of this. There is a danger of "greenwash" though. Greenwash is when organizations give the impression that they have changed their practices to have less impact on the environment but, in fact, they have changed nothing. When mining has extracted all the minerals it economically can, areas can be restored after mine closure. But just bulldozing spoil heaps and adding grass seed is not enough and could be seen as greenwash.

Organizations that work to conserve or preserve biodiversity and the environment may be local or global or both. There are national branches and regional and local offices of the large international organizations. They also fall into two categories depending on their constitution and funding. Governmental organizations (GOs) may be intergovernmental or part of a national government; non-governmental organizations (NGOs) may be international or local and funded by individuals or groups. We shall now look at some of these in more detail.

To research

Which is which?

Consider the logos of some international agencies and organizations in Fig. 6.3 that work to conserve the environment. Work out which organization is which from the logo and then do research to find out:

1 Is it a governmental or non-governmental organization?

2 What are its main aims?

3 How does it accomplish these aims?

Fig. 6.3 Some conservation organization logos

The organizations you have investigated operate in different ways. In their dealings with people, some work at strategic level while others work with local people "in the field". To bring about change, some work conservatively by careful negotiation, while others are more radical and draw attention to issues using the media.

4 Copy Fig. 6.4. Place the organizations on the axes and add others that you know of locally and nationally.

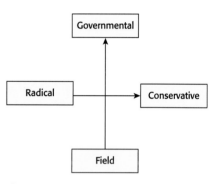

Fig. 6.4

How NGOs and GOs compare

	GO	NGO
Use of media	Media liaison officers prepare and read written statements	Use footage of activities to gain media attention
Speed of response	Considered, slow – depends on consensus often between differing viewpoints	Can be rapid
Political diplomatic constraints	Considerable – often hindered by political disagreement especially if international	Unaffected by political constraints – can even include illegal activity
Enforceability	International agreements and national or regional laws can lead to prosecution	No legal power – use of persuasion and public opinion to put pressure on governments

Table 6.1

The good news

No nation today can afford to ignore the rest of the world. Authoritarian nation states are on the decline and the rise of the information age and freedom of information via the worldwide web, human rights and democracy mean that the individual has more power than ever before. International cooperation was formalized with the United Nations and its huge number of specialist agencies which have a major impact. UNEP set up the Intergovernmental Panel on Climate Change (IPCC) and drove the Montreal Protocol for phasing out the production of ozone-depleting substances. Countries are forming economic coalitions – ASEAN, the EU, the African Union, OPEC to name but a few – and some of these organizations are working for sustainable development and environmental protection. They are complemented by NGOs, some of which started as grassroots movements (locally by individuals) but are now having global impacts, e.g. Greenpeace, WWF. Institutional inertia has been a block to change but the power of people wanting change and voting for politicians who say they will deliver it, may be able to stop the swing of the pendulum and slow down and even reverse degradation of our planet by human activities.

The IUCN

The **IUCN** is the International Union for Conservation of Nature, also sometimes called the World Conservation Union. It is an international agency, was founded in 1948 (when the Universal Declaration on Human Rights was adopted by the UN General Assembly) and brings together 83 states, 110 government agencies, more than 800 NGOs, and some 10 000 scientists and experts from 181 countries in partnership.

The IUCN's mission is to influence, encourage and assist societies throughout the world to conserve the integrity and diversity of nature and to ensure that any use of natural resources is equitable and ecologically sustainable.

It monitors the state of the world's species through its **Red List of Threatened Species** (see page 125) and supports the Millennium Ecosystem Assessment as well as educating the public, advising governments, and assessing new World Heritage sites.

Timeline of sustainability and biodiversity dates

Year	Events
1961	World Wildlife Fund set up by IUCN and Julian Huxley
1966	Species Survival Commission published Red Data Lists
1973	Convention on the International Trade in Endangered Species of Wild Fauna and Flora (CITES)
1980	World Conservation Strategy
1980	Brandt Commission published – beginnings of sustainable development
1982	UN World Charter for Nature
1987	Brundtland Commission on Our Common Future – first defined sustainable development
1991	IUCN Caring for the Earth: A Strategy for Sustainable Living
1992	UN Earth Summit Rio de Janeiro produced Agenda 21, Convention on Biological Diversity and the Rio Declaration Earth Council Global Biodiversity Strategy
2000	UN Millennium Summit and the MDGs (millennium development goals)
2002	UN World Summit on Sustainable Development held in Johannesburg
2005	UN World Summit, New York

The World Conservation Strategy (WCS) and subsequent milestones

The WCS was published by IUCN, UNEP and WWF in 1980. It was a ground-breaking achievement which presented a joined-up and integrated approach to conservation for the first time. Its aims were to:

- maintain essential ecological processes and life support systems
- preserve genetic diversity.

It called for international, national and regional efforts to balance development with conservation of the world's living resources. Many countries adopted the WCS and developed their own strategies for addressing national issues.

In 1982, the General Assembly of the UN adopted the UN World Charter for Nature. According to the UN its principles were:

- Nature shall be respected and its essential processes shall not be impaired.
- The genetic viability on the Earth shall not be compromised; the population levels of all life forms, wild and domesticated, must be at least sufficient for their survival, and to this end necessary habitats shall be safeguarded.
- All areas of the Earth, both land and sea, shall be subject to these principles of conservation; special protection shall be given to unique areas, to representative samples of all the different types of ecosystems and to the habitats of rare or endangered species.
- Ecosystems and organisms, as well as the land, marine and atmospheric resources that are utilized by man, shall be managed to achieve and maintain optimum sustainable productivity, but not in such a way as to endanger the integrity of those other ecosystems or species with which they coexist.
- Nature shall be secured against degradation caused by warfare or other hostile activities.

In 1991, the update to the WCS, *Caring for the Earth: A Strategy for Sustainable Living* was launched in 65 countries. It stated the benefits of sustainable use of natural resources, and the benefits of sharing resources more equally among the world population.

At the UN Earth Summit in Rio in 1992, 179 heads of government signed their countries up to Agenda 21, a blueprint for sustainable development into the 21st century.

Agenda 21 is a guide for individuals, businesses and governments in making choices for development that help society and the environment. Agenda 21 is a huge document, with 900 pages. Essentially it considers what we should do as individuals, local councils, nations and the world as a whole to make the Earth a fit place for life in this century and beyond. It includes "making decisions for sustainable development" and states that it is necessary to understand the links between the environment and economic activities. This will force us to make development choices that will be economically efficient, socially equitable and responsible, and environmentally sound. It looks at:

1 Social and economic dimensions – developing countries; poverty; consumption patterns; population; health; human settlements; integrating environment and development.

2 Conservation and management of resources – atmosphere; land; forests; deserts; mountains; agriculture; biodiversity; biotechnology; oceans; fresh water; toxic chemicals; hazardous, radioactive and solid waste and sewage.

3 Strengthening the role of major groups – women; children and youth; indigenous peoples; non-governmental organizations; local authorities; workers; business and industry; farmers; scientists and technologists.

4 Means of implementation – finance; technology transfer; science; education; capacity-building; international institutions; legal measures; information.

In the Rio Summit, a treaty called the Convention on Biological Diversity was also agreed. The aim of the convention was to assist countries in integrating biodiversity into their national planning. The three main objectives were:

- the conservation of biological variation
- the sustainable use of its components
- and the equitable sharing of the benefits arising out of the utilization of genetic resources.

The UN Johannesburg Summit on sustainable development in 2002 was intended to consolidate the Rio one but little action came out of its deliberations. However the 2000 UN Millennium Summit was the largest-ever gathering of world leaders who agreed to a set of time-bound and measurable goals for combating poverty, hunger, disease, illiteracy, environmental degradation and discrimination against women. Now known as the **Millennium Development Goals** (MDGs; see also page 188), the aim was that they are to be achieved by 2015. The follow-up 2005 World Summit in New York outlined a series of global priorities for action, and recommended that each country prepare its own national strategy for the conservation of natural resources for long-term human welfare. National conservation strategies have been written as a result of these international meetings and integrating conservation with development has a much higher priority now than it had a few decades ago.

IUCN Red Lists

The Red List of Threatened Species is a collection of objective lists of species under varying levels of threat to their survival. The data is gathered scientifically on a global scale and species under threat are put into one of the Red List categories which assess the relative danger of extinction. There are Red Lists of 'extinct' and 'extinct in the wild' species. The lists are regularly updated and inform government policies on trade in endangered species and conservation measures.

Strengths and weaknesses of species-based conservation

The CITES agreement

CITES (the Convention on International Trade in Endangered Species of Wild Fauna and Flora) is an international agreement between governments set up to protect the many species which were becoming endangered because of international trade. Its aim is to ensure that international trade in specimens of wild animals and plants does not

To research

Focus on the WWF and Greenpeace

Go to www.panda.org and www.greenpeace.org/international/. Find out:

1 What WWF stands for.

2 Why do you think WWF has a panda as its logo?

3 For both WWF and Greenpeace, find out:
  a Who they are.
  b What they do.
  c Where they work.
  d How they work.

4 Find the nearest WWF project to where you live.

5 a What are Greenpeace's main **biodiversity** campaigns?

  b Choose one and identify one aspect (e.g. whales). Describe an action Greenpeace has taken to draw international attention to the issue.

threaten their survival. Governments sign up to CITES voluntarily and have to write their own national laws to support its aims. It is a great example of what can be done voluntarily to conserve species. About 5000 animal species and 28 000 plant species are on its lists and, since 1975, it has been one of the most effective international wildlife conservation agreements in the world. Threatened species from elephants to turtles, orchids to mahogany are protected by CITES.

CITES has dramatically reduced the trade in endangered species, whether live animal imports (e.g. tortoises) or animal parts (e.g. elephant tusk, rhino horn).

The species are grouped in the CITES Appendices according to how threatened they are by international trade.

Appendix I: species cannot be traded internationally as they are threatened with extinction.

Appendix II: species can be traded internationally but within strict regulations ensuring sustainability.

Appendix III: a species included at the request of a country which then needs the cooperation of other countries to help prevent illegal exploitation.

The appendices include some whole groups, such as primates, cetaceans (whales, dolphins and porpoises), sea turtles, parrots, corals, cacti and orchids, and many separate species or populations of species.

Captive breeding and zoos

It might seem a good idea to retain examples of species in zoos or aquaria and so keep the species from becoming extinct. But, of course, we cannot keep captive every species there is and we would have to breed them to keep the species alive, or keep their DNA in the hope that one day we can recreate the organism. What is more, reintroducing a captive-bred species into a habitat where it has been wiped out is not a simple process.

Fig. 6.5 Captive gorilla in London Zoo

Programmes to reintroduce populations or establish new ones are expensive and difficult. A few are successful, e.g. for the condor in areas of California, USA, Przewalski's horse in Mongolia and black-footed ferret in Wyoming, USA. Most are not. Criticisms of programmes are that they waste money, are unnecessary, poorly run or unethical. But they can represent the best hope of saving a species that is extinct in the wild or in severe decline. Providing incentives to local people to support a reintroduction programme is often useful. Released animals may need extra feeding or care or even to be recaptured if their food supply runs out or they are in danger of dying. But the successful programmes are worthwhile. The reintroduction of the golden lion tamarin to the remaining Atlantic coast forest of Brazil has been a rallying point for local people, and reintroducing the Arabian oryx in Oman has given employment to the Bedouin locally. However, in some cases the animal has become used to humans. Orphan orang utans have to be taught by humans how to climb and socialize with each other. It takes patience and perseverance to get them to live in the wild after years in captivity and many do not make the jump successfully.

Rare plants can be reintroduced to the wild more easily, after raising seedlings in controlled conditions. But these may be dug up by collectors once planted out, or outcompeted by other plants or eaten by herbivores.

Sometimes it is impossible to reintroduce a species to its native habitat because the habitat has gone. Then the only way to keep that species may be in captivity – animals in zoos, game farms and aquaria, plants in arboretums, botanic gardens and seed banks.

Zoos have had a bad press and sometimes this is justified when animals are kept in close confinement in small cages or treated with cruelty. Most zoos are open to the public and so need to keep the megafauna (giraffe, elephant, hippo, rhino, polar bear) as these are what people pay to see, rather than the beetles or worms. The best zoos look after the animals very well, many have educational centres and breeding programmes and exchange animals with each other to widen the gene pools for reproductive success. Aquaria may keep the large cetaceans – killer whale and porpoises – and many fish and invertebrate species.

Botanical gardens and seed banks

The Royal Botanic Gardens at Kew, London, has a seed bank containing 25 000 plant species (10% of the world's total). About 10% of these are threatened in the wild. Around the world are some 1500 botanical gardens and they not only grow plants but also identify and classify them and carry out research, education and conservation.

In seed banks seeds are stored, frozen and dry, for many years. Seed banks are gene banks for the world's plant species and an insurance policy for the future. Up to 100 000 plant species are in danger of extinction, and by preserving their seeds for future use, the seed bank helps preserve the genetic variation of a species. For many crop plants just a few varieties are widely grown and the seed bank can maintain many more varieties of the species. There are seed banks around the world, holding national and international collections. A major concern for some people is who owns the seeds in the seed bank. As they need high levels of technology to maintain them, they tend to be in MEDCs yet may contain seeds from many countries. A seed bank in Svalbard, Norway, funded by the Global Crop Diversity Trust and the Norwegian government, has been built within the permafrost so needs no power to keep the seeds frozen. This is the ultimate safety net against the loss of other seed banks through disaster or civil strife.

To think about

While we can watch stunning wildlife documentaries on DVD, they are not the same as watching an elephant feed from three metres away. If you believe that animals and plants are there for humans to enjoy, then zoos, aquaria and botanical gardens provide you with a service.

1 Discuss whether keeping animals for humans to look at is a pointless exercise or has a value. If it has a value, what is it?

2 Do we have the right to capture and cage other species even if we treat them well?

3 If there is a choice between allowing a species to become extinct in the wild or keeping the last few individuals in a zoo, which is right?

To research

Choose an animal species that is threatened, by browsing the Red Lists or WWF websites. Your selected species must be in the CR, EN or VU Red List categories. Find the following information and prepare and give a presentation to your class including a one-page information sheet. Your presentation should include:

- Images of your chosen species (correctly cited)
- Its Red List category and CITES appendix position (I, II or III)

- A map of its global distribution
- Its estimated current population size and how this has changed over time
- Threats facing the species (ecological, economic and social)
- Conservation strategies including how local people and government actions are involved.

Designing protected areas

Many protected areas or nature reserves were set up in the past on land that no one else wanted. It may have been poor agricultural land, land not near areas of high human population density or land that was degraded in some way. The haphazard nature of this meant that early reserves may not have been large enough or been inappropriate to the needs of the species they were aiming to protect.

UNESCO's **Man and the Biosphere Programme** (MAB), started in 1970, created a world network of international reserves which now has over 500 reserves in over 100 countries. The Millennium Development Goals, sustainability and conservation are their aims.

Questions that conservationists now ask themselves when planning a protected area are:

- How large should it be to protect the species? Are there species that need protection in the middle of a large reserve?
- How many individuals of an endangered species must be protected?
- Is it better to have one larger or many smaller reserves? What about the edge effects?
- What is the best shape?
- If there are several reserves, how close should they be to each other?
- Should they be joined by corridors or separate?

The large or small debate is known as the SLOSS debate (single large or several small). A large reserve will contain sufficient numbers of a large wide-ranging species (e.g. top carnivores), minimize edge effects and may provide more habitats for more species. But several smaller reserves that are well placed may be able to provide a greater range of habitats and more populations of a rare species than would one large block. With several reserves, the danger of a natural or human-made disaster (fire, flood, disease) wiping out the reserve and its inhabitants is reduced as some reserves may escape the damage. Small reserves near cities can also be used for education and so further long-term goals of conservation through education.

The consensus is that large or small will depend on the size and requirements of the species that the reserve is there to protect. Sometimes only small is possible, and better than nothing. But, whatever the size, edge effects should be minimized so that the circumference-to-area ratio is low. Edge effects occur at **ecotones** (where two habitats meet and there is a change near the boundary). More species are present in ecotones as species from each habitat occur along with some opportunists so there is increased predation and competition. There is a change in abiotic factors as well, e.g. more wind or precipitation. Long thin reserves have a large edge effect; a circular one has the least. But, in practice, the shape is determined by what is available and parks are irregular in area. Dividing up the area with fences, pylons, roads, railways or farming should be avoided if possible as this fragments the habitat. But sometimes it is easier for governments to put road and rail links across national parks as this generates less opposition than plans to use privately owned land.

Corridors, strips of protected land, may link reserves. These allow individuals to move from reserve to reserve and so increase the size of the gene pool or allow seasonal migration. This approach worked well in Costa Rica where two parks were linked by a 7700 ha corridor several kilometres wide – at least 35 species of bird use this corridor to fly from park to park. But corridors have disadvantages as a disease in one reserve may be spread to the other, and it may be easier for poachers and hunters to kill the animals in the corridor, which is harder to protect than the reserves. Exotic or invasive species may also gain entry to a reserve via the corridors.

The MAB reserves have, if possible, a buffer zone. This is a zone around the core reserve which is transitional. Some farming, extraction of natural resources, e.g. selective logging, and experimental research can go on in the buffer zone. The core reserve is undisturbed and species that cannot tolerate disturbance should be safer there.

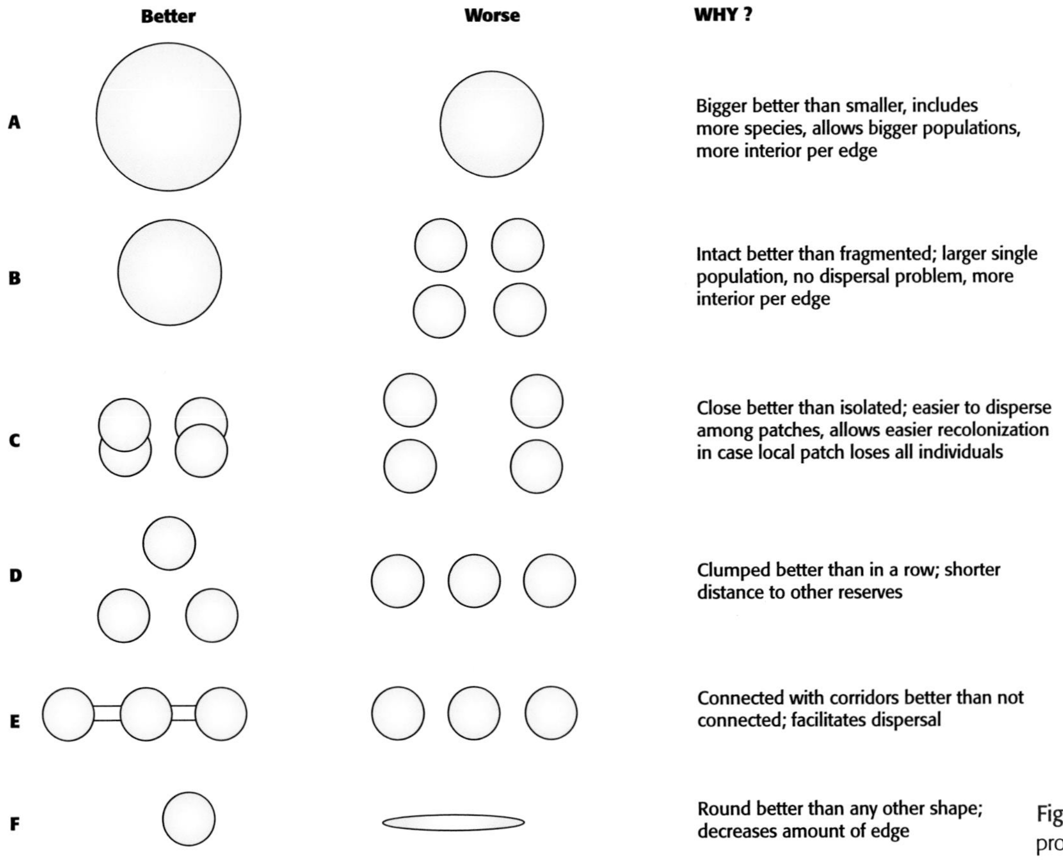

Better	Worse	WHY ?
A		Bigger better than smaller, includes more species, allows bigger populations, more interior per edge
B		Intact better than fragmented; larger single population, no dispersal problem, more interior per edge
C		Close better than isolated; easier to disperse among patches, allows easier recolonization in case local patch loses all individuals
D		Clumped better than in a row; shorter distance to other reserves
E		Connected with corridors better than not connected; facilitates dispersal
F		Round better than any other shape; decreases amount of edge

Fig. 6.6 Possible shapes of protected areas

1 What is meant by the word **conservation**?

2 Here are four ways to conserve species on the edge of extinction. Explain how this can be done with named examples. *You may need to look up some on the Internet.*

a Legislation and conserving habitats
b Zoos and captive breeding
c Botanic gardens and seed banks
d Nature reserves

Conservation in action

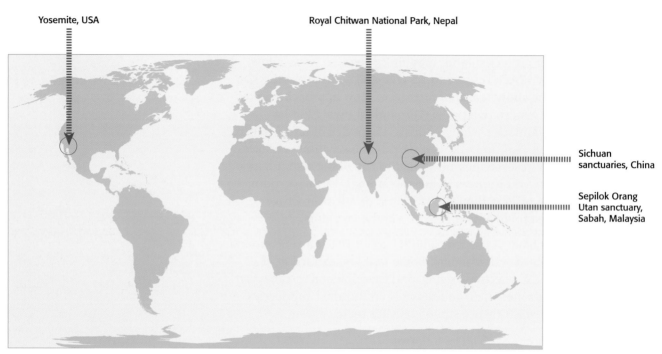

Yosemite, USA

Royal Chitwan National Park, Nepal

Sichuan sanctuaries, China

Sepilok Orang Utan sanctuary, Sabah, Malaysia

Fig. 6.7 World map with location of case studies

Case study 1

Royal Chitwan National Park, Nepal

Fig. 6.8 Rhino in Chitwan National Park, Nepal

This national park is in southern Nepal in flat lowlands near the border with India. It is a habitat for species of tiger, rhino, elephant, dolphin, crocodile and lizard. All have been hunted by big game hunters in the past but are now a major tourist attraction. The human population of the park and surrounding areas has increased dramatically since anti-malarial drugs became available. The forest was cleared to grow food, and animals were poached and killed. Populations of tigers and rhino became very low, estimated at 30 and 200 respectively. A moratorium on hunting was introduced, people were relocated from the park area and it was declared a national park in 1973.

Tiger and rhino numbers have increased but so has the human population size. People and domestic livestock are killed by tiger or rhino and crops are damaged. There are tight restrictions on using the park's resources for collecting firewood, fishing and hunting.

As tourist facilities increase, there are more jobs for local people but food and household goods prices have also increased and there is evidence that local people became poorer as a result of the park. Now, some revenue from the park goes to local villagers but there is still pressure on both human and animal populations.

Case study 2

Sichuan giant panda sanctuaries

In Sichaun province, China, are some 900 000 hectares of national reserves which are habitat for the giant panda as well as the red panda and snow and clouded leopards. They have been made a World Heritage site and it is thought that about 1600 giant pandas live there. The forests are species-rich with up to 6000 plant species present. But the pandas' habitat is shrinking as people fell the bamboo forests and degrade the habitats. Some pandas live in strips of bamboo only one kilometre wide.

In the 1960s, errors were made in trying to protect the pandas and many were caged in hopes of them breeding in captivity but they did not do so successfully. Captive breeding is now more successful as the pandas are housed in larger areas. Since human populations were moved out of reserves and laws on gun use tightened, panda numbers may have increased but they are not yet out of trouble.

The specialist diet of the panda, living almost entirely on bamboo, means its habitat is limited. If the bamboo goes, they starve. But they cannot digest the cellulose in bamboo effectively so have to eat huge quantities per day to derive enough energy from it.

The earthquake in June 2008 destroyed most of the infrastructure and buildings in the panda sanctuaries and over 100 people died. The effect on the pandas is not yet known, however some held in captivity were also killed.

Fig. 6.9 Giant pandas in the wild in China

Case study 3

The Sepilok orang utan rehabilitation centre, Sabah, Malaysia

Fig. 6.10 Orphan orang utan in Sepilok, Sabah, Malaysia

This sanctuary was established in 1964 to return orphaned orang utans back to the wild. The mothers of these orphans are either shot by loggers and poachers or die when the tree in which they are hiding is felled. Often the baby survives the fall. Many are taken as pets or sold on in markets. Once they are too big to be cute, they are a problem in households.

Although run by the national government, the centre has to raise some funds so education of the public as well as research and rehabilitation go on there. Some animals are "sacrificed" to the need of visitors to see and be close to the orang utans. Once they are too used to humans, it is almost impossible to reintroduce them to the wild as they keep returning to the centre for food.

The habitat of the orang utan is lost as the forest is felled for timber. Mining or forest fires also destroy it. But the biggest loss of primary rainforest in Sabah is to oil palm plantations. Orang utans are critically endangered.

Case study 4

Yosemite National Park, USA

Fig. 6.11 Yosemite National Park

Yosemite National Park is in California and covers 260 000 hectares. It was designated as a scenic area in 1864 by US President Abraham Lincoln and this was the first time that a government had set aside land to protect it and allow people to enjoy it. It became a national park in 1890 after John Muir led the battle to save it. There are many habitats in Yosemite from chaparral to alpine regions. Within these diverse ecosystems are 400 species of vertebrates and 1400 of plants. Black bears and bobcats are the large animals that people come to see there (or to avoid) but, even in this protected area, species have become extinct and others are threatened. This is partly due to pressures of visitors – over 4 million a year – but the loss of the natural fires, introduction of non-native species, air pollution and habitat fragmentation all affect the park. Although visitors are carefully educated and managed in the park, so many feet on the paths and campsites no doubt have an impact.

Key words

conservation	endangered
preservation	vulnerable
biorights	threatened
CITES	protected areas
UNEP	corridors
NGO	fragmentation
WCS	island biogeography
Red Lists	captive breeding
WWF	seed banks
Greenpeace	habitat conservation
extinct	species conservation

To do

The four case studies mentioned above are in different parts of the world.

1 What do they all have in common?

2 Is it easier to protect an area if there is a big furry animal that lives in it? If so, why?

3 Write your own case study on a protected area that you know.

Key points

- Greenhouse gases (GHGs) maintain the temperature on Earth at a level suitable for life.
- Human activities increase the amount of GHGs in the atmosphere.
- Increased GHGs may lead to warming of the atmosphere and climate change.
- Increased GHGs may alter weather patterns, shift biomes and increase sea levels.
- Feedback mechanisms are very complex and changes may lead to positive or negative feedback.
- Governments and individuals can act to reduce GHG emissions.
- There is still much controversy and varying perceptions about global warming.

"An increase of two or three degrees wouldn't be so bad for a northern country like Russia. We could spend less on fur coats, and the grain harvest would go up."

Vladimir Putin, Russian President, October 2003

The controversy

The only environmental issue to have caused as much debate and discussion as climate change was probably human population growth. But that is, in some ways, clearer to deal with. We can count how many we are, more or less, and can see a direct effect of more people wanting to use more resources from a finite stock.

Climate change and global warming have become very emotive issues involving national and international politics, global economics and the fate of national economies all bound up with scientific debate about the evidence and cause and effect. Added to this are the questions of whether millions or billions of people will suffer, whether there will be losers and winners if climate shifts to a new equilibrium, and whether the power bases of different nations will be affected. So you can begin to see what a complex issue this is.

We talked about environmental viewpoints in Chapter 2. Your viewpoint certainly colours how you interpret the evidence on climate change, as technocentrists and ecocentrists clash on the question of what we should do or can do to mitigate the effects that we are seeing.

While there are facts that are not in debate – that there is a greenhouse effect, that greenhouse gas emissions are increasing due to human activities and are probably increasing the greenhouse effect, that there has been a recent pattern of increased average global temperature – there is not total agreement over the cause of the rise in temperature nor over what we should be doing about it.

The vast majority of scientists working in this field accept the correlation between increased greenhouse gas emissions and increased temperature, causing climate change and different weather patterns. But a significant minority question the cause and effect, some citing the Earth's rotational wobble, sunspot activity or that increased temperature is causing increased greenhouse gases, not the other way around.

All agree that the feedback mechanisms are very complex in such a complex system as the Earth and that our models, though much improved, may not model the climate exactly.

What should and could be done in response to climate change is a very big question. Possible actions may be preventive (e.g. reduce fossil fuel burning) or lessen the impact of climate change (e.g. build flood barriers) or we may take no action. There is a time lag and some inertia between making a decision to act and the actions being carried out, especially when they may involve large lifestyle changes.

The IPCC is the Intergovernmental Panel on Climate Change. It was set up by UNEP (UN Environment Programme) and the World Meterological Organization in 1988 to look into this major issue. Here, we take the view of the IPCC in its fourth assessment report in 2007, that there is "very high confidence (9 out of 10 chance of being correct) that the globally averaged net effect of human activities since 1750 has been one of warming".

The greenhouse effect

First, let us make clear that the greenhouse effect is a good thing for life on Earth. Indeed, without it, there would be no life on Earth as we know it at all. The nearest planets to the Earth are Mars with a surface temperature of $-53°C$ and Venus with a surface temperature of $+450°C$. Earth's average surface temperature is a comfortable 15°C so just right for life and 33°C warmer than it would be without the greenhouse effect. The effect is caused by gases in the atmosphere reducing heat losses by radiation back into space. They trap heat energy that is reflected from the Earth's surface and reradiate it – some back to space and some back to the Earth. A common misconception is that this works in the same way as a glass greenhouse. But greenhouses work by reducing convection currents in the air inside the greenhouse which is warmer than air outside. Greenhouse gases absorb infrared radiation radiated from the Earth's surface and pass this as heat to the other atmospheric gases.

Incoming solar radiation is mostly made up of visible light, ultraviolet light and infrared heat. This passes through the atmosphere of Earth mostly unaffected (although the longer wavelength infrared is absorbed by the atmosphere). About 45% of the incoming light is absorbed, scattered or reflected by the atmosphere and clouds before it reaches the surface of the Earth. Of the 55% of incoming solar radiation that reaches the Earth's surface, 4% is reflected and 51% absorbed. This is used in several processes including photosynthesis, heating the ground and seas, and evaporation. It is released back to the environment and the atmosphere as longer wavelength infrared energy – heat energy (see Fig. 7.2). If we had no atmosphere with greenhouse gases in it, this would go straight back into space and the temperature on Earth would fall drastically every night. But some of

"Sadly, it's much easier to create a desert than a forest."

James Lovelock

this heat is absorbed by gases in the atmosphere which re-emit it as heat energy back to Earth.

There are many gases that can do this but the main greenhouse gases are water vapour, carbon dioxide and methane. Only molecules with two or more bonds joining the atoms can absorb and re-emit energy in this way. Nitrogen and oxygen make up most of the atmosphere but are not greenhouse gases.

The enhanced greenhouse effect

As humans increase emissions of some greenhouse gases, the greenhouse effect is exaggerated or enhanced. Most climate scientists believe that this is causing global warming and climate change.

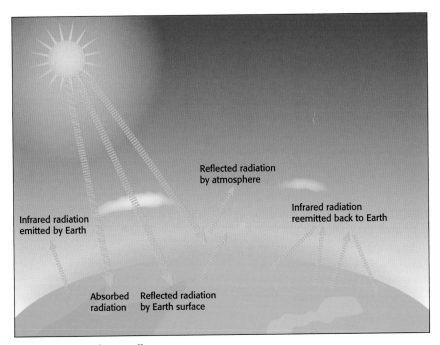

Fig. 7.1 The greenhouse effect

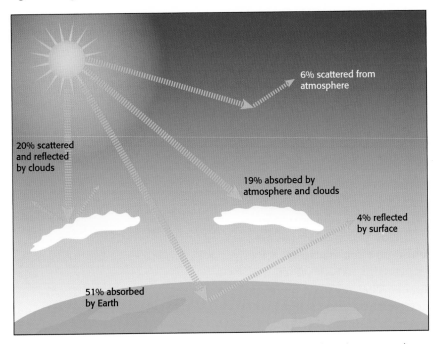

Fig. 7.2 The greenhouse effect showing percentages absorbed, reflected or scattered

Greenhouse gases

There is very little carbon dioxide in the atmosphere (0.038% of the total gases) but it is increasing as are other gases, some of which are greenhouse gases, through the activities of humans (anthropogenic activities). The list of greenhouse gases includes not only carbon dioxide, water vapour and methane but also chlorofluorocarbons (CFCs and HCFCs), nitrous oxide (nitrogen(I) oxide) and ozone.

There are three points that may be confusing when reading about or reviewing statistics on climate change.

1 The role of ozone and CFCs.

2 The role of water vapour.

3 Whether figures refer to total greenhouse gas effects or the enhanced (anthropogenic) greenhouse effect.

Ozone forms a layer in the stratosphere that absorbs much of the ultraviolet radiation from the sun. CFCs are chemicals made by humans that coincidentally break down ozone when they reach the stratosphere but act as greenhouse gases in the troposphere. Ozone is also a greenhouse gas in the troposphere but it acts in the stratosphere to cool the Earth (see Fig. 4.2, page 69). There is no direct link between global warming and ozone depletion but, because the climate is so complex, there are indirect links.

When data is presented, consider whether the contribution of water vapour is included or excluded. Water vapour has the largest effect on trapping heat energy so is the most potent greenhouse gas but it is not usually listed because it varies so much in concentration and is constantly condensing to water, snow and ice so stops acting as a greenhouse gas. Somewhere in the region of 36–66% of the greenhouse effect is due to water vapour. The IPCC, and most scientists, omit water vapour from their calculations but the IPCC work on the figure of a 50% contribution by water vapour. Clouds may contribute up to 25% of the greenhouse effect (depending on the type of cloud and its altitude) and other greenhouse gases cause the rest, with carbon dioxide having the largest effect.

Also remember that most greenhouse gases in the atmosphere are there through natural processes (except CFCs which are human-made) and it is the increase in these due to anthropogenic activities that is of concern. Carbon dioxide concentration may be higher now than at any time during the last 160 000 years; the recent rapid rate of increase of 30 ppm (parts per million) in 20 years is unprecedented and is due to human activities.

CFCs are human-made chemicals so are not present in the atmosphere as a result of natural processes. There are many types, e.g. CFC11 and CFC12 as well as HCFCs. Although their concentration in the atmosphere is measured in parts per trillion (10^{12}), they have a large contribution to the enhanced greenhouse effect because each molecule has a high **global warming potential** (GWP) which may be thousands of times that of carbon dioxide. That means a molecule of a CFC is up to 10 000 times more effective at trapping longwave radiation than a molecule of carbon dioxide, which has a GWP of 1.

"We're in a giant car heading toward a brick wall and everyone's arguing over where they're going to sit."

David Suzuki

Greenhouse gas	Pre-industrial concentration (ppm)	Present concentration (ppm)	Global warming potential	% contribution to enhanced greenhouse effect	Atmospheric life time / years
Carbon dioxide, CO_2	270	379	1	50–60	50–200
Methane, CH_4	0.7	1.774	21	20	12
Nitrous oxide, N_2O	0.27	0.31	206	4–6	140
CFC11	0	0.00025	3500	14 (all CFCs)	45
Ozone	not known	variable	2000	variable	not known

Table 7.1 Major greenhouse gases

Radiative forcing

You may see the term "radiative forcing" with respect to greenhouse gases. This means the change in the balance between radiation coming into the atmosphere and radiation going out of the atmosphere. If it is positive, the Earth's surface is warmed up; if negative, it is cooled. The IPCC report plots this in 2005 compared with a zero radiative forcing pre-1750. Solar radiative forcing (i.e. how much hotter the sun is) is measured since 1850.

While greenhouse gases are positive in Fig. 7.3, other factors are negative. Stratospheric ozone has a cooling effect; tropospheric ozone a positive one. Aerosols in the atmosphere cool the surface of the Earth as they block sunlight (global dimming) (see page 154); water, bare soil and dark crops have a warming effect as they absorb heat while ice and snow reflect more heat (have a higher albedo) and cool the surface. The net anthropogenic effect may not depend simply on the result of adding all the values as some effects counteract others, hence the error bar is the largest.

A positive radiative forcing tends on average to warm the surface of the Earth, and negative forcing tends on average to cool the surface. The figure shows estimates of the globally and annually averaged anthropogenic radiative forcing (in Wm^{-2}) due to changes in concentrations of greenhouse gases and aerosols from pre-industrial times to present day and to natural changes in solar output from 1850 to the present. The height of the rectangular bar indicates a mid-range estimate of the forcing and the error bars show the uncertainty range. Note: forcing associated with stratospheric aerosols resulting from volcanic eruptions is not shown because it is very variable over this time period. Halocarbons include all CFCs and related greenhouse gas compounds.

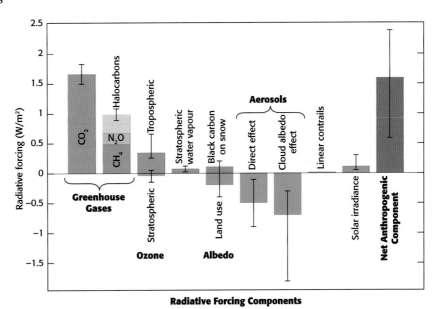

Fig. 7.3 Global average radiative forcing estimates and ranges in 2005 for anthropogenic greenhouse gases and other important agents and mechanisms

Methane as a greenhouse gas

Methane is a simple hydrocarbon, CH_4. Since about 1950, the concentration of methane in the atmosphere has increased by about 1% per year due to human activities. About 60% of the methane in the air is from human activity and 15% of this from cattle. We also use methane as a fuel. Natural gas is methane; biogas digestors produce methane which is used for cooking or heating. (See also Fig. 4.26, page 84)

Sources of methane

Fig. 7.4 Cows produce methane

There are 1.3 billion **cattle** in the world and they are ruminants with bacteria in their stomachs that break down the cellulose in the grass that they eat. These bacteria live in anaerobic conditions and release methane as a waste product. It comes out of both ends of the cattle and amounts to 100 million tonnes of methane per year. Each methane molecule is 21 times more effective than one of carbon dioxide at absorbing and radiating heat energy and methane contributes about 20% of the anthropogenic greenhouse gases. If we could capture this methane, we could use it as a fuel. Some scientists are trying to reduce the amount that cows produce by feeding them special diets higher in sugar levels or taking the bacteria in kangaroo stomachs and putting them in cow stomachs.

Waste tips in more developed countries give off methane as waste food decomposes in the tip in anaerobic conditions. There can be so much methane that it is tapped and piped away to be used to generate electricity or for heating. But much is not captured and is released into the atmosphere.

Rice paddy fields cover 1.5 million km² of land and rice is a staple crop. The fields release up to 100 million tonnes of methane per year due to anaerobic respiration by bacteria in the soil. But they only release methane when flooded which is about one third of the year and the rest of the time may act as a sink for methane, absorbing it.

Natural sources of methane

Swamps and bogs: 5 million km² of bogs and marshes release methane.

Termites may produce 5% of atmospheric methane as bacteria in their guts release it as they break down cellulose.

The tundra: The bogs and swamps of the tundra contain much methane produced by decomposition in waterlogged soils. But this methane is locked up as it is frozen in the permafrost. There is evidence that the permafrost is melting and that some methane is being released. In some parts of Siberia or Northern Canada, you can dig a hole and set fire to the methane. As the permafrost melts, it releases more methane which causes more warming – an example of positive feedback. In Arctic seas is another source of methane, also locked up as methyl hydrates in clathrates which are molecular cages of water that trap methane molecules within them. These are only stable when frozen and under high pressure at the bottom of the seas. There may be up to 10×10^9 tonnes of methane in these structures and companies are already trying to mine them. But it is very dangerous work as the methyl hydrates can bubble up to the surface and sink any ships in the area.

The carbon cycle

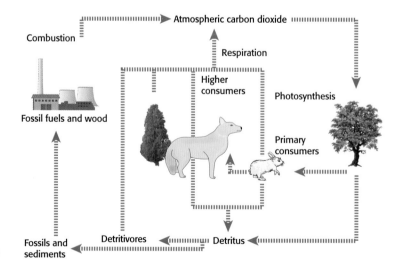

Fig. 7.5 The carbon cycle

Reminder: Movement of matter, such as nutrients, through an ecosystem is very different to the movement of energy. Energy is transferred from the sun, through food webs and is eventually lost to space as heat. Nutrients and matter are finite and are recycled and reused (via the decomposer food chain). Organisms die and are decomposed and nutrients are released, eventually becoming parts of living things again, when they are taken up by plants. These are the **biogeochemical cycles**.

Most carbon in Earth is locked up or fixed into solid forms as sedimentary rocks and fossil fuels. There is also a sink in the biosphere – living plants and animals and the oceans where carbon is dissolved or locked up as carbonates in the shells of marine organisms. The remaining carbon is carbon dioxide in the atmosphere. If the rocks, organisms or oceans take in more carbon dioxide from the atmosphere than they release, they are called **carbon sinks**. If they release more carbon dioxide than they take in, they are **carbon sources**.

All fossil fuels contain carbon because they are fossilized life forms. Life on Earth is based on carbon. Carbon is an essential element in living systems, providing the chemical framework to form molecules that make up living organisms. The molecules of organic compounds are built from chains of carbon atoms to which atoms of the other elements (mainly hydrogen, oxygen, nitrogen and sulfur) are attached. Carbon makes up around 0.038% of the atmosphere as carbon dioxide, and is present in the oceans as carbonate and bicarbonates and in rocks such as limestone and coal. Most carbon on Earth is stored in rocks and sediments (limestone, chalk, fossil fuels) and this is locked up for millions of years.

The **carbon cycle**, in which carbon circulates through living and non-living systems, occurs in the ecosphere. Here carbon is found in four main storages: the soil, living things (biomass), the oceans and the atmosphere. Carbon not in the atmosphere is stored in carbon dioxide sinks (soil, biomass and oceans) as complex organic molecules or dissolved in seawater.

Carbon cycles between living (biotic) and non-living (abiotic) chemical cycles: it is fixed by photosynthesis and released back to the atmosphere through respiration. Carbon is also released back to the atmosphere through combustion of fossil fuels and biomass. When dead organisms decompose, when they respire and when fossil fuels are burned, the carbon is oxidized to carbon dioxide and this, water vapour and heat are released. By photosynthesis, plants recapture this carbon – carbon fixation – and lock it up in their bodies for a time as glucose or other organic molecules.

When plants are harvested and cut down for food, firewood or processing, the carbon is also released again to the atmosphere. As we burn fossil fuels and cut down trees, we are increasing the amount of carbon in the atmosphere and changing the balance of the carbon cycle. Carbon can remain locked in the biotic or abiotic cycle for long periods of time, i.e. in the wood of trees or as coal and oil. Human activity has disrupted the balance of the global carbon cycle (carbon budget) through increased combustion, land use changes and deforestation.

The carbon budget

The amount of carbon on Earth is finite and we have a rough idea of where it goes. The diagram of the carbon cycle in Fig. 7.6 shows carbon sinks (storages) and flows in Gt C.

Our annual current global emissions from burning fossil fuels are about 5.5 Gt C. About 20% of this is from burning natural gas, 40% from burning coal and the other 40% from burning oil. Another 1.6 Gt C are added through deforestation. So 7.1 Gt C enter the atmosphere each year. Only about 2.4–3.2 Gt C of this stay in the atmosphere. Some is taken up by living things. Diffusion of carbon dioxide into the oceans and uptake by oceanic phytoplankton accounts for 2.4 Gt C. New growth forests fix about 0.5 Gt C a year. But this still leaves between 1 and 1.8 Gt C – a large amount – unaccounted for. We are not sure where it goes because of the complexity of the system. The proportion of carbon in Gt C in other reservoirs are: atmosphere 750; standing biomass 650; soils 1500; oceans 1720. Since the pre-industrial period, we have added 200 Gt C to the atmosphere.

Questions

1 Draw your own diagram of the carbon cycle. Include the processes of photosynthesis, respiration, feeding, death and faecal loss, fossilization, combustion.

2 Label the biotic and abiotic phases on your diagram.

3 Add Gt C data to the storages and flows.

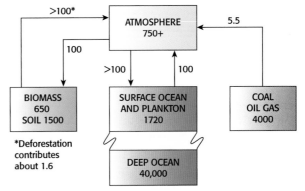

Fig. 7.6 The carbon cycle with flow values in gigatonnes of carbon per year

To do

Anthropogenic sources of greenhouse gases
Copy and complete the table of GHC sources due to human activities.

Greenhouse gas	Sources due to human activities
Carbon dioxide	
Methane	
Ozone	
Nitrous oxide	
CFCs	

What is climate change and what will happen?

Changes in the climate can be seen in different ways. It may be in changed temperatures or rainfall patterns, more severe storms, icesheet thinning or thickening, and sea level rises. It may not be a steady process – global cooling was recorded in the 1970s and, according to some measurements, since the year 2000. That was possibly due to sunspot variations, global dimming or dust in the atmosphere from volcanic activity, but the general trend is for warming over the last century.

There are five ways in which the climate can change over time due to a change in conditions on Earth – a change in the forcing, e.g. of greenhouse gas levels.

(a) There may be a direct relationship – more forcing leads to proportionately more climate change.

(b) There may be a buffering action in which forcing increases but climate change does not follow in a linear way – it is resistant to change.

(c) Climate change may respond slowly at first but then accelerate until it reaches a new equilibrium.

(d) The climate may make no response to forcing changes but then be tipped over a threshold and change rapidly until a new, much higher equilibrium is reached.

(e) In addition to the threshold change, it may then get stuck at the new equilibrium even when the forcing decreases until it then tips over a new threshold and falls rapidly. These threshold changes could occur in just a few decades.

Figure 7.7 shows these scenarios. Our problem is that we do not know which one we are living in.

A way of visualizing this is to imagine the climate is a car that you (the forcing mechanism) are pushing uphill. In (a) you push steadily uphill. In (b) you push with the same force but the car moves more slowly – it has more resistance. In (c) you reach a part of the hill with a shallower gradient and the car moves more easily before the hill gets steeper again. In (d) you push until the car reaches the edge of a cliff and then falls over it. Can you explain (e) in this analogy?

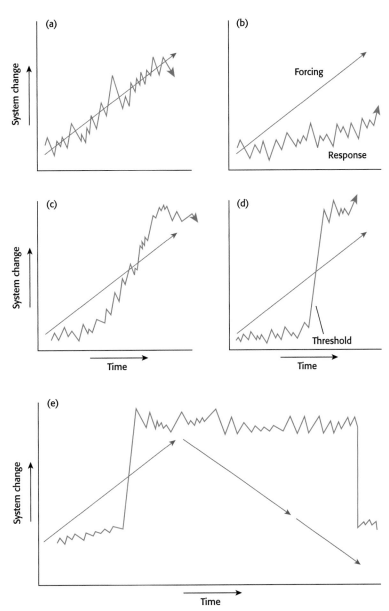

Fig. 7.7 Possible climate change system responses to a forcing mechanism

Where is the evidence?

Past climate changes

The climate of Earth has always changed. It is unstable and has fluctuated greatly in the past. Temperature and precipitation are the main factors that determine climate and biome distribution (see Chapter 3). We cannot measure precipitation in the distant past but we can measure temperature both directly and indirectly by proxy. We can also measure atmospheric gas concentration in bubbles trapped in ice.

In the geological timescale, average temperature on Earth was 20°C in the Early Carboniferous period 350 MYA (million years ago) and this cooled to 12°C later in the Carboniferous, slightly lower than our 15°C today. When the temperature was 20°C, the carbon dioxide concentration in the atmosphere was probably about 1500 ppm (parts per million) and this decreased to about 350 ppm when the average temperature was 12°C. Today there are about 380 ppm carbon dioxide in the atmosphere, less than in the previous 600 million years except during the Carboniferous. But how are these data obtained? Various direct and indirect measurements are taken on sediments from the period which contain fossilized animal shells, but it is hard to say how accurate they are.

1 What is the relationship between carbon dioxide and temperature in the graph in Fig. 7.8?

2 What other factors besides atmospheric carbon may influence Earth temperatures?

3 Can we rely on the data collection methods used?

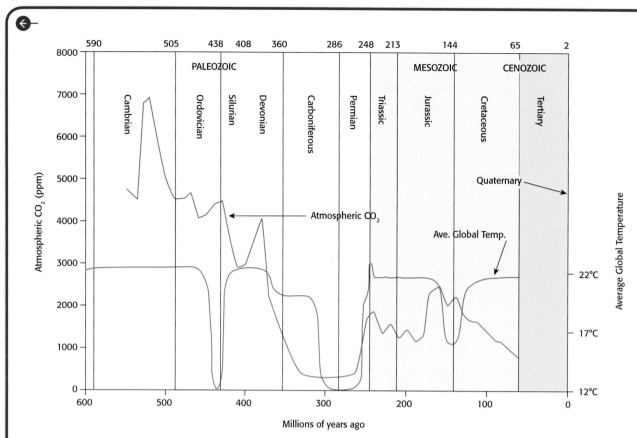

Fig. 7.8 Global temperature and atmospheric carbon dioxide over geological time

More recent climate change

Ice cores (cylinders of snow and ice extracted using a hollow tube) have been taken from the Antarctic and Greenland ice caps. The Vostok core in the Antarctic retrieved ice from a depth where it was laid down 420 000 years ago. A more recent core went to 720 000-year-old ice. The bubbles of air trapped in the ice can be analysed to tell us what the climate was like at the time the ice froze. The proportions of different isotopes of hydrogen and oxygen give an indication of the climate then and levels of gases can be measured. The age of the ice can be calculated by ice rings in the top layers (rather like tree rings showing summer and winter) and by the dust from volcanic eruptions lower down. Dating gets less accurate the deeper you go. From the ice cores, a picture of carbon dioxide and temperature over time can be built up (see Fig. 7.9a and b).

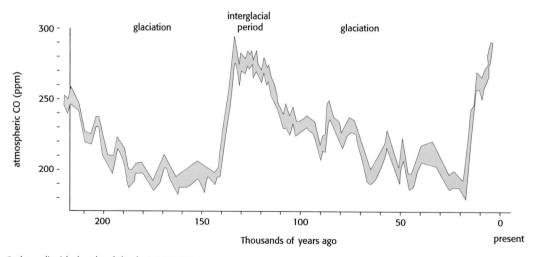

Fig. 7.9a Carbon dioxide levels of the last 240 000 years

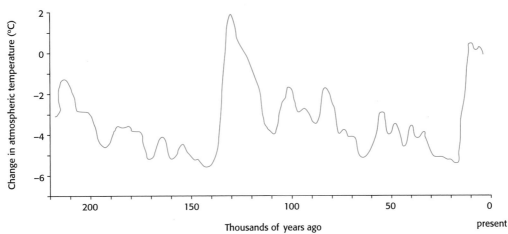

Fig. 7.9b Temperature levels of the last 240 000 years

4 What conclusion can you draw from Fig. 7.9a and b?

Since we started burning large amounts of fossil fuels, humans have added carbon dioxide to the atmosphere in addition to the natural amounts in the carbon cycle. Ice core records show carbon dioxide levels have risen from about 270 ppm before 1750 to 379 ppm by volume today. Although carbon dioxide levels were far higher than this (Fig. 7.8) in geologic time, the recent increase has been due to human activities and adds up to 2.7 Gt C per year to the atmosphere.

In Hawaii in the Pacific, atmospheric carbon dioxide has been measured since 1958 (see Fig. 7.10).

5 Describe and explain the trend in Fig. 7.10.

6 What is the percentage increase in carbon dioxide levels between 1960 and 2007?

7 Why is there an annual cycle? (*Hint*: think about which hemisphere has the most land mass and then when plants photosynthesize.)

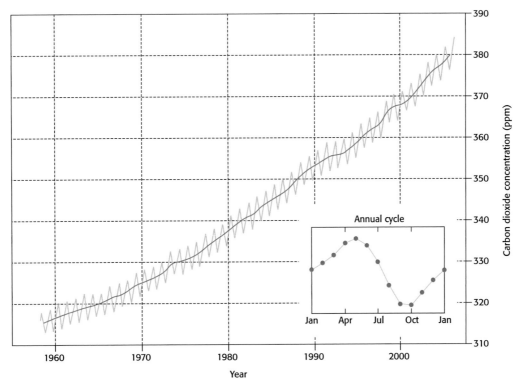

Fig. 7.10 Atmospheric carbon dioxide in ppm 1958–2007, measured at Mauna Loa, Hawaii

8 The temperature of the Earth since 1850 is shown in Fig. 7.11. Although this has fluctuated, there is a trend. What is it?

9 What would you have said in answer to question 8, in 1910 or 1950?

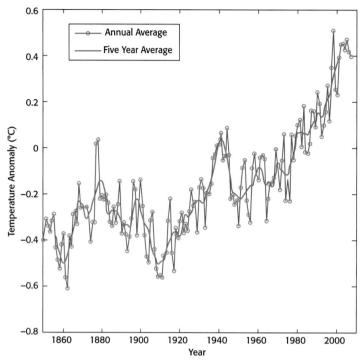

Fig. 7.11 Average annual temperature 1850–2007

More recently, over the last 100 years, sea level changes have been recorded as well. But these are difficult to measure as the land is not static but also moves up and down slowly. What measurements do show, however, is that sea levels have risen and sea ice thickness at the poles has decreased.

10 Sea temperature has also been recorded. Figure 7.12 shows average surface sea temperature over the last 25 years. What is the trend?

11 In Fig. 7.12, what happened to temperature when Pinatubo (a volcano in the Philippines) erupted? Explain this. Research what happens in El Nino years. Explain this (see pages 211–12).

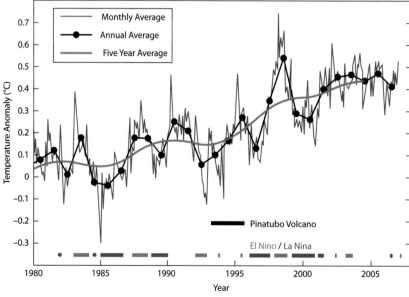

Fig. 7.12 Sea surface temperature 1980–2007

What is the consensus view?

There is evidence that greenhouse gases concentrations have increased since the Industrial Revolution and human activity has caused this. Climate is changing. Some of that may be due to natural climatic variation (sunspot activity, wobble of the Earth), some due to atmospheric forcings. But the climate and weather patterns are very, very complex and some parts respond quickly, e.g. the atmosphere, while others do so very slowly, e.g. the deep oceans.

The IPCC is made up of hundreds of scientists from around the world who research and regularly report on climate change. The fourth assessment report of the IPCC in 2007 used language that qualified some statements as scientists are wary of stating certainty; further experiment could falsify a theory, e.g. "all swans are white" works as an hypothesis until you see a black swan in Australia. However, they used terms such as "extremely likely" (>95% confidence of an outcome), and "very high confidence" (9 out of 10 chance of being correct) so went as far as possible in stating their case.

A summary of the IPCC report 2007 is that there is very high confidence that the effect of human activity since 1750 has been one of warming the Earth. In more detail, the report says:

- Carbon dioxide level predominately from fossil fuel burning and land use change has increased from pre-industrial levels of 280 ppm to 379 ppm in 2005 and is increasing 1.9 ppm per year. Methane and nitrous oxide levels have increased due to agriculture.
- Warming of the planet is now unequivocal. Air and sea temperatures are higher, ice and snow are melting and mean sea level is rising.
- Warming in the last 100 years has caused about a 0.74°C increase in global average temperature.
- Average Arctic temperatures increased at almost twice the global average rate in the last century.
- Global dimming has reduced the warming to some extent.
- Glaciers and ice sheets have melted and this water and the thermal expansion of oceans has very likely led to a sea level rise of 3.1 mm per year in the last few years. But Antarctic sea ice has not changed.
- Hurricane intensity has increased, which is expected as sea temperatures rise.

The IPCC report predicts:

- If carbon dioxide levels double, the temperature will go up by 3°C.
- Dry regions will get drier, droughts worse and wet regions wetter with more floods.
- Coasts will erode more and corals will bleach more.
- Ecosystems will degrade and be less able to act as carbon sinks.
- Food production will increase as long as the temperature does not rise more than 3°C.

Effects of climate change

Global average temperature has increased by 0.8°C, most of that since 1980. What are the effects?

1 Effects on oceans and sea levels

Sea levels are rising. This is because water expands as it heats up and ice melting on land slips off the land and into the sea, increasing the volume of sea water. The Greenland and the Antarctic ice sheets are on land and are thinning. This and the thermal expansion of the seas will mean that sea levels rise more. By how much is not clear but predictions are becoming more accurate as climate modelling improves with more variables entered into the programs. An increase of between 1.5 and 4.5°C could mean a sea level rise of 15–95 cm (IPCC data). But that is assuming a proportional relationship. If there is a threshold and this is exceeded, sea levels could rise by many metres. Up to 40 nations will be affected. Low-lying states (e.g. Bangladesh, the Maldives, the Netherlands) would lose land area; some, e.g. Tuvalu, would disappear completely.

The oceans absorb carbon dioxide and this makes them slightly acidic. They have become slightly more acidic by 0.1 pH as they have absorbed about half the carbon produced by anthropogenic activities. This may affect marine organisms, particularly corals. But as they warm, they absorb less carbon dioxide.

2 Effects on polar ice caps

Melting of land ice in Antarctica and Greenland will cause sea levels to rise as it flows into the oceans. Melting of the floating ice cap of the Arctic will not increase the volume of water as ice has the same displacement as liquid water. But glaciers are melting into the seas and these will increase the volume of water. The Greenland ice sheet could melt completely and slow down or stop the North Atlantic Drift current by diluting the salt water. If the North Atlantic Drift current and the Gulf Stream slow or even shut down, the climate of the UK and Scandinavia would be much colder (see page 210–11).

Melting in the Arctic could open up trade routes, make travel in the region easier and allow exploitation of undersea minerals and fossil fuel reserves.

Methane clathrate is a form of ice under the Arctic ocean floor that traps methane. If this were to melt and reach the surface, the release of methane might trigger a rapid increase in temperature.

3 Effects on glaciers

In the Little Ice Age between about 1550 and 1850, glaciers increased in size. They then decreased (except for the period 1950–80 when global dimming possibly masked global warming) and have continued to decrease in size. Some have melted completely. Loss of glacier ice leads to flooding and landslides. Glacier summer melt provides a fresh water supply to people living below the glacier and this has provided water to many major Asian rivers (Ganges, Brahmaputra, Indus, Yellow, Yangtze) which are fed by the Himalayan glaciers.

4 Effects on weather patterns

More heat means more energy in the climate and so the weather will be more violent and sporadic with bigger storms and droughts. Global precipitation may increase by up to 15%. This will cause more soil erosion and lack of water will mean more irrigation and consequent salinization. There were more hurricanes in 2007 and some were more violent (e.g. hurricane Katrina). There is evidence that severe weather and more extreme rainfall or droughts are occurring. Monsoon rains fail more often than they used to.

5 Effects on food production

Warmer temperatures should increase the rate of biochemical reactions so photosynthesis should increase. But respiration will increase too so there may be no increase in NPP. In Europe, the crop growing season has expanded.

But if biomes shift away from the equator, there will be winners and losers. It very much depends on the fertility of the soils as well. If production shifts northwards from the Ukraine with its rich black earth soils to Siberia with its thinner, less fertile soils, NPP will fall. Various predictions state huge ranges of changes from −70% to +11%. There are just too many variables to be certain. But what is certain is that some crop pests will spread to higher latitudes as they will not be killed by cold winters.

In the seas, a small increase in temperature can kill plankton, the basis of many marine food webs.

Heatwaves and drought kill livestock.

6 Effects on biodiversity and ecosystems

Melting of the tundra permafrost would also release methane which is trapped in the frozen soils. In Alaska, Canada and Russia, permafrost is melting and houses built on it are shifting as it thaws.

Animals can move to cooler regions but plants cannot. The distribution of plants can shift as they disperse seeds which germinate and grow in more favourable habitats. But this happens slowly at about 1 km per year and perhaps too slowly to stop them becoming extinct. Species in alpine or tundra regions have nowhere to go, neither up nor towards higher latitudes. Polar species could become extinct in the wild.

Birds and butterflies have already shifted their ranges to higher latitudes.

Plants are breaking their winter dormancy earlier.

Loss of glaciers, decreased salinity of marine waters and changes to ocean currents alter habitats.

If droughts increase, then wildfires are more likely to wipe out other species or at least habitats for animals.

An increase in the temperature of fresh and salt water may kill sensitive species, and national parks and reserves could find their animals dying and the park boundaries static.

Indonesian forest fires have set peat bogs alight, which have burned continually for years. The amount of carbon released by these adds significantly to carbon in the atmosphere.

Pine forests in British Columbia are being devastated by pine beetle which is not being killed off by previously cold winters which have become milder.

Corals are very sensitive to increased sea temperature. An increase of 1°C can cause coral bleaching as the mutualistic algae in the corals are expelled and the coral dies. Corals are the basis for many food webs. If the corals die, the ecosystem dies.

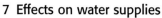

7 Effects on water supplies

Increased evaporation rates may cause some rivers and lakes to dry up. Without a water supply, human populations would have to move away. The UN says that 2.4 billion people live in the river basins fed by the Himalayas and their water supply is reducing. In Europe and North America, glaciers are also in retreat.

8 Effects on human health

Heatwaves killed many in Europe in 2006. These may increase. Insect disease vectors will spread to more regions as the less cold winters will not kill them. Malaria, yellow fever and dengue fever could spread to higher latitudes. Algal blooms may be more common as seas and lakes warm and some are toxic (red tides) and can kill humans. In a wetter climate, fungal disease will increase; in a drier climate, dust increases leading to asthma and chest infections.

Warmer temperatures in higher latitudes would reduce the number of people dying from the cold each year and reduce heating bills for households.

9 Effects on human migration

If people cannot grow food or find water, they will move to regions where they can. Global migration of millions of environmental refugees is quite possible and this would have implications for nation states, services and economic and security policies. The IPCC estimates 150 million refugees from climate change in 2050.

10 Effects on national economies

Some economies would suffer if water supplies decrease or drought occurs. Others would gain if it became easier to exploit mineral reserves (e.g. tar sands of Canada and Siberia) that would have been frozen in the permafrost or under ice sheets. If rivers do not freeze, hydroelectric power generation is possible at higher latitudes.

The Northwest Passage could be a sea route for shipping from the Atlantic Ocean to the Pacific Ocean via the seas of the Arctic Ocean north of Canada. Many explorers tried to find this route in the northern summers but were stopped by sea ice. In 2007, the passage was navigable for the first time in recorded history.

Overall, there will be gains and losses for national economies. Agricultural production may rise in higher latitudes but fall in the tropics. Africa will probably lose food production and rainfall. Northern Darfur in Sudan has seen desertification on a massive scale already as many millions of hectares become desert.

To put a monetary value on this is difficult but the Stern Report (from the former chief economist of the World Bank) suggested in 2006 that 1% of global GDP should now go to mitigating the effects of climate change to save up to 20% of global GDP in a recession later.

How fast could climate change happen? It is happening now and there will be more changes in your lifetime.

Feedback mechanisms and climate change

Feedback (see also p. 79) is the return of part of the output from a system as input, so as to affect succeeding outputs. There are two kinds of feedback:

- **Negative feedback** is feedback that tends to dampen down, reduce or counteract any deviation from an equilibrium, and promotes stability. For example, increased evaporation in tropical latitudes leads to increased snowfall on the polar ice caps, which reduces the mean global temperature.
- **Positive feedback** is feedback that amplifies or increases change; it leads to exponential deviation away from an equilibrium. For example, increased thawing of permafrost leads to an increase in methane levels, which increases the mean global temperature.

While climate modelling keeps improving, some feedbacks are so complex that we cannot be sure of the results.

To do

Table 7.2 lists changes that may have positive and negative feedbacks.

1　Draw diagrams to show these feedbacks, labelling storages, flows, inputs and outputs.

Also look at Fig. 4.21 on page 81 and Review box on page 82 to see some of these feedbacks.

2　Look at the effects of climate change again and put them in diagrammatic form to include the effects, positive and negative changes and feedback mechanisms.

	Positive feedback or amplified change	Negative feedback or dampened down change
Oceans	Oceans are a carbon sink containing 50 times the amount of carbon as the atmosphere. They release more carbon dioxide to the atmosphere as they warm up as warm liquids hold less gas. Stalling of the North Atlantic Drift could reduce transfer of heat to the north and increase temperatures dramatically. Huge amounts of methane are frozen in methane clathrates in the ocean sediments. If these are released, the volume of methane in the atmosphere will increase dramatically.	Oceans absorb more carbon dioxide in warmer water as phytoplankton photosynthesize faster, producing more phytoplankton that absorb more carbon dioxide so dampening global warming.
Clouds	More evaporation leads to more clouds which trap more heat.	More evaporation leads to more clouds which reflect more heat.
	In the dark, clouds keep heat in, in the light they reflect it. But it depends what type of cloud as well. Cirrus (high, thin) clouds have a warming effect, low, thick ones a cooling one.	
Pollution	At night, cloud formation is increased by aerosols; clouds act as insulation, trapping heat. More clouds, more heat trapped. Black soot falling on ice decreases albedo, increasing heat absorption, increasing temperature and melting.	Aerosols from pollution, particularly sulfates, form condensation nuclei and more clouds form. These reflect heat and increase albedo, reducing warming in the day.
Polar ice	Ice has a high albedo – reflects heat and light. When it melts, the sea or land have a lower albedo and absorb more heat and more ice melts.	Warmer air carries more water vapour so more rainfall, some of which will be snow so more snow, more reflection, lower temperatures, more snow and ice. Possibly the next ice age.
Forests	Forests are cut down and burnt. Less carbon is absorbed. More CO_2 in the atmosphere so higher temperatures. Forests die due to high temperature and may catch fire, more CO_2 released, temperature rises.	Forests act as a carbon sink, removing carbon dioxide from the atmosphere, so temperature rise decreases.
Tundra	As temperatures rise, permafrost melts, releasing CO_2 which is trapped in the frozen soil. Methane is also released.	

Table 7.2 Possible feedback mechanisms in climate change

Carbon emissions from fossil fuel burning

Although figures vary from source to source, there is general
agreement about the amount of carbon dioxide released to the
atmosphere by burning fossil fuels. Carbon dioxide is responsible for
two-thirds of the anthropogenic (enhanced) greenhouse effect. Most
of our efforts are going to reduce these emissions of carbon dioxide,
which may not look much in parts per million but amount to four
tonnes per year per capita on average worldwide. Since publication,
the data have dated rapidly as Asian economic growth has
continued. China is almost certainly now the largest emitter, having
overtaken the United States in carbon emissions. Look up the latest
information on this.

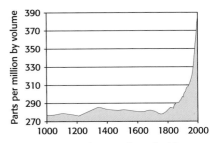

Fig. 7.13 Atmospheric carbon dioxide
levels 1000–2000

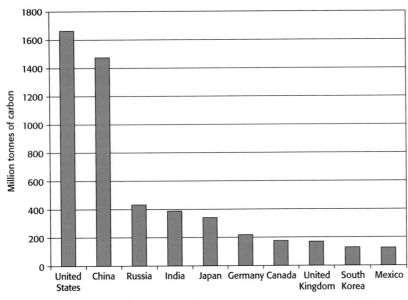

Fig. 7.14 Carbon emissions from fossil fuel burning in the top ten countries, 2006

Who produces what carbon emissions?

Country	Emissions	Share of global total
	Million tonnes of carbon	Percent
United States	1656	19.8
China	1480	17.7
Russia	437	5.2
India	391	4.7
Japan	342	4.1
Germany	221	2.6
Canada	177	2.1
United Kingdom	171	2.0
South Korea	130	1.6
Mexico	123	1.5
All other countries	3249	38.8
Global total	**8379**	**100.0**

Table 7.3 Carbon emissions from fossil fuel burning in the top ten countries, 2006

Reminder: 1 Gt is 1 billion
tonnes (10^9 tonnes) and 1
Gt C (carbon) corresponds to
3.67 Gt CO_2 (carbon dioxide).

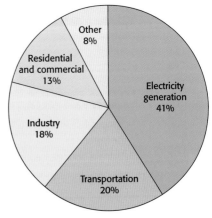

Fig. 7.15 Contributions of sectors to global
carbon dioxide emissions in 2004

Country	Emissions per person
	Tonnes of carbon
Qatar	22.4
United Arab Emirates	13.3
Kuwait	10.4
Singapore	9.2
USA	5.5
Canada	5.4
Norway	5.3
Australia	4.5
Kazakhstan	4.1
Saudi Arabia	3.9
World average	**1.3**

Table 7.4 Carbon emissions from fossil fuel burning per person for top ten countries and world, 2006

According to the Earth Policy Institute, carbon emissions from fossil fuel burning was 8.38 Gt C in 2006, 20% above the 2000 level and running at an increase per year of about 3.1%. Five countries produce over half the fossil fuel-related carbon dioxide emissions; China and the US produce the most and China may now have overtaken the US as the biggest emitter. The rapid growth in emissions in China and India reflect their rapid industrialization and large size. But per capita emissions do not relate to the size of the country. Singapore and some Gulf states have the highest per capita emissions, followed by the US with per capita emissions five times that of China.

The United Nations calculates that an average air-conditioner in Florida is responsible for more carbon dioxide every year than a person in Cambodia is in a lifetime, and that a dishwashing machine in Europe annually emits as much as three Ethiopians. But a higher per capita emission level does not necessarily mean a higher standard of living.

Deforestation also contributes to carbon emissions and Indonesia would be the highest emitting country of greenhouse gases if emissions due to deforestation and forest fires were included.

Strategies to alleviate climate change

There are three strategies that we can adopt on this issue: do nothing, wait and see, or take precautions now. Science cannot give us 100% certainty on the issue of global warming nor predict with total accuracy what will happen. What it can do is collect data and provide evidence. How that evidence is interpreted and extrapolated will depend on individual viewpoints, scientific consensus, economics and politics.

A minority of scientists and others do not accept that global warming and climate change is a problem for human activity and development on Earth. They may take the "do nothing" and business-as-usual approach, saying that we may forfeit economic development, and progress out of poverty for many, by reacting to a non-threat. Alternatively, they say that warming is a good thing and technology can manage the effects.

151

The danger in the "wait and see" strategy is that it takes a long time for actions to have results. To move the global economy away from a fossil fuel base is a long, slow process and the possible disruption of national economies incurred may not be necessary. But it is possible that we will reach the tipping point, when our actions will have little effect as positive feedback mechanisms change the climate to a new equilibrium which could be 8°C warmer than it is now. So better safe than sorry, perhaps.

The precautionary strategy is the majority choice. Act now in case. Even if we find out that fossil fuel burning is not causing global warming, we know these fuels will run out and it makes sense to clean up the Earth and find alternative fuel sources now, before we run out. What we are seeing in national policies and international targets are precautions – carbon emission reduction, carbon offset and lifestyle changes – against increased climate change.

These precautions can be divided into three categories: international commitments, national actions and personal lifestyle changes.

International action: a timeline of agreements and commitments for action

Year	Events
1979	First World Climate Conference. Climate change officially recognized as a serious problem needing an international response when evidence of increasing carbon dioxide levels established.
1988	**Intergovernmental Panel on Climate Change (IPCC)** established by United Nations Environment Programme (UNEP) and the World Meteorological Organization (WMO). The IPCC is a collaborative activity comprising over 2000 climate scientists worldwide. Its main activity is to provide at regular intervals an assessment of the state of knowledge on climate change
1990	**First IPCC Report** on Climate Change. The Report confirms that *climate change is a reality and is supported by scientific data.*
1992	UN Earth Summit, Rio de Janeiro (United Nations Conference on Environment and Development). **United Nations Framework Convention on Climate Change (UNFCCC)** signed by 154 governments.
1995	First UNFCCC conference. Governments recognized that voluntary commitments were inadequate and work started to draft a protocol for adoption at the third Conference of Parties in 1997. **Second IPCC Report** concludes that *the balance of evidence suggests a discernible human influence on the global climate.*
1997	**Kyoto Protocol** signed by some 160 nations at third UNFCCC conference. The Protocol calls for the first ever legally binding commitments to reduce carbon dioxide and five other greenhouse gas emissions to 5.2% below 1990 levels before 2012. The US signed but has not ratified the protocol.
2001	**Third IPCC Report** states that *anthropogenic emissions will raise global mean temperature by 5.8°C by 2050.*
2004	**Kyoto Protocol is still ineffective**. For the Kyoto Protocol to be effective at least 55 countries have to ratify (fully adopt the commitments) and there must be enough annex I (developed) countries who together are accountable for more than 55% of the emissions according to the 1990 levels. However the percentage of annex I countries is only 37.5%.
2005	**Kyoto Protocol goes into effect**, signed by major industrial nations except US. Work to retard emissions accelerates in Japan, Western Europe, US regional governments and corporations.
2007	**Fourth IPCC Report** warns that serious effects of warming have become evident; cost of reducing emissions would be far less than the damage they will cause. In December 2007 UN climate conference in Bali agreed on the **Bali road map** which set the framework for negotiations on a long-term agreement on emissions cuts and recognized the urgency. However, it did not specify emission reduction levels.
2008	**Bangkok Climate Change talks** in April and **Poznan, Poland, Climate Change conference** in December, negotiations continued towards an international agreement to be concluded in Copenhagen at the end of 2009 to take effect in 2012.

Management strategies can be looked at using the pollution management model in Table 7.5.

Strategy for reducing global emissions	Example of action
Altering the human activity producing pollution	Improve efficiency of energy production Energy-efficient light bulbs and appliances Electric and hybrid vehicles Replace fossil fuel, especially coal, use by using renewable or nuclear energy Reduce overall demand for energy and electricity by being more efficient and using less Insulate and cool buildings more effectively Use less private or any transport; cycle and walk more Fly less often Reduce deforestation Sustainable agriculture Slow population growth Adopt carbon taxes and remove fossil fuel subsidy Set national limits on greenhouse gas production and a carbon credit system Low emissions zones in cities
Regulating and reducing the pollutants at the point of emission	Capture carbon dioxide using carbon capture technologies and store it underground Reduce methane emission by cows with diet/fewer cows Replace CFCs Add iron to oceans to remove more carbon dioxide from the air Plant more trees or fast-growing plants that act as carbon sinks
Clean-up and restoration	Place mirrors in space to reflect sunlight away from the Earth Add sulfate particulates to the atmosphere to make more clouds Develop crops that need less water Use more efficient irrigation Relocate people away from regions that could be flooded by sea level rise Stop building on flood-susceptible land Stockpile foods Move nature reserves to higher latitudes as biomes shift

Table 7.5 Clean-up actions on climate change

To research

1 Find out what major climate change meetings there have been since 2007 and what the outcomes are.

2 Find out:
   a What is your own government policy on climate change and carbon emissions?
   b What alternative energy sources is the country in which you live developing?
   c What are the advantages and disadvantages of these?

3 List the possible ways that countries could reduce their carbon emissions.

4 Research your own carbon footprint (www.carbonfootprint.com is a good place to start).
   a Calculate your carbon footprint. (This measures your carbon use in tonnes of carbon dioxide, not hectares as in ecological footprints.)
   b List as many ways as you can of reducing your own carbon footprint size.
   c How many of these will you do?

5 Which of the strategies you have listed would be the most effective in reducing carbon dioxide emissions and why? Consider whether they need people to co-operate, if they reduce your quality of life, if the technology is available, how easy they are to do. Are the strategies ecocentric or technocentric?

To think about

Carbon offset and carbon emissions trading

As part of the Kyoto Protocol in which 163 countries agreed to aim to limit their greenhouse gas production, the concept of carbon emissions trading evolved. In this scheme, countries that go over their quota (set by international agencies) on carbon dioxide emissions can buy carbon credits from countries which do not meet their quotas. In this way they still produce carbon dioxide but, globally, the limits are still met. You can imagine how complex this system is to operate. Monitoring carbon emissions for an industry is hard enough but for a country it is very difficult. Which country owns the emissions from an international flight or container ship – where it started or ends up, or the country where the airline is based? Who sets the quotas? While a market has grown up for trading carbon emission permits, it is a volatile one. The EU emissions trading scheme, for example, has seen the value of carbon credits fall due to an overestimate of the allocation required when it started. The scheme does not encourage industries or countries to reduce their emissions either if they can buy permission to continue emitting. Carbon emissions trading is an alternative to a carbon tax where organizations are taxed for polluting, i.e. for releasing carbon dioxide.

1 Which do you think is the better option, carbon emissions trading or carbon tax?

There are voluntary schemes now to offset carbon emissions for individuals and companies.

Book a plane ticket now and you will be asked by many airlines if you want to pay to offset the carbon emissions that you are causing by flying on a plane. If you agree to this, the money should go to a company that invests it in a scheme that reduces carbon emissions. This is usually in a renewable energy scheme such as wind turbines, tree planting or hydroelectric power generation. Although small, the market is increasing as environmentally aware individuals invest. But some schemes have minimal impact on global carbon emissions as they must invest in a scheme that would otherwise not have happened. Just taking the money and planting trees which would still have been planted is not recapturing any more carbon dioxide.

2 To become "carbon neutral" is a goal that some talk about. Do you think this is possible for an individual? What actions would you take in your home and in your lifestyle to try to achieve this?

Personal carbon allowances (PCAs) are another idea. This would mean we are all issued with an allowance for carbon emissions. If we travel a lot or live in energy-inefficient homes, we would either have to change our lifestyles or buy more PCAs on the open market from people who produce less carbon dioxide.

3 What lifestyle changes would you be prepared to make if PCAs become mandatory?

4 How do you think different societies would react to this?

Global dimming

Global dimming is a reduction in solar radiation reaching the surface of the Earth. It was first noticed in the 1950s when scientists in Israel and Australia measured pan evaporation rates (evaporation of water from a pan). In the 1980s, more research in Switzerland, Germany and the USSR also found that the incoming radiation was less than it had been. At the time, most were skeptical of the results which showed a reduction in the rates because of a reduction in sunlight and so a cooler Earth at a time when it conflicted with evidence that the Earth was getting warmer. When aircraft were grounded for a few days in the US after 9/11, the absence of contrails (vapour trails from aircraft) produced a sharp rise in the range of the Earth's surface temperature.

Fig. 7.16 Contrails

Other experiments measuring sunlight levels over the Maldives showed that particulates in the atmosphere were causing global dimming. The small pollutant particles of mostly sulfate aerosols in the clouds both block sunlight from reaching the Earth's surface and reflect it back into space. These act as nuclei around which water droplets form and the clouds then reflect more sunlight back into space. Other particles come from volcanic eruptions, dust storms and incomplete burning of fossil fuels which produce black carbon or soot. It appears that global dimming affects weather patterns to the extent of shifting the monsoon rains and causing the long-term drought in 1970s and 1980s in sub-Saharan Africa.

The drop in temperature caused by global dimming was about 2–3% from 1960 to 1990 but no drop has been recorded since then. The amount varies around the Earth with more in the temperate zones of the Northern Hemisphere and may be masking the full increase of global warming. There is some evidence that, as we clean up our atmospheric pollutants, global dimming decreases and we see the full effect of global warming.

While global warming would increase temperatures and give more energy to the water cycle, decreased energy input would slow it down and make the climate more humid with less rain.

It has been suggested that we could control global dimming and so mitigate the effects of global warming. Putting sulfur in jet fuel would cause it to produce sulfates in flight and so cool the Earth. But the danger in this is that we would be creating more pollution, not less, and may find ourselves in a cycle of increased warming and increased pollution to counteract it. Perhaps not the best idea?

Viewpoints on climate change

Viewpoint 1
George Monbiot is an author, journalist and environmental campaigner based in Wales.

In his book, *Heat: How to Stop the Planet Burning*, Monbiot shows how our carbon emissions can be reduced by 90% by 2030 and without bringing civilization to an end. He argues that this is technologically possible with a will from governments but that industrial leaders and politicians do not make this a priority.

Excerpts from *Heat: How to Stop the Planet Burning* by George Monbiot, 2006

"When you get your 80% cut, what will this country look like?"

"A very poor third-world country."

"When this civilization falls, when the Romans, whoever they are this time round, have finally left and the new dark ages begin, this will be one of the first luxuries to go. The old folk crouching by their peat fires will tell their disbelieving grandchildren of standing naked mid-winter under jet streams of hot clean water, of lozenges of scented soaps and of viscous amber and vermilion liquids they rubbed into their hair to make it glossy and more voluminous than it really was, and of thick white towels as big as togas, waiting on warming racks."

Fossil fuels helped us fight wars of a horror never contemplated before, but they also reduced the need for war. For the first time in human history, indeed for the first time in biological history, there was a surplus of available energy. We could survive without having to fight someone for the energy we needed. Our freedoms, our comforts, our prosperity are all the products of fossil fuel, whose combustion is also responsible for climate change. Ours are the most fortunate generations that have ever lived. Ours might also be the most fortunate generations that ever will. We inhabit the brief historical interlude between ecological constraint and ecological catastrophe.

If carbon dioxide released from the burning of fossil fuels reaches a certain concentration in the atmosphere – 430 parts per million parts of air – the likely result is two degrees of warming. Two degrees centigrade is the point beyond which certain major ecosystems begin collapsing. Having, until then, absorbed carbon dioxide, they begin to release it. This means that 2° inevitably leads to 3°. This in turn triggers further collapses, releasing more carbon and pushing the temperature 4–5° above pre-industrial levels: a point at which the survival of certain human populations is called into question. Beyond 2° of warming, in other words, climate change is out of our hands: there is nothing we can do to prevent it from accelerating. The only means, Forrest argues, by which we can be fairly certain that the temperature does not rise to this point is for the rich nations to cut their greenhouse gas emissions by 90% by 2030. This is the task whose feasibility *Heat* attempts to demonstrate.

After looking at what the impacts of unrestrained climate change might be, and at why we have been so slow to respond to the threat, I begin my search for solutions within my own home. I show how years of terrible building, feeble regulations and political cowardice have left us with houses scarcely able to perform their principle function, which is keeping the weather out. I look at the means by which our existing homes could be redeemed and better ones could be built, and discover what the physical and economic limits of energy efficiency might be.

What is the point of cycling into town when the rest of the world is thundering past in monster trucks? By refusing to own a car, I have simply given up my road space to someone who drives a hungrier model than I would have bought. Why pay for double-glazing when the supermarkets are heating the pavement with the hot air blowers above their doors? Why bother installing an energy-efficient lightbulb when a man in Lanarkshire boasts of attaching 1.2 million Christmas lights to his house? (Mr Danny Meikle told journalists that he needs two industrial meters to measure the electricity he uses. One year his display melted the power cable supplying his village. The name of the village – which proves, I think, that there is a God – is Coalburn.)

Viewpoint 2

The second viewpoint is expressed by Al Gore. He presented his lecture on climate change around the world before making the film *An Inconvenient Truth*. In 2007, he and the IPCC won the Nobel Peace Prize for their actions. While the science in the film is basically sound, some hyperbole was used in the figures presented.

The following extract is from the website http://www.climatecrisis. net/thescience/.

An Inconvenient Truth by Al Gore

The vast majority of scientists agree that global warming is real, it's already happening and that it is the result of our activities and not a natural occurrence. (This is according to the IPCC, this era of global warming "is unlikely to be entirely natural in origin" and "the balance of evidence suggests a discernible human influence of the global climate.") The evidence is overwhelming and undeniable.

We're already seeing changes. Glaciers are melting, plants and animals are being forced from their habitat, and the number of severe storms and droughts is increasing. The number of Category 4 and 5 hurricanes has almost doubled in the last 30 years. Malaria has spread to higher altitudes in places like the Colombian Andes, 7000 feet above sea level. The flow of ice from glaciers in Greenland has more than doubled over the past decade. At least 279 species of plants and animals are already responding to global warming, moving closer to the poles. [All these statements have citations on the website as do the ones below.]

If the warming continues, we can expect catastrophic consequences. Deaths from global warming will double in just 25 years – to 300 000 people a year. Global sea levels could rise by more than 20 feet with the loss of shelf ice in Greenland and Antarctica, devastating coastal areas worldwide. Heatwaves will be more frequent and more intense. Droughts and wildfires will occur more often. The Arctic Ocean could be ice-free in summer by 2050. More than a million species worldwide could be driven to extinction by 2050.

There is no doubt we can solve this problem. In fact, we have a moral obligation to do so. Small changes to your daily routine can add up to big differences in helping to stop global warming. The time to come together to solve this problem is now – TAKE ACTION.

Al Gore's claims have been reviewed by respectable journals.
http://news.nationalgeographic.com/news/2006/05/060524-global-warming.html
http://www.newscientist.com/blog/environment/2007/10/al-gores-inconvenient-truth.html
Read these articles and summarize the arguments in 500 words.

Climate modelling

We all want to know what the weather will be and we base our predictions on past experience (what it was like last June) and authority (the experts who tell us their informed predictions). We have been trying to model the climate using general circulation models (GCMs) since the first computers. The early ones had few inputs and were not very accurate predictors but we now have atmosphere-ocean GCMs (AOGCMs) which split the Earth into subregions and consider the inputs of the atmosphere, oceans, ice sheets, land and biosphere. This includes the effect that humans are having on forests (deforestation). The latest AOGCMs quite accurately reflect past climate change and have predicted changes with various concentrations of greenhouse gases. These are reported by the IPCC.

To do

1   Watch the film *An Inconvenient Truth*. Watch *The great global warming swindle*. Watch *Global Dimming* (a BBC Horizon documentary) or the NOVA programme *Dimming the Sun*.
2   Read peer-reviewed articles and books on the issue of climate change. (There is a lot of propaganda out there. Make sure you are reading a reliable source.)
3   Form an opinion based on evidence and decide if you will take any actions to reduce your carbon emissions.

What next on global warming and climate change?

Your parents or grandparents were probably concerned about the Cold War, nuclear proliferation and deterrent, the Second World War or human population growth. Until the 1980s, we were not utilizing more than the Earth could provide and were not all really aware of, or chose to ignore, climate change as an issue affecting all life on Earth. Now and for the rest of your lifetime, the climate change debate will not go away as it is a moral as well as an economic issue and will affect your lifestyle and choices.

The moral dilemmas involve your personal choices (whether to fly or not, to drive or not, what type of car, to live in a city or the countryside) as these all have an effect on your personal carbon emissions. There is an ethical question for governments of MEDCs whose people mostly have a high standard of living – why should they deny this to those in LEDCs by asking them to decrease their economic development rate? In the economic model we have, LEDCs are aiming to increase their GDP (gross domestic product) and so improve the lives of their people and would be unwilling to reduce growth rates.

The cost of alleviating climate change has been given a value of 2% of the world's GDP. Should governments all vote to give 2% of their GDP towards this and so not towards something else? We have and can develop the technology to make cleaner energy production systems but we do not have the collective will to put efforts into this as long as we can burn fossil fuels. Perhaps this is because democratic politics is short-termist – until the next elections, we do not plan for 50 or 100 years but for less than five.

Humans are resourceful and climate change will not make the human species extinct. But it will alter your lifestyle choices and possibilities and you should understand the science and the politics behind decision-making on these issues.

Review

1 Explain what is meant by the greenhouse effect.

2 Why is the greenhouse effect normally of benefit to the Earth and its organisms?

3 List four greenhouse gases in order of importance.

4 Explain the sources of these four greenhouse gases.

5 Describe the increase in the levels of carbon dioxide, i.e. how much has it increased, when did the increase begin and what are the main causes of the increase in carbon dioxide level.

6 State whether the levels of the other three greenhouse gases have increased since 1850 and, if so, why?

7 Explain how at least four human activities such as deforestation, burning fossil fuels, rice and cattle farming and use of CFCs add to greenhouse gases.

8 Discuss four ways in which global emission of greenhouse gases can be reduced (include conservation of energy (including carbon tax) and use of alternative energy sources).

9 What is global warming and how is it related to the greenhouse effect?

10 Why might the following processes occur due to global warming:
   a thermal expansion of the oceans,
   b melting of the polar ice caps,
   c increased evaporation in tropical latitudes leading to increased snowfall on the polar ice caps, triggering a new ice age,
   d the effect of air pollutants (aerosols) in reflecting radiation, thus offsetting the warming trends?

11 How may global warming affect the planetary distribution of biomes?

12 If the distribution of biomes changes what impacts might this have on global agriculture?

13 What international measures have been taken to combat global warming? How successful have these measures been?

14 List five ways in which emissions of greenhouse gases can be reduced in your local community.

15 Discuss the complexities and uncertainties surrounding the issue of climate change.

Key words

greenhouse gas
GHG
greenhouse gas emissions
anthropogenic
enhanced greenhouse effect
radiative forcing
clathrates
carbon dioxide/carbon sink

carbon budget
Gt C
ice cores
IPCC
carbon emissions
carbon offset
carbon taxes

8 Population dynamics

Key points

- A given area can only support a certain size of population.
- Population numbers may either crash or reach an equilibrium around the carrying capacity. The human population is growing at an exponential rate overall.
- The population of LEDCs is 80% of world population and this is growing faster than MEDCs.
- Crude birth rate, death rate, fertility, doubling time and natural increase rate are all indicators of population change.
- Population pyramids and the demographic transition model show how populations change over time.
- These models help in predicting population changes.

"It is well to remember that the entire universe, with one trifling exception, is composed of others."

John Andrew Holmes Jr

Population changes

Over time the numbers of individuals within a population change. If we were to collect a few bacterial cells, place them in a suitable supply of nutrients and then, under a microscope, count the number of cells every hour, we would find that there would be many more bacteria at the end of a 24-hour period than at the start. Bacteria can reproduce asexually by splitting in two (binary fission) so, if you start with one bacterium, there will be 2, 4, 8, 16, 32, 64, etc. if there are no limiting factors slowing growth. This is called **exponential** or **geometric growth**.

The human population has also been growing at an exponential rate.

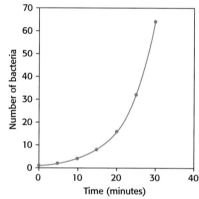

Fig. 8.1 Exponential growth in a bacterial population over time

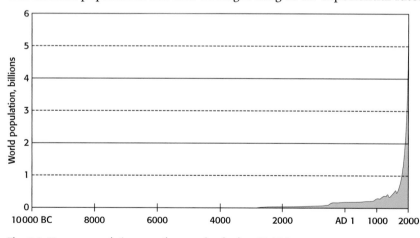

Fig. 8.2 Human population growth curve for the last 12 000 years

Limiting factors

Exponential growth only occurs when a species lives under optimal conditions, with enough food, water and space. However, limiting factors often reduce or stop population growth. There are two types of limiting factors: density-dependent limiting factors and density-independent limiting factors.

In general, **density-dependent limiting factors** are biotic. Their effects increase with increasing population size. Density-dependent limiting factors act as **negative feedback** mechanisms and lead to stability or regulation of the population. Density-dependent limiting factors can be further divided into internal and external factors.

Internal factors act within a species. Examples include limited food supply, limited availability of territories, and density-dependent fertility. A limited food supply will lead to competition between population members. The stronger, more efficient, more aggressive individuals will get a larger share of the food. The less successful individuals get less food and will have a hard time surviving and reproducing. Territorial animals can suffer from a lack of suitable territories. The weaker animals will not have a territory and may therefore starve or be unable to find a partner and reproduce. Some species show a reduced fertility and high population densities. An example is the gerbil, a common pet. The intraspecific competition between seagulls (page 63) is an example of a density-dependent limiting factor: the effect depends on the population density (low density, small effect; high density, large effect). This is mainly associated with pressure for food, nutrients or space.

External factors act between different species. Two examples are predation and disease. When the population size of a prey animal increases, predators will find their prey more easily. This will reduce the population size of the prey animal. Furthermore, the predators will be producing more offspring, which also need to feed on the prey animals. Finally, the population size of the prey animal will decrease, which will lead to a decrease in the predator population, e.g. the Canadian lynx and snowshoe hare (pages 81–2). Diseases tend to be density dependent. At high population densities, diseases can spread very fast and strongly affect population size. The Black Death was a world pandemic for humans in the 1300–1400s. It probably killed one-sixth of the world's human population (75 million – equivalent to 1 billion people today) and spread via rats among a crowded population that was short of food.

Density-independent limiting factors tend to be abiotic. Their effects are not related to population density. Examples include weather (short-term effects like a storm), climate (long-term weather conditions like a cold winter or a dry summer), volcanic eruptions and floods. Note that these limiting factors are not part of a feedback system, as their effect is by no means related to population size. Perhaps wars are density-independent limiting factors for humans.

These two types of limiting factors give rise to two different growth patterns: S-curves and J-curves.

S-curves

S-curves (see Fig. 8.3) start with exponential growth. However, above a certain population size, the growth rate slows down gradually, finally resulting in a population of constant size. The growth is slowed down more in larger populations, so this pattern is consistent with density-dependent limiting factors.

Fig. 8.3 illustrates this for a colony of yeast grown in a constant but limited supply of nutrient. During the first few days the colony grows slowly as it starts to multiply (lag phase), then it starts to grow very rapidly as the multiplying colony has a plentiful nutrient supply (exponential phase). Eventually the population size stabilizes as only a set number of yeast cells can exploit the limited resources (stationary phase). Any more yeast cells and there is not enough food to go around. The numbers stabilize at the **carrying capacity** of the environment which is the maximum number or load of individuals that an environment can carry or support.

The maximum population size is called **carrying capacity** (K) of the ecosystem. The area between the exponential growth curve and the S-curve is called **environmental resistance**. This is any factor that limits the increase in numbers of the population (e.g. lack of food, space, light, predation or disease).

J-curves

J-curves (see Fig. 8.4) show a "boom and bust" pattern. The population grows exponentially at first and then suddenly collapses. These collapses are called **diebacks**. Often the population exceeds the carrying capacity before the collapse occurs (**overshoot**). Therefore, the carrying capacity is best defined as the maximum number of individuals of a species that can be supported by a particular ecosystem on a long-term basis. It is important to include "long-term basis" as the carrying capacity can be exceeded in the short term. It seems likely that the human race is overshooting its carrying capacity at the moment.

The J-curve does not show the gradual slowdown of population growth with increasing population size. So, most likely, a density-independent limiting factor is causing the population decrease.

A J-shaped population growth curve is typical of microbes, invertebrates, fish and small mammals.

S- and J-curves are idealized curves. In practice, both density-dependent and density-independent limiting factors act on the same population and the resulting population growth curve normally looks like a combination of an S- and a J-curve.

The growth rate of the human population is slowing, as is predicted for us as we reach the carrying capacity of the environment. The rate of growth peaked at 2.1% per year in 1965–70 and is now 1.3% and falling, but falling more in some regions (most MEDCs; see p. 167) than others.

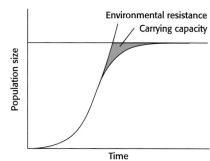

Fig. 8.3 S-shaped growth curve of a population

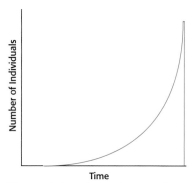

Fig. 8.4 J-shaped growth curve of a population

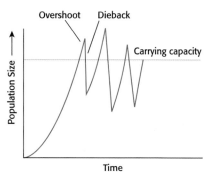

Fig. 8.5 Fluctuation of population size around the carrying capacity

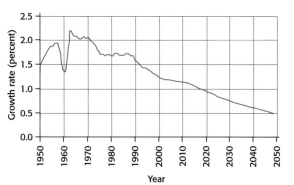

Fig. 8.6 World human population growth rates 1950–2050

Prior to the 1800s, the world population was increasing only slowly due to environmental resistance (both density-dependent and density-independent factors): diseases (smallpox, measles, scarlet fever with high infant mortality), epidemics (typhoid, cholera which killed adults), famine and natural catastrophes. High birth rates were balanced by high mortality rates (especially of children). The population explosion occurred after the Industrial Revolution because of improved medicine, education and nutrition.

K- and r-strategists: reproductive strategies

Species can be roughly divided into **K- and r-strategists or K- and r-selected species**. K and r are two variables that determine the shape of the population growth curve, where r is the growth rate of the population and K is the carrying capacity of the environment.

Different species vary in the amount of time and energy they use to raise their offspring. There are two extremes. Species like humans and other large mammals have small numbers of offspring but invest large amounts of energy in parental care, and most offspring survive. These are K-strategists. K-strategists are good competitors. Population sizes of K-strategists are usually close to the carrying capacity, hence their name. In stable, climax ecosystems, K-strategists outcompete r-strategists.

In contrast, invertebrates and fish use lots of energy in the production of vast numbers of eggs but none in raising the young after hatching. They lay their eggs and leave them forever. They are r-strategists. The r-strategists reproduce quickly, are able to colonize new habitats rapidly and make opportunistic use of short-lived resources. Because of their fast reproductive and growth rates, they may exceed the carrying capacity, with a population crash as a result. They predominate in unstable ecosystems.

r-strategist	K-strategist
short life	long life
rapid growth	slower growth
early maturity	late maturity
many small offspring	fewer large offspring
little parental care or protection	high parental care and protection
little investment in individual offspring	high investment in individual offspring
adapted to unstable environment	adapted to stable environment
pioneers, colonizers	later stages of succession
niche generalists	niche specialists
prey	predators
regulated mainly by external factors	regulated mainly by internal factors
lower trophic level	higher trophic level
Examples: annual plants, flour beetles, bacteria	trees, albatrosses, humans

Table 8.1 Typical characteristics of r- and K-strategists

It is important to appreciate that K- and r-strategists are the extremes of a continuum of reproductive strategies and many species show a mixture of these characteristics.

Survivorship curves

A survivorship curve shows the fate of a group of individuals of a species. Three hypothetical survivorship curves are shown in Fig. 8.7. Note that the vertical axis is logarithmic.

Curve I is typical for *K*-strategists. Because of their small number of large offspring and parental care, mortality at low age is small and most of the individuals reach reproductive age. They live for most of their lifespan.

Curve III is typical for *r*-strategists. High mortality occurs in the very early stages of the species' life cycle.

Curve II is rather rare. It represents species that have an equal chance of dying at any age. It occurs for example in the hydrozoan *Hydra* and some species of birds.

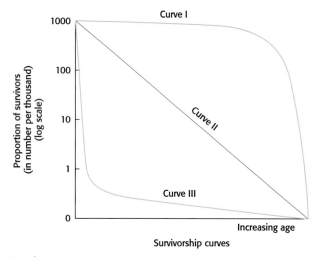

Fig. 8.7 Survivorship curves

Test yourself

1 List the characteristics of *r*-strategists and *K*-strategists and give three examples of each type.
2 Explain whether *K*- or *r*-strategists would be more likely to be regulated by density-independent limiting factors, e.g. the weather.
3 Describe and explain the shape of the survivorship curve for *K*-selected species.

Look at Fig. 8.8, the diagram of a theoretical survival model of a small bird population.

4 What is the lifespan of these birds?
5 What is the potential population size?
6 How many survivors after the first year?
7 What is the percentage mortality at end of year 1?
8 What is the percentage mortality at end of year 2?
9 What is the percentage mortality at end of year 3?
10 What would the survivorship curve for these birds look like?

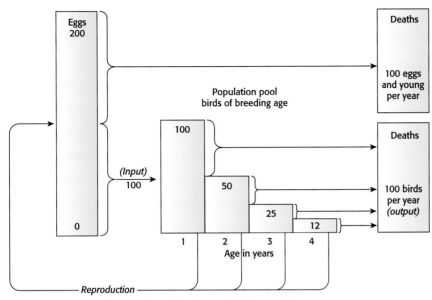

Fig. 8.8 A theoretical survival model of a small bird population

Setting the scene on human populations

Enter "world population clock" into a search engine and you will find a number of websites that give an estimate of the human population on Earth. In August 2008 the US Census Bureau figure was **6 720 225 893**.

What is the population on the day you are reading this? The global birth rate is about 6 per second and death rate is about 3 per second. So in the time it takes you to read this, how many extra humans are there on Earth? Each year about 90 million people are born at current growth rates. Our growth rate is phenomenal. It took about 100 years for the population to double from 1.25 billion to 2.5 billion, from 1850 to 1950. In 1990, there were 5 billion, double the population in 40 years. Predictions are that, even with slowing growth rates, it will double again within another 100 years. Of the 6.6 billion humans alive today, about half live in poverty.

The US Census Bureau has made this prediction for human population growth (Table 8.2):

Year	Population (billions)
2010	6.8
2020	7.6
2030	8.3
2040	8.9
2050	9.4

Table 8.2

Inequalities of life

If we could reduce the world's population to a village of precisely 100 people, with all existing human ratios remaining the same, the demographics would look something like this.

The village would have 60 Asians, 14 Africans, 12 Europeans, 8 Latin Americans, 5 from the USA and Canada, and 1 from the South Pacific.

51 would be male; 49 would be female

82 would be non-white; 18 white

67 would be non-Christian; 33 would be Christian

80 would live in substandard housing

67 would be unable to read

50 would be malnourished and 1 dying of starvation

33 would be without access to a safe water supply

39 would lack access to improved sanitation

24 would not have any electricity (and of the 76 that do have electricity, most would only use it for light at night)

7 people would have access to the Internet

1 would have a college education

1 would have HIV

2 would be near birth; 1 near death

5 would control 32% of the entire world's wealth; all 5 would be US citizens

33 would be receiving – and attempting to live on – only 3% of the income of "the village".

These statistics come from the *State of the Village Report* by Donella H. Meadows, published in 1990 as *Who lives in the Global Village?* and updated in 2005. It is controversial as some people think she was biased in the use of statistics and some of these are inaccurate. Examples of this are:

- Male:female ratio is 1.05:1
- Almost 80% of the world's population is now literate
- There is less than 1/6th of the world's population malnourished
- About 3% of the world's population will have a college education
- About 9% will now own a computer
- The US controls no more than 30% of the world's wealth.

However, it is a stark demonstration of the haves and have-nots.

Do you think this is an example of propaganda for a particular viewpoint or does it make valid points about the unequal distribution of wealth and goods on Earth – or is it both?

Are there too many humans alive on the Earth today? Have we exceeded the carrying capacity of the Earth? Are we heading for a population crash? The difficulty we have in trying to answer these questions is that humans are able to manipulate the environment. We can increase the carrying capacity locally, for example by living in large cities, or living in regions that traditionally cannot grow enough food for the population by using technology to improve productivity. In this chapter, we look at ideas concerning **demographics –** the study of the dynamics of population change.

Too many people in one geographical area, using local resources beyond a sustainable level and producing waste at a rate that does not allow assimilation back into the environment without causing damage, could be viewed as an environmental problem that is directly associated with the size of the population. Too many families dependent on a single spring may lead not only to the depletion of the spring but also to a depletion of other resources around the spring, e.g. firewood, food plants and animals. The contamination of waterways, coastlines and terrestrial habitats by pollution as urban centres expand, or the development of virgin land for housing, could be driven as much by the forces of the marketplace as it is by human population expansion in general.

LEDCs and MEDCs

The **Human Development Index** (**HDI**) has been adopted by the UN Development Programme as a measure of the "well-being" of a country. It combines measures of life expectancy, standards of living, education and gross domestic product (GDP) per capita into one value. It is used to rank countries. Iceland, Norway and Canada have been at the top of this list in recent years.

Countries are also grouped into more and less economically developed, based on their industrial development and GDP. More economically developed countries (MEDCs) are industrialized nations with high GDPs. Their population is relatively rich and individuals are unlikely to starve through poverty. They have a relatively high level of resource use per capita (per person) and relatively low population growth rates. Less economically developed countries (LEDCs) are less industrialized or have hardly any industry at all. They may have raw materials (natural capital) but this tends to be exported and processed in MEDCs. The population has a lower GDP and higher poverty rates. More people are poor with low standards of living. Most LEDCs have high population growth rates. MEDCs include most countries in Europe and the EU (European Union), North America, South Africa, Israel and Japan. LEDCs tend to be most of the countries outside Europe and North America. Various other terms are used to describe the differences in industrialization and wealth of countries. "Developed" and "developing" have been used but MEDC and LEDC have replaced these as some LEDCs are in economic decline and are called failed states or least developed countries.

The terms first, second, third and fourth world used to refer to technologically advanced democracies, Communist states within the USSR, economically underdeveloped countries and stateless nations, respectively. Newly industrialized countries (NICs) have accelerated their industrial development and have increased GDP, often accompanied by massive foreign investment, population migration to the cities to provide a workforce, free trade and increased civil rights. At the moment, the following countries are considered to be NICs: China, India, South Africa, Malaysia, Thailand, the Philippines, Turkey, Mexico and Brazil.

Population and the environment

Human population affects the environment:

- more people require more resources
- more people produce more waste
- the more people there are, the greater the impact they have
- control population increase and you control resource demand
- therefore population control and sustainability are intrinsically linked.

Measures of population changes

The four main factors that affect population size of organisms are birth rate, death rate, immigration and emigration.

Measures of population change are crude birth rate, crude death rate, doubling rate and natural increase rate.

To think about

Perhaps it would be possible to apply population ecology models to human populations. There are a range of models available that explore population dynamics and the interrelationship between a species' population size and other environmental parameters, resources and internal controls.

Could we use a predator-prey model to examine the dynamics of human population change over time? What would limit the use of this model?

Crude birth rate (CBR) is the number of births (or natality) per thousand individuals in a population per year. The crude birth rate for the world is 20.3 per 1000 per year.

Crude death rate (CDR) is the number of deaths (or mortality) per thousand individuals in a population per year. The crude death rate for the world is about 9.6 per 1000 per year.

Crude birth and death rates are calculated by dividing the number of births or deaths by the population size and multiplying by 1000. Write these out as formulae.

Natural increase rate (NIR) is (crude birth rate – crude death rate)/10. This gives the natural increase rate as a percentage. It excludes the effects of migration.

Doubling rate is the time in years that it takes for a population to double in size.

An NIR of 1% will make a population double in size in 70 years. This is worth remembering.

The doubling time for a population is 70 divided by the NIR.

To do

1 Calculate the population density, crude birth rate, crude death rate and natural increase rate from the data provided in Table 8.3. Put the data into your own table.

2 **Population density** is the number of people per unit area of land. Calculate the population density in number per km².

Region	Popn 10⁶	Land area km² × 10⁶	Births 10⁶	Deaths 10⁶	Crude birth rate	Crude death rate	Natural increase rate	Popn density
World	6000	131	121.0	55.8				
Asia	3500	31	88.2	29.4				
India	1000	3	29.0	10.0				
Africa	730	29	30.7	10.0				
Tanzania	30	0.9	1.3	0.4				
Europe	730	22.7	8.5	8.2				
Switzerland	7	0.04	0.09	0.07				
N America	460	21.8	9.3	3.6				
USA	270	9.6	4.3	2.4				

Table 8.3

3 Describe and explain the differences in the data for the three regions Asia, India and Africa.

Population growth can be defined in terms of birth rate, doubling time and fertility rate.

Total fertility rate is the average number of children each woman has over her lifetime.

A fertility rate higher than 2.0 results in population increase, while lower than 2.0 leads to population decrease, because the two parents should be replaced by two children in order to maintain a stable population. Migration is not taken into consideration.

Fertility rate is the number of births per thousand women of child-bearing age. In reality, **replacement fertility** ranges from 2.03 in MEDCs to 2.16 in LEDCs because of infant and childhood mortality.

The fertility rate is sometimes considered a synonym for the birth rate. However, there is a difference in expression. Birth rate is expressed as a percentage: births per 1000, or hundred (%) of the total population, not of each woman. However, in practice, **crude birth rate (CBR)** is used which is the number of births per thousand individuals, male and female, young and old.

Human population growth

Demography is the study of the statistical characteristics of human populations, e.g. total size, age and sex composition, and changes over time with variations in birth and death rates.

Carrying capacity is the maximum number of a species or "load" that can be sustainably supported by a given environment, i.e. without destroying the stock (see also pp. 63 and 162). Populations remain stable when the **death rate** and the **birth rate** are equal and so there is no net gain or reduction in population size.

There are numerous examples of the impact of resource failure and the consequence on human population. Modern history is littered with examples of the direct and indirect effects of famines and droughts across the Earth. Consider the ones about which you have heard.

Size of population alone is not the only factor responsible for our impact on our resource base and our impact on the environment in which we live. We need to also consider the wealth of a population, resource desire and resource need (or use). Many population impact models function on the assumption that all individuals (or all populations of a similar size) have the same resource needs and thus have the same impact environmentally (based on resource use and waste associated with exploiting a resource). However, individual resource use (and population resource use) is a dynamic principle. Resource use varies in time and space. MEDCs and LEDCs demonstrate contrasting resource use per head; urban and rural populations demonstrate varying resource use profiles. Young people have different resource needs than the elderly. Amazonian Indians have different resource needs than Parisians. Yet all these groups may have an impact, though the impact will vary in scale, type and severity, and may not necessarily be linearly related to population size.

About 20% of us live in MEDCs, 80% in LEDCs. The proportion in MEDCs is falling as birth rates are higher in LEDCs and sometimes population growth rate is negative in some MEDC countries (e.g. Italy, Germany).

To do

1 From Fig. 8.9, calculate the percentage of the world's population living in LEDCs and MEDCs in 1950 and predicted in 2150.
2 What is plotted on the y-axis in Fig. 8.10?
3 Why are the numbers not increasing arithmetically (e.g. 2, 4, 6, 8)?
4 Do these lines show exponential growth?
5 Estimate the percentage of the world population living in Asia in 2005 from Fig. 8.11.

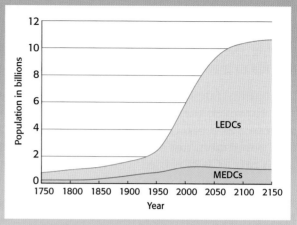

Fig. 8.9 Human world population growth 1750–2150 for MEDCs and LEDCs

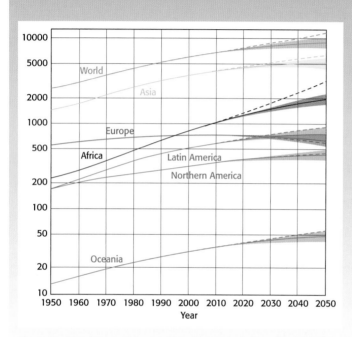

Fig. 8.10 Population of the world and its regions (in millions) from 1950 to 2050 (predicted). Solid line: medium variant. Shaded region: low to high variant. Dashed line: constant-fertility variant.

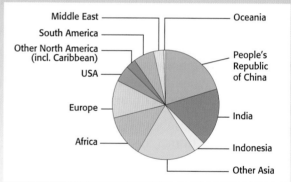

Fig. 8.11 Distribution of world population in 2005

To think about

(1) Population growth benefits the human race because more people have more ideas, are inventive and work. Necessity is the mother of inventions. If we run out of living space, we build high-rise blocks of flats; if we need more food, we intensify production.

(2) Economic growth is a good thing because it means more people can have safe drinking water, sanitation, collection of refuse.

(3) Economic growth is a bad thing as it means we use more resources – fossil fuels, chemicals.

(4) Human population growth is levelling off so we shall reach the carrying capacity of the environment and numbers will fluctuate around it. We can clean things up then.

(5) Individuals suffer less from disease as medicine improves so we should save as many lives as possible.

(6) Contraception is against nature so we should procreate naturally.

(7) Any attempt to limit population growth will have to be made in LEDCs and this could be construed as an attempt by the "haves" to limit the "have nots" and so safeguard their own lifestyle.

(8) Increasing affluence gives people the ability to clean up their immediate environment but at a cost to the distant environment – e.g. factories move to LEDCs.

(9) Increased population size in LEDCs leads to subdivision of farms for children until they are

too small to support the family, encroachment on forest to gain more land, migration to cities or MEDCs, illicit trade, poaching, etc. as unemployed people become desperate.

Question

Do you agree with these statements? Label each viewpoint expressed as technocentric, cornucopian, environmental manager, or ecocentric.

To do

1 Write a definition of **exponential growth**.
2 Exponential growth is characterized by increasingly short doubling times. Doubling time is the number of years it would take to double the size of a population at a particular rate (%) of growth. For example with a 2% growth rate or **natural increase rate**, the population doubling time would be about 35 years, with 4% natural increase rate a population will double in about 17 years. How long would it take for a population to double if the natural increase rate was 1%?
3 Copy and complete Table 8.4 with doubling times for the global population (in billions).

Date	Population	Doubling time	Date	Population	Doubling time
1500	0.5	1500		5.0	
1800	1.0	300		6.0	
1927	2.0			7.0	
	3.0			8.0	
	4.0			9.0	

Table 8.4

4 What do you notice about the changing doubling times?

Population growth and food shortages

There are two main theories relating to population growth and food supply, from Malthus and Boserup.

Malthusian theory

Thomas Malthus was an English clergyman and economist who lived from 1766 to 1834. In his text *An essay on the principle of population,* published in 1798, Malthus expressed a pessimistic view of the dangers of overpopulation and claimed that food supply was the main limit to population growth. Malthus believed that the human population increases geometrically (i.e. 2, 4, 8, 16, 32, etc.) whereas food supplies can grow only arithmetically (i.e. 2, 4, 6, 8, 10, 12, etc.), being limited by available new land. Malthus added that the "laws of nature" dictate that a population can never increase beyond the food supplies necessary to support it.

According to Malthus, population increase is limited by certain "checks". These prevent numbers of people increasing beyond the optimum population, which the available resources cannot support. As long as fertile land is available, Malthus believed that there would be more than enough food to feed a growing population. However, as population and the demands for food increase, there is a greater pressure to farm more intensively and cultivate poorer, more marginal land. According to Malthus, though, food production can only increase to a certain level determined by the productive capacity of the land and existing levels of technology.

Beyond the ceiling where land is used to its fullest extent, overcultivation and, ultimately, soil erosion occurs, contributing to a general decline in food production. This is known as the law of diminishing returns where, even with higher levels of technology, only a small increase in yield will eventually occur. These marginal returns ultimately serve as a check to population growth. Malthus did acknowledge that increases in food output would be possible with new methods in food production, but he still maintained that limited food supply would eventually take place and so limit the population.

Neo-Malthusians agree with Malthus' arguments and believe that we are now seeing the limits to growth as the increase in food production is slowing. The Club of Rome, an NGO, is neo-Malthusian.

Limitations of Malthusian theory

Anti-Malthusians criticize the theory as being too simplistic. A shortage of food is just one possible explanation for the slowing in population growth. Malthus' reasoning ignores the reality that it is actually only the poor who go hungry. Poverty results from the poor distribution of resources, not physical limits on production. Except on a global scale, the world's community is not "closed" but does not enjoy a fair and even distribution of food supplies. Even so, Malthus could not possibly have foreseen the spectacular changes in farming technology which mean we can produce enough food from an area the size of a football pitch to supply 1000 people for a year, i.e. there is enough land to feed the whole human population. Thus evidence of the last two centuries contradicts the Malthusian notion of food supply increasing only arithmetically. Rather than food shortage, food surpluses exist and agricultural production increases. In 1992, European surpluses reached 26 million tonnes and there are indications that this trend will continue, contrary to Malthusian theory. There were 7 million people in Britain when Malthus lived. Now there are 60 million and most have a high standard of living and enough food – though some is imported. This model is repeated in MEDCs which import food from across the world. Globalization is something Malthus could not have expected.

Boserup's theory

In 1965, Ester Boserup, a Danish economist, asserted that an increase in population would stimulate technologists to increase food production (the optimistic and technocentric view). Boserup suggested that any rise in population will increase the demand for food and so act as an incentive to change agrarian technology and produce more food. We can sum up Boserup's theory by the sentence "necessity is the mother of invention".

Boserup's ideas were based on her research into various land use systems, ranging from extensive shifting cultivation in the tropical rainforests to more intensive multiple cropping, as in South-East Asia. Her theory suggests that, as the population increases, agriculture moves into higher stages of intensity through innovation and the introduction of new farming methods. The conclusion arising from Boserup's theory is that population growth naturally leads to development.

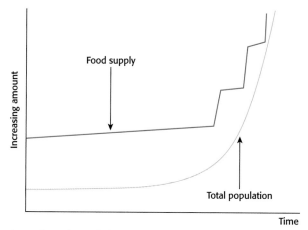

Fig. 8.12 Food supply and population curves

Limitations of Boserup's theory

Like Malthus, Boserup's idea is based on the assumption of a "closed" community. In reality, except at a global scale, communities are not "closed" because constant immigration and emigration are common features. It has therefore been very difficult to test Boserup's ideas. This is because emigration usually occurs in areas of overpopulation to relieve the population pressure, which, according to Boserup's theory, then leads to technological innovation.

Overpopulation can lead to unsuitable farming practices that may degrade the land; for example, population pressure may be responsible for desertification in the Sahel. From this it is clear that certain types of fragile environment cannot support excessive numbers of people. In such cases, population pressure does not always lead to technological innovation and development.

Applications of Malthus and Boserup

There is evidence to suggest that the ideas of both Boserup and Malthus may be appropriate at different scales. On a global level the growing suffering and famine in some LEDCs today may reinforce Malthusian ideas. On the other hand, at a national scale, some governments have been motivated by increasing population to develop their resources and so meet growing demands.

Both Malthus and Boserup can be right because Malthus refers to the environmental limits while Boserup refers to cultural and technological issues.

To do

1 Read the description of the theories of Malthus and Boserup and summarize their models in a table like Table 8.5.

	Malthus	**Boserup**
Model diagram		
Main ideas		
Limitations		
Applications		

Table 8.5

2 Look at Fig. 8.13, a graph of food supply in India.

  a According to the Indian National Commission on Population, the population of India was about 439 million in 1961, 548 million in 1971, 665 million in 1981, 846 million in 1991, 1012 million in 2001 and estimated to be 1179 million in 2009. As the population of India has increased what happens to the **per capita** food supply?

  b Sketch the graph and add a third line to show increase in the Indian human population.

  c Whose theory is represented by this data? Explain your reasoning.

3 The pattern of human growth is not uniform, with most growth currently taking place in LEDCs. Use the following data to construct population growth curves for MEDCs and LEDCs from 1800 until 2100. Plot the LEDCs above the MEDCs (the units for population are 10^9).

Date	**MEDCs**	**LEDCs**
1800	0.3	0.7
1850	0.4	0.8
1900	0.6	1.1
1950	0.8	1.7
1960	0.9	2.1
1970	1.0	2.8
1980	1.1	3.3
1990	1.2	4.1
2000	1.3	5.0
2025	1.4	7.2
2050	1.4	8.0
2100	1.4	8.0

Table 8.6

4 The values in Table 8.6 for the next century are only estimates. What will be the most important social factor that will determine human population size?

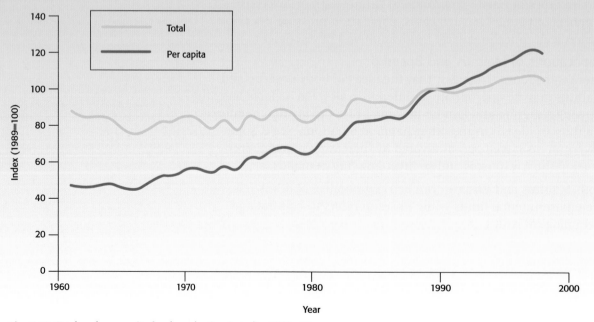

Fig. 8.13 Total and per capita food production in India, 1961–98

What will be the future world population?

Three billion women will decide the world population in 2050.
A fertility rate of 2.0 means that a couple replace themselves, and do not add to the population. In this scenario the population will increase from 6 billion now to 10.8 billion. If every second woman decides to have three rather than two children, a fertility rate of 2.5, the population will rise to 27 billion by 2150. If, however, every second woman decides to have only one child instead of two, a fertility rate of 1.5, the world population will sink to 3.6 billion. The total world fertility rate is now about 3.0; 1.7 in MEDCs, and averaging 3.4 (but up to 6.0) in LEDCs. Fertility rate is falling although the population size continues to increase. The UN has calculated estimates for population change based on fertility rates stabilizing at 2.6 (high), 2.1 (medium/replacement level) and 1.6 (low).

When is a country overpopulated?

If the optimum population is defined as being the level at which the population produces the highest economic return per capita, using all available resources, then some countries may have a higher optimum population density than others. The UK and Netherlands have high population densities but can support this population with a high living standard. Brazil with two people per km^2 in the north is overpopulated as resources are much scarcer. The richer countries can import goods and services from elsewhere to support their populations.

Why do people have large families?

It appears that the decision to have children is not correlated with GNP of a country nor personal wealth. Some reasons may be:

1 High infant and childhood mortality: one child dies every three seconds – 26 500 per day – due to malnutrition and disease, according to UNICEF. It is an insurance to have more children than you may need so that some of them are likely to reach adulthood.

2 Security in old age: the tradition in the family is that children will take care of their parents. The more children, the more secure the parents, and the less the burden for each child. If there is no social welfare network, children look after their parents.

3 Children are an economic asset in agricultural societies. They work on the land as soon as they are able. More children mean more help but more children need feeding. In MEDCs, children are dependent on their parents during their education and take longer to contribute to society.

4 Status of women: the traditional position of women is that they are subordinate to men. In many countries, they are deprived of most rights, like owning property, having their own career, becoming educated. Instead they do most of the agricultural work, and are considered worthy only for making children and their social status depends on the number of their children, particularly boys. Breaking down such barriers of discrimination (social or religious), allowing girls to be educated and capable of gaining status outside the context of bearing children has probably contributed more than anything toward the very low fertility of MEDCs.

5 Unavailability of contraception: in MEDCs contraception is the prime way of reducing fertility. In LEDCs, many women would like to use contraception but they are too poor to afford contraceptives or they cannot obtain them.

The ways to reduce family size are to:

1 **Provide education** even if only basic literacy to children and adults.

2 **Improve health** by improving nutrition, basic hygiene to prevent the spread of diseases, and simple medicines and vaccines.

3 **Provide contraception** and family planning counselling. Improved health reduces infant and adult mortality and encourages smaller families as more children survive into adulthood.

4 **Increase family income** by small-scale projects, e.g. the Grameen Bank in Bangladesh gives small loans to the poor to start up small enterprises. This "micro-lending" has been very successful.

5 **Improve resource management** by encouraging local people to take part in reforestation projects or prevent erosion through soil conservation measures. Large projects in LEDCs often do not work. Major projects like building dams for hydroelectric power, or road-building, cost an LEDC which is then in debt (so-called Third World debt) and force the population into cash cropping (e.g. tobacco, oil palm).

Population pyramids

These pyramids show how many individuals are alive in different age groups (five-year cohorts) in a country for any given year. They also show how many are male and female. Population numbers are on the *x* axis and age groups on the *y* axis.

The overall shape of these pyramids can be sketched as in Fig. 8.14 and indicate the stage of development in a country at a particular time. LEDCs tend to have expanding populations so the pyramids are wide at the bottom; MEDCs tend to have stationary or contracting pyramids as birth rates fall and individuals live longer overall.

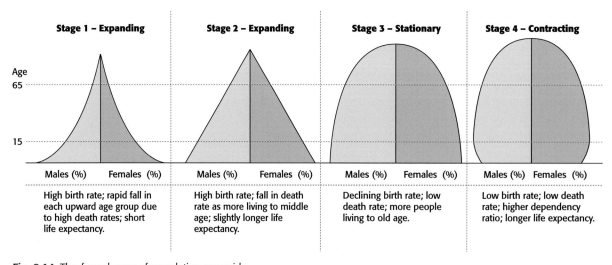

Fig. 8.14 The four shapes of population pyramids

To do

Population pyramids

All pyramids are taken from the US Census Bureau website (www.census. gov/ipc/www/idb/pyramids.html). Have a look at this site as it has dynamic pyramids which change over time.

Population or age/sex pyramids show the distribution of individuals in a population, by sex and age. They contain a lot of information.

1   List as full sentences, five pieces of information that you can find in the population pyramid of Afghanistan in 2000 (Fig. 8.15). e.g. In Afghanistan in 2000 there were approximately 2.2 million males under the age of 5 and 2.1 million females.

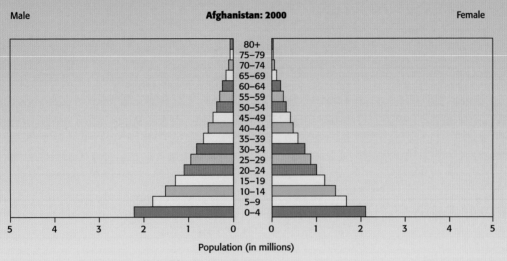

Fig. 8.15 Population pyramid of Afghanistan in 2000

2   What changes to the population are there in the predicted pyramid of 2025 (Fig. 8.16) compared with 2000?

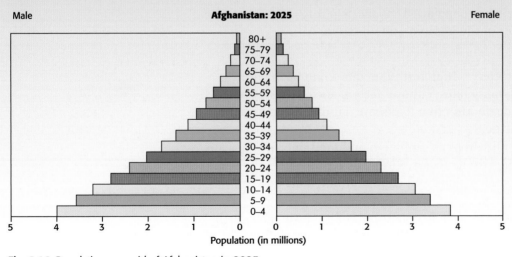

Fig. 8.16 Population pyramid of Afghanistan in 2025

3   In Fig. 8.16, imagine two horizontal lines at 15 and 65 years. What do the three bands then formed represent?

Population pyramids can indicate political and social changes. China used the concept of an optimum population size to try to stabilize its population at 1.2 billion by the year 2000 and reduce the population to a government-set level of 700 million by the end of the century.

4 Explain the decrease in population younger than 45 in China in 2007 (Fig. 8.17). Think of at least two reasons.

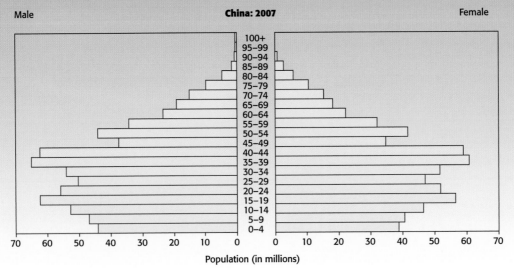

Fig. 8.17 Population pyramid of China in 2007

Demographic transition model

The demographic transition model (DTM) describes the pattern of decline in mortality and fertility (natality) of a country as a result of social and economic development. Demographic transition can be described as a five-stage population model (Fig. 8.18), which can be linked to the stages of the sigmoid growth curve.

The stages are:

1 Pre-industrial society – High birth rate due to no birth control; high infant mortality rates; cultural factors encouraging large families. High death rates due to disease, famine, poor hygiene and little medicine.

2 LEDC – Death rate drops as sanitation and food improve, disease is reduced so lifespan increases. Birth rate is still high so population expands rapidly and child mortality falls due to improved medicine.

3 Wealthier LEDC – As a country becomes more developed, birth rates also fall due to access to contraception, improved health care, education, emancipation of women. Population begins to level off and desire for material goods and low infant death rates mean that people have smaller families.

4 MEDC – Low birth and death rates. Industrialized countries. Stable population sizes.

5 MEDC – Population may not be replaced as fertility rate is low. Problems of aging workforce.

As a model, the DTM explains changes in some countries but not others. China and Brazil have passed through the stages very quickly. Some sub-Saharan countries or those affected by war or civil unrest do not follow the model. It has been criticized as extrapolating the European model worldwide.

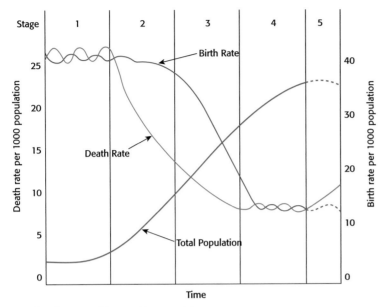

Fig. 8.18 The demographic transition model

The controversial thing about this model is that it is based on change in several industrialized countries yet it suggests that all countries go through these stages. Like all models, it has limitations. These are:

The initial model was without the fifth stage, which has only become clear in recent years when countries such as Germany and Sweden began to experience population decline.

The fall in the death rate has not always been as steep as the steepest part of the death rate curve suggests, as movement from the countryside to cities has created large urban slum areas which have poor or no sanitation and consequent high death rates of the young and infirm. Deaths from AIDS-related diseases may also affect this.

The fall in the birth rate assumes availability of contraception and that religious practices allow for this. It also assumes increasing education and increased literacy rates among women. This is not always the case.

Some countries have compressed the timescale of these changes. The Asian "Tiger economies" of Malaysia, Singapore and Hong Kong, for example, have leapt to industrialized status without going through this sequence in the same time period as others.

This is a eurocentric model and assumes that all countries will become industrialized. This may not be the case in some "failed states", for example.

To do

1 Copy and complete Table 8.7 by filling in the characteristics of each pyramid.

Stage	1 Expanding	2 Expanding	3 Stationary	4 Contracting
Birth rate				
Death rate				
Life expectancy				
Population growth rate				
Stage of DTM (see page 180)				
Example				

Table 8.7

2 For each pyramid in Fig. 8.19, identify the stage. You might like to look up the pyramid for your own country, if not included, and do the same. Comment on the birth rate, fertility, death rate, life expectancy, gender differences and stage of development of the country.

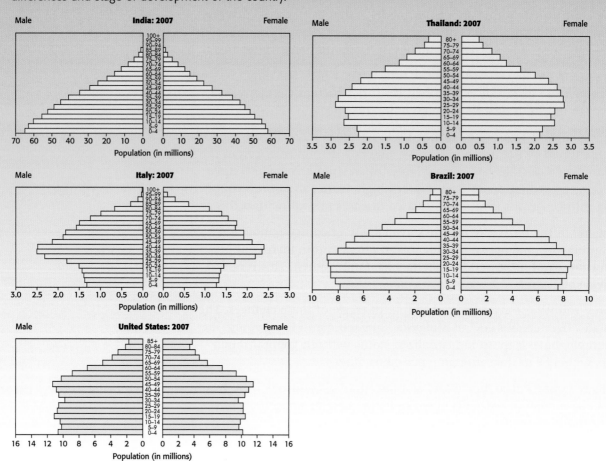

Fig. 8.19 Population pyramids for five countries in 2007

To think about

Models may be very generalized and simple to use, or so complex that they are difficult to use. They should present the significant factors without extra detail that may confuse us. They should be useful in helping us to make predictions and make sense of the real world. How far does the DTM help us or does it hinder our understanding of population change which is far more complex than this model suggests?

To research

Using computer models to predict population change

Go to the UN Economic and Social Affairs office website (http://esa.un.org/unpp/index.asp?panel=1).

1 Using the basic data and median variant, find the data for the following:

  a World population in 2050.

  b More developed regions population in 2050.

  c Less developed regions population in 2050.

  d Calculate the percentage of total world population that each region makes in 2050.

  e Repeat steps 1–4 for the year 2010.

  f What does this tell you?

2 Now go to the detailed data in panel 2.

  a Look up and make a note of fertility rates for the same regions and same years.

  b Which region has the highest fertility rate and why?

  c What has happened to birth rates of these regions over time and what does this mean?

  d What has happened to death rates of these regions over time and what does this mean?

  e What has happened to life expectancy in these regions?

Using the same website:

3 Pick two variables that you think are important when trying to explain population growth, justify why they are important and what they show over the period 1950–2050.

4 What are the problems with using computer models to predict population expansion? How valid are they and what are they useful for?

5 For the country in which you live or where you hold nationality, examine the change in population between 1950 and 2050.

The AIDS epidemic – Africa, an orphaned continent

Information taken from UNAIDS, a joint UN agency on AIDS research and relief

Worldwide 33.2 million people live with the HIV virus. In 2007, 2.5 million were infected with the virus and 2.1 million died from AIDS-related illnesses. Three-quarters of these were in sub-Saharan Africa. While the total numbers seem to be levelling off, that is still a great many people.

Sub-Saharan Africa is still the most affected region and 22.5 million live with the virus there. 61% of sufferers are women. South Africa is the worst affected country.

Most people who die from AIDS-related illnesses are the wage earners of a family and those who are poor cannot afford antiretroviral drug treatments. Although the multinational drug companies have agreed to reduce the prices of these drugs in Africa, they are still out of the reach of all but a few. The effect is to leave a generation of orphan children, looked after by their grandparents.

Questions

1   Describe the population pyramid for South Africa in 2000 (see Fig. 8.20).
2   What type of pyramid is this?
3   Explain why it changes for 2025.
4   Explain the factors that influence the pyramid in 2050. (Remember to include economic factors.)

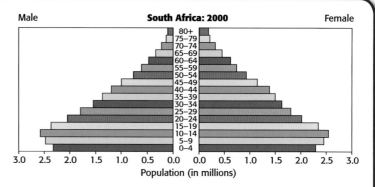

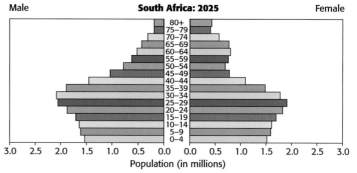

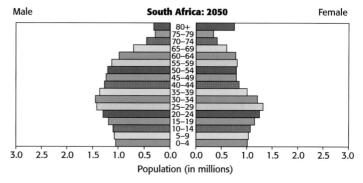

Fig. 8.20 Population pyramids for South Africa 2000, 2025 and 2050

The greying of Europe

Europe's population growth rate is falling and its population is ageing (going grey). Who will be the workers who support the older population? Immigration may help but immigrants also grow old and need more state support.

So the combination of decreased replacement rate and fewer workers means it is not looking good for Europe. Within the geographical boundaries of Europe are some 728 million people (with 495 million of these in the 27 EU member states). In 1900, Europeans made up 25% of the world population; by 2050, they will be 7% according to UN projections. This is because the growth rate in other countries is far higher than that of Europe. By 2025, the population over 65 will be 22.4%. In some countries, fertility rates are lower than in others. Italy has a fertility rate of 1.2 children per woman,

compared with the EU average of 1.52. This is not enough to replace the population yet some see immigration as bringing major social problems as well. The UK anticipates more immigration, up to 135 000 per year for the next 25 years. But services such as education, medical care and social support are not increasing to meet the demand. Immigration occurs across the EU, mostly from east to west but also into the EU from LEDCs.

Debate about immigration soon becomes political with terms like "illegal immigrant", "economic migrant" and "bogus asylum seeker", yet migration has always occurred for a host of economic, social and political reasons. It is inevitable. The challenge for politicians will be to intervene to change retirement age, modify the pension systems, improve productivity and stimulate worker mobility.

Influencing human population growth

National and international policies influence human population growth. If you pay more tax to have more children or even lose your job, you may decide to have a smaller family. Agricultural development, improved public health and sanitation, etc. may lower death rates and stimulate rapid population growth without significantly affecting fertility. Parents in subsistence communities may be dependent on their children for support in their later years and this may create an incentive to have many children. Policies that stimulate economic growth may reduce birth rates as a result of increased access to education about methods of birth control. Urbanization may also be a factor in reducing crude birth rates. Policies directed toward the education of women, and enabling women to have greater personal and economic independence, may be the most effective in reducing fertility and therefore population pressures.

While we may be able to count how many people are alive, what age they are and where they live, and even predict changes in the future, we do not really know how many people and other species the Earth can support. All evidence we have at the moment is that we are using the Earth's resources unsustainably, but are we inventive enough to either live within our means or find ways to increase productivity? Time will tell.

To do

Research population policies in one of India, China, Iran, Columbia, Brazil, Singapore or your own country. Write a short case study on this and exchange it with your classmates.

Key words

exponential growth	*K*-selected
geometric growth	*r*-strategist
logistic growth	*r*-selected
limiting factor	survivorship curves
density-dependent factor	human carrying capacity
internal factor	technological development
external factor	import
density-independent factor	CBR
S-curve	CDR
carrying capacity	NIR
environmental resistance	doubling time
J-curve	total fertility
dieback	fertility rate
overshoot	age-sex pyramids
K-strategist	demographic transition model

9 Resources – natural capital

Key points

- Resources (or natural capital) are goods or services that have some value to humans.
- We can exploit resources which then produce a yield (or natural income).
- The three types of resource are renewable, non-renewable and replenishable.
- The value of a resource can be different in different cultures, and can change over time as technology and economic development progress.
- Sustainability relies on living within only the natural income generated by natural capital and not using the capital.
- Sustainable development is development that meets current needs without compromising the ability of future generations to meet their needs and without degrading the environment.
- The human carrying capacity can be increased by use of technology, reducing energy use or recycling.
- The ecological footprint of a population is the land area required to sustainably support it and can be calculated.
- Population growth and ecological footprint size are influenced by stage of development of a country and by its worldview.

> "*The ultimate test of a moral society is the kind of world that it leaves to its children.*"
>
> Dietrich Bonhoeffer, German theologian

Natural capital and natural income

In the past, economists spoke of capital as the products of manufacturing, human-made goods, and separated these from land and labour. But we now recognize that capital includes natural resources that have value to us, e.g. trees, soil, water, living organisms and ores bearing minerals, and also includes services that support life, e.g. flood and erosion protection provided by forests. So the term **natural capital** is now used to describe these goods or services that are not manufactured but have value to humans. We may be able to process these to add value to them, e.g. mine tin or uranium, turn trees into timber, but they are still natural capital.

Natural capital may also be processes, e.g. photosynthesis that provides oxygen for life forms to respire, the water cycle or other processes that maintain healthy ecosystems. With this view, we can then count natural resources and processes as capital which can be improved or degraded and given a value – we can begin to give values to ecosystems. We use the terms resource and natural capital interchangeably here.

Just as capital yields income in terms of economics, natural capital yields **natural income** (yield or harvest or services) – factories produce objects, cherry trees produce cherries, and the water cycle provides us with fresh water. The measure of the true wealth of a country must include its natural capital, e.g. how many mineral resources, forests, rivers, it has. In general MEDCs add value to natural income by manufacturing goods from it and LEDCs may have greater unprocessed natural capital. The World Bank now calculates the wealth of a country by including the rate of extraction of natural resources and the ecological damage caused by this, including carbon dioxide emissions. The depletion of natural resources at unsustainable levels, i.e. providing more natural income than is sustainable, and efforts to conserve these resources are often the source of conflict within and between political parties and countries.

Renewable, non-renewable and replenishable

Natural capital can be classified into three groups: renewable, non-renewable and replenishable.

Renewable resources are living resources that can replace or restock themselves – they can grow. Timber, for example, can be harvested, and more trees can be planted to be harvested in the future, as long as the environment is managed in a sustainable fashion. Usually, renewable natural capital has solar radiation via photosynthesis as its primary source of energy. Solar radiation itself may also be viewed as a renewable resource as it is produced over such a long timescale. But renewable natural capital can run out if the standing stock (how much is there) is harvested unsustainably, i.e. more is taken than can be replaced by the natural growth rate. Then, it will eventually run out.

Fig. 9.1 A forest – a renewable resource

Fig. 9.2 A coal stockpile – a non-renewable resource

Non-renewable resources are resources that exist in finite amounts on Earth and are not renewed or replaced after they have been used or depleted (or only over a long timescale – normally geological scales). Non-renewable resources include minerals and fossil fuels. As the resource is used, natural capital or stocks are depleted. New sources of stock or alternatives need to be found.

Replenishable resources are arguably a middle ground between renewable and non-renewable resources. These resources are replaceable, however they tend to be replaced over a time period that does not allow them to be viewed as renewable. A good example is groundwater. Groundwater can be used as a resource and depleted as it is used. Normally depletion rates are magnitudes larger than recharge rates; therefore the natural capital is depleted in the same way as natural oil and gas are depleted (wells run dry). However the wells will recharge over time with "new groundwater" whereas natural gas or oil wells will not. Thus these reserves replenish. The difference between replenishable and renewable is in the rate of resource use against resource replacement. Soil is another example of replenishable natural capital.

Depending on your source of drinking water, where you live and the annual rainfall, water may be considered a renewable resource (high rainfall regions where most rain is collected and used for drinking) or a replenishable resource (drier regions where underground aquifers refill slowly at rates longer than an average human lifetime).

Recyclable resources – the fourth element

Iron ore is a non-renewable resource. Once the ore has been mined and processed it cannot be replaced. However from iron ore we produce iron. Iron can be cast into numerous forms and represents a significant commodity within modern societies. About 90% of a car is made from iron or iron-derived products – steel. However, steel and iron can also be recycled. Old or damaged cars can be broken down. Their parts can be used to replace parts in other cars or their parts can be remanufactured into new metal objects. Therefore iron ore is non-renewable but the iron extracted from the ore becomes a renewable resource. The same goes for aluminium.

Exploiting the poles

The Arctic and Antarctic are perhaps the last wildernesses on Earth and are beautiful. Their ecosystems are fragile and contain much biodiversity found nowhere else. Any disturbance takes a long time to recover as growth is slow because temperature is limiting. On land, water is also limiting as it is frozen for much of the year and so unavailable to plants.

The Arctic

Until recently humans could not exploit the resources of the Arctic on a large scale as the seas are frozen for all but a few months of the year and conditions are harsh. But there are mineral riches locked under the Arctic Ocean and surrounding land masses, especially hydrocarbons.

The world's oil supply comes from many countries. To have a national source of oil is a desire for many countries which would then not be dependent on importing oil. Some 40% of oil comes from and is exported by OPEC (Organization of the Petroleum Exporting Countries; 13 LEDCs whose economies rely on oil exports) and prices and supply are controlled. North America produces about 19%, with the remaining dispersed fairly evenly across a number of other countries. The price of a barrel of crude oil reached over US$100 in March 2008 while it hovered below $30 a barrel for much of the 1990s.

With climate change causing the Arctic to warm up, there are more ice-free days. High oil prices mean that reserves that were uneconomic

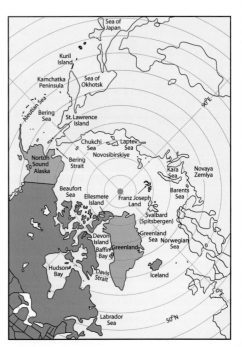

Fig. 9.3 Map of the Arctic

to extract are no longer so, and the Arctic could be the next goldmine or environmental disaster, depending on your environmental worldview. At 2008 prices, the estimated value of the Arctic's minerals is US$1.5-2 trillion. There are crude oil reserves under Northwestern Siberia and Alberta, Canada. There is also oil right under the North Pole. Humans have the technology to extract this oil. Why would we not?

Who owns the Arctic?

There is no land at the North Pole, it is ice floating on water. Under international convention, the United Nations Convention on the Law of the Sea, a state can claim a 200 nautical mile zone and beyond that up to 150 nautical miles of rights on the seabed. So the owning state may fish or exploit the minerals exclusively in this zone and other countries may not. This distance is not measured from the border or edge of a country but from the edge of the continental shelf, which may be some distance away from the border of the country under the sea.

In August 2007, a Russian submarine expedition planted a Russian flag on the seabed at the North Pole, two miles under the Arctic ice cap. They claimed that the seabed under the pole, called the Lomonosov Ridge, is an extension of Russia's continental shelf and thus Russian territory.

Eight countries—Canada, Denmark, Finland, Iceland, Norway, Russia, Sweden and the United States—have Arctic Ocean coastlines, and Denmark has sent its own scientific expeditions to study the opposite end of the Lomonosov ridge to see if they can prove it is part of Greenland, which is a Danish territory.

The Antarctic

Antarctica is a continent of which 98% is covered in ice and snow. In Antarctica, no large mineral or oil reserves have been found. But humans exploit the continent through tourism, fishing, sealing and whaling. About 10 000 tourists visit Antarctica each year and this is rising. No one country owns Antarctica but seven had staked territorial claims. The Antarctic Treaty was signed in 1959 by 12 countries which agreed to keep all land south of latitude 60°S free of nuclear tests and nuclear waste, for peaceful purposes, preserve the environment and not to dispute territory. It was perhaps the first step in recognition of international responsibility for the environment, and surprisingly as the original signatories included the US, UK and USSR who signed it in the middle of the Cold War. In 1991, the treaty was strengthened to include prevention of marine pollution, cleaning up of sites and no commercial mineral extraction. Sealing has annual limits and commercial whaling is now tightly regulated. Fishing is less of a success story with overfishing of species which is hard to regulate in the seas around Antarctica.

There is so much ice on Antarctica, that it is approximately 61% of all fresh water on Earth. If all this melted, it would add 70 m in height to the world's oceans. It appears that the ice is melting and some large ice sheets are "calving" or breaking up and slipping away from the land. Over three weeks in 2002, a huge ice shelf, over 3000 square km and 220 m deep, Larsen B, broke up and floated out to sea. But in other areas, the ice is getting thicker.

Questions

1 Discuss the reasons for the absence of an Arctic Treaty.
2 Suggest who should own the oceans.
3 Identify who should regulate human exploitation of the oceans.

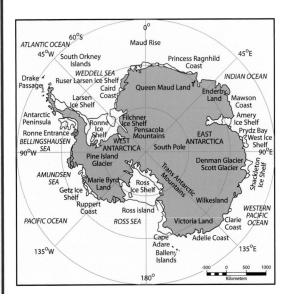

Fig. 9.4 Map of the Antarctic

Fig. 9.5 The breakup of the Larsen B ice shelf in 2002

Sustainability

Sustainability means living within the means of nature, on the "interest" or sustainable natural income generated by natural capital. But sustainability is a word that means different things to different people. Economists have a different view from environmentalists about what sustainable means. The word sustainable is often used as an adjective in front of words such as resource, development and population.

Any society that supports itself in part by depleting essential forms of natural capital is unsustainable. There is a finite amount of materials on Earth and we are using much of it unsustainably – living on the capital as well as the interest. According to UN data showing how humanity has overshot its sustainable level of resource exploitation, in 1961 we used 49% of the Earth's resources. By 2001, it was 121%. This demand is due to the level of overall consumption and per capita consumption and is more in some parts of the world than others. Overall, we are in ecological overshoot at the moment, living on natural capital.

You may wonder why this continues if we all know it to be so. It is perhaps due to many factors including inertia, when changing what we do seems too difficult; the result of the "tragedy of the commons". This is when many individuals who are acting in their own self-interest to harvest a resource may destroy the long-term future of that resource so there is none for anyone. It may be obvious that this will happen, but each individual benefits from taking the resource in the short term so they continue to do so. For example, hunting an endangered species may result in its extinction but if your family are starving and it is the only source of food, you will probably hunt it.

Some people think that the real worth of natural capital is about the same as the value of the gross world product (total global output) – about US$65 trillion per year, yet we are only just beginning to give economic value to soil, water and clean air and to measure the cost of loss of biodiversity.

To think about

The Millennium Development Goals

The Millennium Development Goals (MDGs) are eight goals to be achieved by 2015 that respond to the world's main development challenges. The MDGs are drawn from the actions and targets contained in the **Millennium Declaration** that was adopted by 189 nations and signed by 147 heads of state and governments during the **UN Millennium Summit** in September 2000.

Goal 1: Eradicate extreme poverty and hunger.
Goal 2: Achieve universal primary education.
Goal 3: Promote gender equality and empower women.
Goal 4: Reduce child mortality.
Goal 5: Improve maternal health.
Goal 6: Combat HIV/AIDS, malaria and other diseases.
Goal 7: Ensure environmental sustainability.
Goal 8: Develop a Global Partnership for Development.

1 Are we on target to reach these goals? Research what actions have been taken since 2000. (Try searching the Internet for Millenium Development Goals BBC and you should find some BBC webpages with an update.)
2 Do you think these goals are attainable, or too ambitious?

Dynamic nature of a resource

The importance of resources varies over time. A resource available today may not be a resource in the future. A resource available in the past may not be a resource today, or it may not have the resource value it previously had. Technocentrists believe that new discoveries will provide new solutions to old problems; for example, hydrogen fuel cells replacing hydrocarbon-based fuel, or harvesting algae as a food source. Arrowheads made from flint rocks are no longer in demand. Uranium is in demand as raw material for nuclear power by fission but may not be if we could harness the energy of nuclear fusion – the hydrogen economy.

Resources can be valued in several ways:
- Economic – having marketable goods or services, e.g. timber and food.
- Ecological – providing life-support services, e.g. water storage and gas exchange by forests.
- Scientific/technological – useful for applications, e.g. genetic, medicinal.
- Intrinsic – having cultural, esthetic, spiritual or philosophical (e.g. moral) values.

Whether a resource can be sustainably used is what we need to know. We may think that agriculture is sustainable as crops are eaten and then more are planted, but it is only sustainable if the soil fertility and structure are maintained and the environment is not degraded overall. If biodiversity is lost due to agriculture, is it sustainable?

Slash and burn agriculture (shifting cultivation) or sporadic logging in virgin forest are all sustainable as long as the environment has time to recover. Are we currently giving it enough time to recover?

To do

1 Review the resources of a rainforest using the values list above. Identify goods and services provided by a rainforest for humans by making a table with two columns – goods and services.
2 Make a list of resources that you are using (you can do this by first writing down what objects you are using and then thinking which resources are required to manufacture these objects).
3 Do all humans use the same types and amounts of resources? Explain, using examples.

4 If wood became so scarce that we could not use it for construction of houses any more, what could we use to replace wood?
5 We all use oil and oil products (fuel oil, diesel oil, chemicals, plastics). Sweden, however, does not have its own oil reserves. Would the carrying capacity of Sweden change, if it could not import oil, and what changes might occur in energy production?

To think about

Urbanization

The drift from the countryside to urban life started long ago and has continued (Table 9.1). According to the UN DESA (Department of Economic and Social Affairs), by 2010 the balance of urban to rural population worldwide will be 50:50 and China will be 50% urbanized. Some 4 billion of us will be city dwellers by 2015–20.

Cities are not necessarily unsustainable. There are efficiencies in living in high density populations where transport costs are reduced for commuters and moving resources around, people tend to live in smaller spaces so they use less energy to heat or cool and services are nearby. But cities have to remove their waste and process it, they need a large land area to supply them with food, and they create pollution. Inevitably they encroach on or degrade natural habitats.

Questions

1   Do you live in a city? (If not, select a city near to your home.) What is its population and land area?
2   How much has it grown since 1955?
3   Where does the food sold in the city come from?
4   Where do the wastes (sewage, garbage) go to?

Year		Million people	% of total population	Cities larger than 5 million people	Population of Tokyo (million)
1955	Urban	850	30	11	5
	Rural	1906	70		
1985	Urban	1983	40	28	30
	Rural	2860	60		
2005	Urban	3150	48	50	35.1
	Rural	3314	52		
2015 (predicted)	Urban	3819	52	60	35.4
	Rural	3400	48		

Table 9.1

To think about

Putting a value on the environment

Consider the systems in this list:

Your school or college
Your home
The Amazon rainforest
Lake Superior
Tundra in Siberia
Great Barrier Reef
Tokyo

Your city
Your local park or protected area
The Sahara Desert
San Francisco
Antarctica
Shanghai

Working in groups, put them in an order of increasing (1) environmental value and then (2) economic value. Write down the criteria that your group used. Compare your list with other groups and be ready to justify your decision. If you change your mind, reorder the items to your personal preference and amend your list of criteria. What are the difficulties in assessing the importance of different types of environment and what characteristics need to be taken into consideration when trying to do this? Do you think that environments can have their own intrinsic value?

Valuing the environment

Since the early 1980s, UNEP (UN Environmental Programme) has been using a system of integrated environmental and economic accounting (or socioeconomic environmental assessment – SEEA) to try to value the environment and track resource depletion. If countries would include the cost of degrading their natural resources within their GNP (gross national product), the real cost and health of the nation would be clearer to see.

From the UN Earth Summit in Rio de Janeiro in 1992 came Agenda 21 (see page 124). An undertaking was given that local councils would produce their own plan, a Local Agenda 21 involving consulting with the local community. What does your local Agenda 21 say?

To research

Fairtrade

Fairtrade (www.fairtrade.net) is an NGO charity.

1   What is the vision of Fairtrade?
2   What do they do?
3   What is their logo?

4   Name three products that have the Fairtrade logo.
5   Evaluate the impact of Fairtrade on (a) the producer and (b) the consumer.

To think about

Globalization

Did you know?

- 51 of the world's top 100 economies are corporations.
- Transnational corporations control two-thirds of world trade, 80% of foreign investment and employ just 3% of the world's labour force of 2.5 billion.
- In 2006 it was reported that the American public corporation Wal-mart may be bringing 38 000 people out of poverty per month in China.

Globalization is the concept that every society on Earth is connected and unified into a single functioning entity. The connections are mostly economic but also allow the easy exchange of services as well as goods, information and knowledge.

Globalization has been facilitated by new technologies, air travel and the communication revolution. The World Trade Organization (WTO) controls the rules of this global trade. Everyone can access the global market – if they are connected by email, the Internet or telephone. Ebay for example allows someone in Europe to purchase goods from an individual in the USA.

Global trade is not new. The Ancient Greeks and Romans traded across their world. The Han dynasty in China traded across the Pacific Basin and India. European empires and the Islamic world traded via trade routes around the world. What is new is the speed and scale of the trade and the communication. Since the end of the Second World War, protectionism of markets has decreased and free trade has increased. The World Bank and the International Monetary Fund (IMF) were set up in 1944 and have influenced development and world finance, including third world debt, since then. Some think that globalization only leads to higher profits for the transnational corporations (TNCs) but there is evidence that poverty has decreased in countries with increased global contacts and economies such as China. Ecologically, international agreements on global issues such as climate change or ozone depletion have tended to be easier to conclude with increased globalization. There is a tendency for globalization to westernize some countries.

Globalization is not internationalism. The latter recognizes and celebrates different cultures, languages, societies and traditions. It promotes the unit as the nation state. The former sees the world as a single unit or system, not necessarily recognizing these differences. Such globalization is both a positive and negative force. In one instance it can make us aware of the plight of others on the other side of the globe and in another instance make us aware of what one society has and another does not have.

Many if not most products are now traded on a global scale. They are part of what is referred to as the "global market" and minerals mined in South Africa or Australia are traded and shipped globally.

Discuss some advantages and disadvantages of globalization.

191

Human carrying capacity and the ecological footprint

Human carrying capacity

Carrying capacity is defined as the maximum number or load of individuals that an environment can sustainably carry or support.

By examining carefully the requirements of a given species and the resources available, it should be possible to estimate the carrying capacity of that environment for the species. This is problematic in the case of human populations for a number of reasons.

1 Humans use a far greater range of resources than any other animal. We also replace resources with others if they run out. We may burn coal instead of wood, use solar energy instead of oil, or eat mangoes instead of apples.

2 Depending on our lifestyles, culture and economic situation, our resource use varies from individual to individual, country to country.

3 Developments in technology lead to changes in the resources we use.

4 We import resources from outside our immediate environment.

While importing resources in this way increases the carrying capacity for the local population, it has no influence on the global carrying capacity. It may even reduce carrying capacity by allowing cheaper imports of food and forcing farmers to reduce their costs to compete with imports and so reduce incentives for conservation of the local environment. If the environment becomes degraded, e.g. by soil erosion, the land may become less productive and so not produce food for as many people. All these variables make it practically impossible to make reliable estimates of carrying capacities for human populations.

Ecocentric thinkers may try to reduce their use of non-renewable resources and minimize their use of renewable ones.

Technocentric thinkers may argue that the human carrying capacity can be expanded continuously through technological innovation and development. We shall always grow enough food, have enough water. It is just a matter of being more efficient and inventive. Using the remaining oil twice as efficiently means it lasts twice as long as it would have otherwise. But, given the estimate of human population size in 2050 of 9 billion, efficiencies will have to increase dramatically.

Conventional economists argue that trade and technology increase the carrying capacity. Ecological economists say that this is not so and that technological innovation can only increase the efficiency with which natural capital is used. Increased efficiency, at a particular economic level, may allow load on the ecosystem to increase but carrying capacity is fixed and once reached cannot be sustainably exceeded. The other difficulty with technology is that it may appear to increase productivity (e.g. energy-subsidized intensive agriculture giving higher yields) but this cannot be sustainable and long-term carrying capacity may be reduced (e.g. by soil erosion).

Reuse, recycling, remanufacturing and absolute reductions

Humans can reduce their environmental demands (and thereby increase human carrying capacity) by reuse, recycling, remanufacturing and absolute reductions in energy and material use.

> ### Test yourself
>
> Review your knowledge of carrying capacity.
>
> 1 Draw a sigmoid growth curve for an animal population.
> 2 Label the phases of growth and indicate the carrying capacity.

Reuse: the object is used more than once. Examples include reuse of soft drink bottles (after cleaning) and secondhand cars.

Recycling: the object's material is used again to manufacture a new product. An example of this is the use of plastic bags to make plastic poles for gardens. A second example is the recycling of aluminium. Obtaining aluminium from aluminium ore requires vast amounts of energy. As melting used aluminium to make new objects only takes a fraction of this energy, much energy can be saved by recycling.

Remanufacturing: the object's material is used to make a new object of the same type. An example is the manufacturing of new plastic (PET) bottles from used ones.

Absolute reductions: absolute reduction means that we can simply use fewer resources, e.g. use less energy or less paper. Unfortunately the advantages of reductions in resource use, i.e. increased carrying capacity, are often eroded by population increase.

Limits to human carrying capacity

In 1798, when the human population was about 1 billion, Thomas Malthus, an economist, wrote, "The power of the population is infinitely greater than the power of the Earth to produce subsistence for man." In 1976, when the population was 3.5 billion, environmentalist Paul Ehrlich warned of "famines of unbelievable proportions" and that feeding a population of 6 billion (exceeded in 1999) would be "totally impossible in practice". So far these predictions of disaster have been wrong and human carrying capacity may continue to increase.

Ecological footprints

As we have seen, humans can exceed their local carrying capacity by several means including trade to import resources. Thus **human carrying capacity** can also be viewed as the maximum load (rate of resource harvesting and waste generation) that can be sustained indefinitely without reducing productivity and functioning of ecosystems wherever those ecosystems are. So human carrying capacity depends not only on population size but also on the areas of land that support that population.

The **ecological footprint** of a population is an area of land (and water) that would be required to sustainably provide all of a particular population's resources and assimilate all its wastes (rather than the population that a given area can sustainably support). An ecological footprint is therefore the inverse of carrying capacity and provides a quantitative estimate of human carrying capacity. How does the cartoon in Fig. 9.6 show this?

Two researchers in Canada, Rees and Wackernagel, first published a book on ecological footprints and their calculation in 1996. Since then, the concept has become widely accepted with many website calculators designed to help you measure your footprint.

Fig. 9.6 Our ecological footprint

Personal ecological footprints

Figure 9.7 shows a fair Earthshare for one person. A fair Earthshare is the amount of land each person would get if all the ecologically productive land on Earth were divided evenly among the present world population.

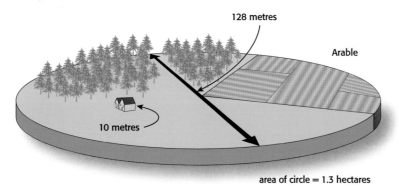

Fig. 9.7 An Earthshare

Do you think your Earthshare is larger? Check it by searching on the Internet for an ecological footprint calculator. www.earthday.net/footprint/ is a good place to start.

On average, a Canadian's ecological footprint is 7.8 hectares or approximately the size of 15 football fields. Only the United States and Australia have larger footprints at 10.3 and 9.0 hectares respectively. To compare, the average person in India has a footprint of 0.8 hectares, China 1.6. In the United Kingdom it is 5.2, in Germany 5.3 and in Switzerland 5.1 hectares.

In 2003, if we all shared equally, there would have been 1.8 hectares available per person or 1.3 if you do not include productive marine areas. Clearly, we are living beyond the Earth's ability to provide for our consumption (see Fig. 9.8).

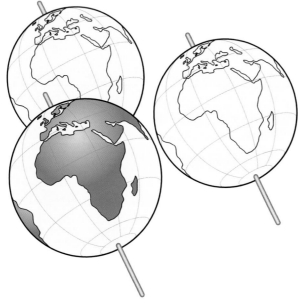

Fig. 9.8 Some suggest that humans are using resources at a rate that is unsustainable and we would need to find two or more Earths to sustain us if we continue to use resources at the current rate

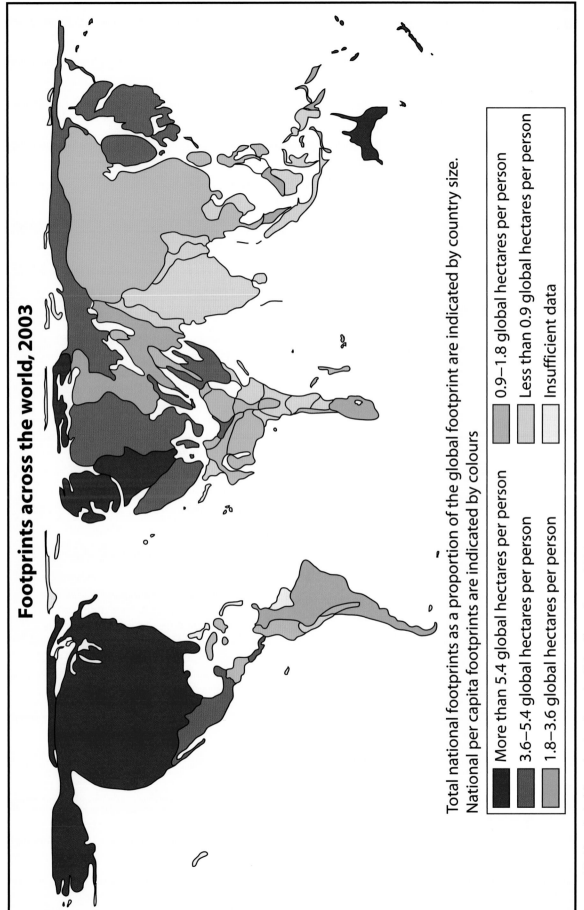

Footprints across the world, 2003

Total national footprints as a proportion of the global footprint are indicated by country size.
National per capita footprints are indicated by colours

■ More than 5.4 global hectares per person	■ 0.9–1.8 global hectares per person
■ 3.6–5.4 global hectares per person	■ Less than 0.9 global hectares per person
■ 1.8–3.6 global hectares per person	□ Insufficient data

Fig. 9.9 World ecological footprint sizes from the *Living Planet Report 2006*

The ecological footprint of a country depends on several factors: its population size and consumption per capita – how many people and how much land each one uses. It includes the cropland and other land that is needed to grow food, grow biofuels, graze animals for meat, produce wood, dig up minerals and the area of land needed to absorb wastes, not just solid waste but waste water, sewage and carbon dioxide. Figure 9.9, taken from the WWF *Living Planet Report* 2006, attempts to show countries as either ecological debtors or creditors. The creditors have smaller footprints than their biocapacity (living capacity or natural resources) and the debtors have larger footprints, represented here by changing the sizes of the countries in proportion. Debtors could be harvesting resources unsustainably in their countries, importing goods or exporting wastes. There is no such thing as "throwing away" on the Earth. There is no "away" in a closed system.

Ecological Footprint FAQs

adapted from http://www.buddycom.com/animal/envirimg/footprint/index.html

Q: Surely there are enough resources to sustain the human population, after all, we keep growing enough food for the increasing population?

A: No. We would need four more planets like Earth to meet the world's consumption habits if everyone consumed resources like the average Canadian. A growing number of people are subscribing to the concept that we are "global citizens" and each deserve a "fair share" of the Earth's resources. By this we mean that each person deserves fair access to acquire material needs and have a high quality of life. Footprint analysts have estimated that our fair Earthshare of resources amounts to **two hectares per person** or less.

Q: What are the top four things we can do to reduce our footprints?

A: Consider the environment during all daily activities.

1 Reduce water and energy consumption and install water- and energy-efficient devices at home and at work.

2 Use alternative modes of transportation (e.g. walking, cycling and public transportation).

3 Think about the food miles and energy subsidy of your food.

4 Practise the 3Rs. Reduce consumption by doing more with less. Eliminate unessential purchases and, when necessary, buy durable locally produced products with little or no packaging. Reuse items as much as possible and donate items no longer used to local charitable organizations. When a product is no longer useful in its current form, recycle it. Consider whether or not the material can be recycled in your local recycling programme when making purchasing decisions and remember that recycling alone is not enough. Buy products that are made with the materials that are collected in your recycling programme and close the loop by buying recycled.

Q: What role will technological development play in reducing our footprints?

A: Technology can help us reduce our footprints and become more sustainable. Many industries are adopting new practices and developing innovative technologies designed to reduce environmental impacts and increase resource efficiency. Examples include renewable energy (e.g. solar and wind power), energy-efficient electronics, and products that are manufactured to be less toxic and recyclable and/or contain recycled content. As consumers, we can support these initiatives by switching to these products and services. However, we cannot rely on technology alone. Even though products and services are becoming more efficient, our growing global population is consuming increasing quantities of these products and services. Therefore it is important that we as individuals reduce our consumption levels by using only what we really need.

To do

Ecological footprints of MEDCs and LEDCs

Data for food consumption are often given in grain equivalents, so that a population with a meat-rich diet would tend to consume a higher grain equivalent than a population that feeds directly on grain. Look at the data in Table 9.2.

1 What does the high per capita grain production in North America suggest about the diet?
2 What does the local grain productivity suggest about the two farming methods in use?
3 Which population is more dependent on fossil fuels? Explain.
4 Why is there a difference in the net carbon dioxide fixation of the two regions?

These, and other factors, will often explain the differences in the ecological footprints of populations in LEDCs and MEDCs.

5 Calculate the per capita ecological footprint (food land and carbon dioxide absorption land only) for each region, using the two respective formulae.

$$\frac{\text{per capita food consumption (kg yr}^{-1})}{\text{mean food production per hectare of local arable land (kg ha}^{-1}\text{ yr}^{-1})}$$

$$\frac{\text{per capita CO}_2\text{ emission (kg C yr}^{-1})}{\text{net carbon fixation per hectare of local natural vegetation (kg C ha}^{-1}\text{ yr}^{-1})}$$

6 State two differences you would expect between the ecological footprint of a city in an LEDC and that of a city in an MEDC.
7 It has been calculated that the ecological footprint of Singapore is 264 times greater than the area of Singapore. Explain what this means.
8 Assume that in a large city with a stable population, the proportion of the population that has a vegetarian diet increases. Explain how this change might affect the city's ecological footprint.

Population from	Per capita grain consumption kg yr^{-1}	Local grain productivity kg ha^{-1} yr^{-1}	Per capita CO$_2$ emissions from fossil fuels kg C yr^{-1}	Net CO$_2$ fixation by local vegetation kg C ha^{-1} yr^{-1}
Africa	300	6000	200	6000
North America	600	300	1500	3000

Table 9.2

To do

Use the information and examples in this chapter to write a summary of human carrying capacity under the title: *"Human ingenuity, reduction of energy and material consumption, technical innovation and population development policies all increase human carrying capacity."*

Consider the following points. Give two good examples of each.

a Define human carrying capacity. List ways in which local human populations can exceed the natural carrying capacity of the area in which they live.

b Define and give examples of reuse, recycling and remanufacturing. How can these lead to an increase in human carrying capacity?

c What is the relation between technological development, resource use, carrying capacity and population growth? What consequences of these can limit population growth?

d How can national population policies decrease population size? What cultural changes can lead to decreased population growth?

Test yourself

The diagrams in Fig. 9.10 represent the area inhabited by, and the ecological footprint of, two human populations, A and B. One population is from an MEDC and the other from an LEDC. The diagrams are drawn to the same scale.

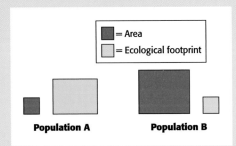

■ = Area
□ = Ecological footprint

Population A   **Population B**

Fig. 9.10

1 Which country is most likely to be an LEDC?

2 Explain your answer.

3 State four pieces of information that would be necessary to calculate the ecological footprint for any human population.

4 Explain two ways in which the latitude of a country might affect the size of an ecological footprint.

5 Which of the populations, A or B, is exceeding the carrying capacity of its local area?

6 Explain your answer.

7 Suggest two ways in which food production of the two populations might differ.

8 Explain how these differences in food production could influence the size of the ecological footprints of these two populations.

Key words

resource	ecological footprint
natural capital	Earthshare
natural income	technological development
renewable	reuse
non-renewable	recycling
replenishable	remanufacturing
sustainability	absolute reductions
sustainable development	

10 Energy resources

Key points

- Society gets its energy from a range of energy sources, non-renewable and renewable.
- We demand more and more energy to run the world economy.
- Different energy sources have advantages and disadvantages.
- Societies may choose energy sources for different reasons: political, cultural, economic, environmental and technological.

"I have no doubt that we will be successful in harnessing the sun's energy... If sunbeams were weapons of war, we would have had solar energy centuries ago."

Sir George Porter

Energy and energy resources

Our energy on Earth comes from our sun. Without it, the planet would be at absolute zero which is -273°C and there would be no life forms. The sun's energy drives the climate, geochemical cycles, photosynthesis, animal life – everything. Humans obtain their energy from the sun via living plants and via fossilized plants in the form of fossil fuels.

Fossil fuels are sources of stored energy from the sun – remains of organic life from millions of years ago, now being extracted and burned by humans. Oil is the economy's largest energy source at the moment, supplying 37% of all the energy we use. Coal is the next largest, supplying 25% and natural gas 23%. So fossil fuels power 80% of the world's economy. The other non-renewable energy source is uranium and this provides nuclear power which gives us just 6% of total energy.

The remaining 15% comes from renewable resources. Burning biomass provides 3% as 3 billion people depend on it for their energy needs for cooking and heating. Hydroelectric power generates 3% of the total and the other sources of renewable energy (solar, wind and wave power, geothermal) 2%.

How much longer for fossil fuels?

Oil 50 years, natural gas 70 years, coal 250 years are common estimates but we do not really know. It depends on our rate of use, other technologies, how efficient we are at using what reserves there are and how good we are at finding and extracting more. What we do know is that they will run out as they are non-renewable resources.

Two-thirds of oil reserves are in the Middle East. Most of the rest are in Russia and some in the USA. The fossil fuel with the most reserves and fastest increase in consumption is coal, mostly because of China's need for power; many power stations in China burn coal.

Humans use more and more energy as there are more of us and, as our wealth increases, we each use more energy if we can. But the use is not evenly distributed. Some countries are richer and use more energy than others. Before the industrial revolution, the energy we used was from wood – biomass. Since then, with the

To do

Draw and label a pie chart of the percentage use of world energy sources in the text.

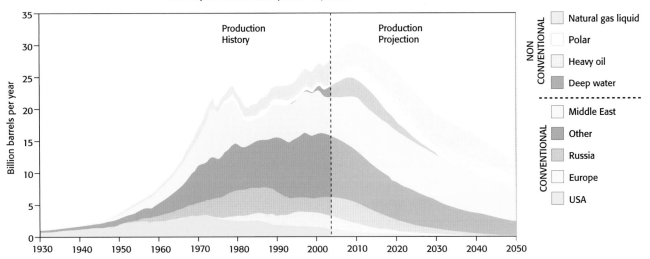

Fig. 10.1 World fossil fuel use and projected use

discovery of fossil fuels, we use far more than that available to us in biomass. We are using the capital stored as fossil fuels formed 300 million years ago. Although it is generally agreed that oil will be exhausted in about 50 years' time, some think it will be as soon as 30–40 years' time – in your lifetimes. We may find more oil but most experts think that we shall not discover more reserves after the next 10–20 years as we shall have looked in most areas by then. If LEDCs use oil at the rate *per capita* of the United States, it will run out in 17 years or sooner. We have reached or are about to reach "peak oil" after which the amount produced will decline. There is more natural gas than oil left but we will use all of that in fewer than 80 years. There is far more coal left to mine and some people say that coal is the energy source of the future. We think there are 1000 billion tonnes of coal left, enough for 150 years. China, in 2007, was building two coal-fired power stations a week and coal's use is growing faster than that of oil and gas. But it is the dirtiest energy source, releasing carbon dioxide and sulfur dioxide when burned. However, we can build power stations with carbon capture and storage (CCS) technology. Since we are going to burn coal anyway, this must be the future – making coal a clean energy source.

Evaluation of energy sources and their advantages and disadvantages

Energy source	From	Advantages	Disadvantages
Non-renewable			
Coal (fossil fuel)	Fossilized plants laid down in the Carboniferous period, mined from seams of coal which are in strata between other types of rock. May be open cast mined (large pits) or by tunnels underground. Burnt to provide heat directly or electricity by burning to turbines in power stations.	Plentiful supply. Easy to transport as a solid. Needs no processing. Relatively cheap to mine and convert to energy by burning. Up to 250 years of coal left.	Non-renewable energy source. Cannot be replaced once used (same for oil and gas). Burning releases carbon dioxide which is a greenhouse gas. Some coals contain up to 10% sulfur. Burning sulfur forms sulfur dioxide which causes acid deposition. Particles of soot from burning coal produce smog and lung disease. Coal mines leave degraded land and pollution. Lower heat of combustion than other fossil fuels, i.e. less energy released per unit mass.

Oil (fossil fuel)	Fossilized plants and micro-organisms that are compressed to a liquid and found in porous rocks. Crude oil is refined by fractional distillation to give a variety of products from lighter jet fuels and petrol to heavier diesel and bitumen. Extracted by oil wells. Many oil fields are under the oceans so extraction is dangerous. Pipes are drilled down to the oil-bearing rocks to pump the oil out. Most of the world economy runs on oil either burnt directly in transport and industry or to generate electricity.	High heat of combustion, many uses. Once found is relatively cheap to mine and to convert into energy.	Only a limited supply. May run out in 20–50 years' time. Like coal, gives off carbon dioxide when burnt. Oil spill danger from tanker accidents. Risk of terrorism in attacking oil pipelines.
Natural gas (fossil fuel)	Methane gas and other hydrocarbons trapped between seams of rock. Extracted by drilling like crude oil. Often found with crude oil. Used directly in homes for domestic heating and cooking.	Highest heat of combustion. Lot of energy gained from it. Ready-made fuel. Relatively cheap form of energy. Cleaner fuel than coal and oil.	Only limited supply of gas but more than oil. About 70 years' worth at current usage rates. Also gives off carbon dioxide but only half as much per unit of energy produced as coal.
Nuclear fission	Uranium is the raw material. This is radioactive and is split in nuclear reactors by bombarding it with neutrons. As it splits into plutonium and other elements, massive amounts of energy are also released. Uranium is mined. Australia has the most known reserves, Canada exports the most, other countries have smaller amounts. About 80 years' worth left to mine at current rates but could be extracted from sea water.	Raw materials are relatively cheap once the reactor is built and can last quite a long time. Small mass of radioactive material produces a huge amount of energy. No carbon dioxide released nor other pollutants (unless there are accidents).	Extraction costs high. Nuclear reactors are expensive to build and run. Nuclear waste is still radioactive and highly toxic. Big question of what to do with it. Needs storage for 1000s of years. May be stored in mine shafts or under the sea. Accidental leakage of radiation can be devastating. Accidents rare but worst nuclear reactor accident at Chernobyl, Ukraine in 1986. Risk of uranium and plutonium being used to make nuclear weapons.
Renewable			
Hydro-electric power (HEP)	Energy harnessed from the movement of water through rivers, lakes and dams to power turbines to generate electricity. Pumped-storage reservoirs power turbines.	High-quality energy output compared with low-quality energy input. Creates water reserves as well as energy supplies. Reservoirs used for recreation, amenity. Safety record good.	Costly to build. Can cause the flooding of surrounding communities and landscapes. Dams have major ecological impacts on local hydrology. Silting of dams. Downstream lack of water (e.g. Nile) and risk of flooding if dam bursts.
Biogas	Decaying organic plant or animal waste are used to produce methane in biogas generators or burnt directly as dung/plant material. More processing can give oils (e.g. oilseed rape, oil palms, sugar cane) which can be used as fuel in vehicles instead of diesel fuel = biofuels.	Cheap and readily available energy source. If the crops are replanted, biogas can be a long-term, sustainable energy source.	May be replacing food crops on a finite crop land and lead to starvation. When burnt, it still gives off atmospheric pollutants, including greenhouse gases. If crops are not replanted, biomass is a non-renewable resource.
Wood	From felling or coppicing trees. Burnt to generate heat and light.	A cheap and readily available source of energy. If the trees are replaced, wood can be a long-term, sustainable energy source.	Low heat of combustion, not much energy released for its mass. When burnt it gives off atmospheric pollutants, including greenhouse gases. If trees are not replanted wood is a non-renewable resource. High cost of transportation as high volume.
Solar - photo-voltaic cells	Conversion of solar radiation into electricity via chemical energy.	Potentially infinite energy supply. Single dwellings can have own electricity supply. Safe to use. Low quality energy converted to high.	Manufacture and implementation of solar panels can be costly. Need sunshine, do not work in the dark.
Solar – passive	Using buildings or panels to capture and store heat.	Minimal cost if properly designed.	
Wind	Wind turbines (modern windmills) turn wind energy into electricity. Can be found singly, but usually many together in wind farms.	Clean energy supply once turbines made. Little maintenance required.	Need the wind to blow. Often windy sites not near highly populated areas. Manufacture and implementation of wind farms can be costly. Noise pollution. Some local people object to on-shore wind farms, arguing that it spoils countryside. Question of whether birds are killed or migration routes disturbed by turbines.

Tidal	The movement of sea water in and out drives turbines. A tidal barrage (a kind of dam) is built across estuaries, forcing water through gaps. In future underwater turbines may be possible out at sea and without dam.	Should be ideal for an island country such as the UK. Potential to generate a lot of energy this way. Tidal barrage can double as bridge, and help prevent flooding.	Construction of barrage is very costly. Only a few estuaries are suitable. Opposed by some environmental groups as having a negative impact on wildlife. May reduce tidal flow and impede flow of sewage out to sea.
Wave	The movement of sea water in and out of a cavity on the shore compresses trapped air, driving a turbine.	Should be ideal for an island country. These are more likely to be small local operations, rather than done on a national scale.	Construction can be costly. May be opposed by local or environmental groups. Storms may damage them.
Geothermal	It is possible to use the heat inside the Earth in volcanic regions. Cold water is pumped into the Earth and comes out as steam. Steam can be used for heating or to power turbines creating electricity.	Potentially infinite energy supply. Is used successfully in some countries, such as New Zealand.	Can be expensive to set up. Only works in areas of volcanic activity. Geothermal activity might calm down, leaving power station redundant. Dangerous underground gases have to be disposed of carefully.

Table 10.1

To do

Select an energy source from Table 10.1. Research information on this source.

Individually or in small groups, present a case to your class about that energy source.

Include in your presentation: what it is, where it is mostly used, how much it is used, whether its use is sustainable, its relative cost, advantages and disadvantages of the resource use, future prospects, reasons why some countries may use this resource although it may not be the cheapest.

Evaluate others' presentations based on:

● Quality of spoken presentation – did they present in a clear, interesting way? Did they simply read it out or did they look at the audience and engage them?
● Quality of content – did you understand the content? Was it clear?
● Quality of associated documentation (handout, slide presentation, data) – did this enhance your understanding of the presentation?

Energy consumption

The world today is in an **energy crisis** yet we continue to use non-renewable fuels at an increasing rate. Oil prices per barrel reached an all-time high in June 2008 of over US$142 per barrel. Are they higher or lower today? We know we use more and more energy and that fossil fuels will run out. We shall have to use energy sources other than fossil fuels. Renewable sources as well as nuclear power are the options. The EU, for example, produces less than 7% of its energy from renewable sources but there is a proposal to increase this to 20% by 2020.

In the future, without fossil fuels, humans will have to gain their energy from other sources unless we revert to a much smaller population only dependent on solar energy directly. Possibilities include the **hydrogen economy** where hydrogen is the fuel that provides energy for transport, industry and electrical generation. There are prototype engines and cars that use hydrogen as their fuel, releasing water as the waste product. But hydrogen is a highly flammable gas and difficult to transport and store.

In **nuclear fusion**, energy can be released by the fusion of two nuclei of light elements (e.g. hydrogen). The sun and stars generate energy by nuclear fusion. In theory, we could generate energy in this way but, in practice, it has only so far been used in hydrogen bombs.

We need to increase our use of renewable energy sources for which we have the technology, to achieve absolute reductions in energy consumption and to increase our efficiency of energy use.

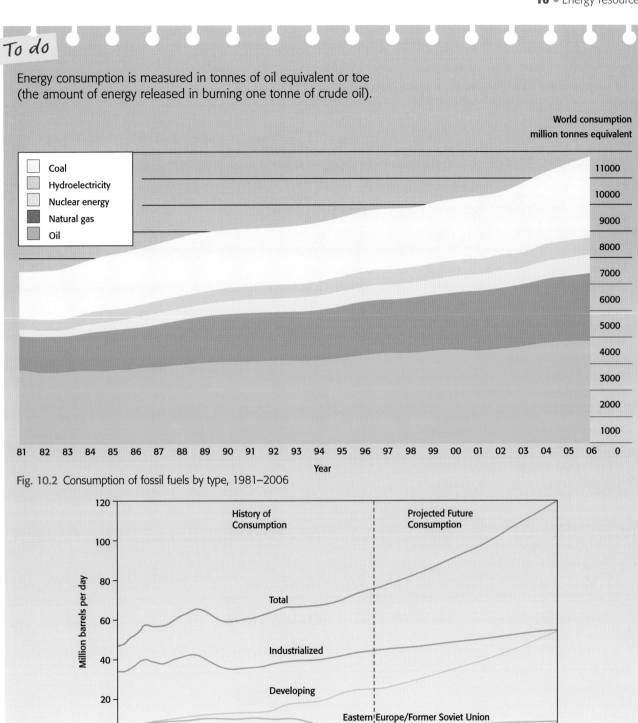

To do

Energy consumption is measured in tonnes of oil equivalent or toe
(the amount of energy released in burning one tonne of crude oil).

Fig. 10.2 Consumption of fossil fuels by type, 1981–2006

Fig. 10.3 World oil consumption by region, 1970–2020

Study the data on energy consumption in Figs 10.2
and 10.3.

1 Describe the lifestyle of someone with the highest
consumption and someone with the lowest. Name
the type of country or regions they might live in.

2 Why has consumption of coal and oil increased
more than that of oil?

3 Suggest reasons for the drop in total and industrialized
countries' oil consumption in the late 1970s.

4 Consider your own use of fossil fuels, list the
products you have and activities you do that rely on
fossil fuel consumption. (Remember, plastics are
derived from oil.)

Renewable energy resources

Theoretically, we could get the energy we need to power the world economy and our own requirements from renewable resources. But, in practice, worldwide we obtain a small percentage at the moment from renewable resources – about 14%.

We know that we must increase the percentage of our energy produced from renewable resources and progress has been and is being made. The EU plans to double its energy supplied from renewables from 6% to 12% by 2010. The US plans similar increases. But investment in research on making renewable energy sources more efficient, e.g. wave and tidal power or solar cells, is small compared with research into finding more oil or gas. Reasons for this may be:

- The TNCs (transnational corporations) and heavy industry are committed to the carbon economy – all machines are made to run on fossil fuels – and the scale of a change is hard to imagine.
- It is cheaper to produce electricity from fossil fuel burning than from most renewable resources at the moment (ignoring the environmental cost).
- Countries are locked into the resource that they currently use, by trade agreements or convenience.
- Some sources are not available, e.g. wave or tidal power are not possible for land-locked countries.
- Some sources are less reliable, e.g. solar energy needs the sun to shine, wind turbines need the wind.

What other reasons can you think of?

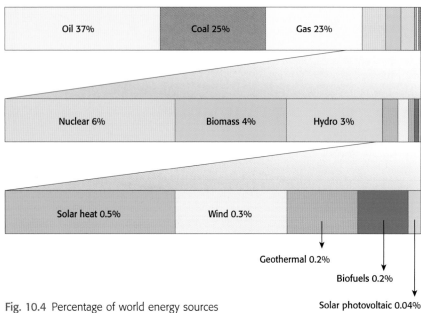

Fig. 10.4 Percentage of world energy sources

To do

Look at the energy sources in Fig. 10.4.
1 State which do not produce carbon dioxide when used to produce energy.

2 Discuss whether any of these energy sources are truly zero carbon emission sources.

Wind power in Denmark

Denmark produces over 18% of its energy requirements from wind energy, converting the kinetic energy in the wind to mechanical energy which drives a wind turbine to produce electricity. Denmark has a higher proportion of its energy from wind than any other country. The reason for this is government intervention. In the 1970s, much of Denmark's energy was from coal-fired power stations but the government wanted to reduce carbon emissions. There was a ban on nuclear power plants and wind power was seen as the solution. Although the wind speeds in Denmark are not particularly high, there are shallow waters offshore where the turbines can be sited and then linked inland. Denmark is connected to the electricity grids of neighbouring countries and can buy electricity from them if the wind drops. This is, of course, one disadvantage of wind power – you need the wind to blow. There is little evidence that wind turbines kill many birds that fly into them. Although some do, most birds fly over or round the turbines and adapt their migration routes. Power lines kill more birds than do wind turbines.

The cost of China's amazing economic growth

A few decades ago, China (the People's Republic of China) was predominantly a country in which peasant farmers were in the majority. Over the last 25 years, it has had the fastest-growing GDP of any country, about 10%, and is now the second largest economy after the USA. Economic development has decreased poverty notably and increased immigration to cities from the countryside. With this rapid development have also come inequalities and industrial pollution but it is working to address these and, as a sign of its opening up to the rest of the world, hosted the Olympic Games in 2008.

Still a communist state, there have been great changes in the state control of the economy. The agricultural communes formed in Chairman Mao's time resulted in decreased food production. Decisions to eliminate the collective farms, increase private enterprise and allow foreign investment over the last 20 years have altered the economy immensely.

With about 20% of the world's population, China has to create 15 million new jobs annually just to employ those joining the workforce. But still, about 300 million people in rural areas live on less than US$1 per day – in poverty. China feeds this 20% of the world's population on 7% of the world's crop lands.

Coal is the biggest energy resource and most coalfields are in the north while industry is concentrated in the southeast. Some 80% of China's energy is from coal-fired power stations with 500 planned to be built. Worldwide about 25% of energy is derived from coal. China is the largest producer and consumer of coal in the world, the largest producer of steel and third largest producer of cars.

This amazing story of industrialization has costs. Due to such rapid growth, environmental legislation has not kept pace and soil erosion, desertification, salinization, water pollution and air pollution are big issues. According to the WHO, seven of the world's ten most polluted cities are in China. People have respiratory problems due to the smog and air pollution. Lead poisoning of children in cities and acid deposition on the countryside in Japan and Korea have increased.

Almost all rivers are polluted and half the population does not have access to clean water. Water scarcity in the north is a problem. Some estimates are that pollution costs up to 10% of GDP in China and it certainly will cost the environment for decades to come. Clean-up is starting but is slow and the task is daunting.

The Worldwatch Institute president has said: "The pace of energy development in China is breathtaking. No other large country has ever built an economy at this speed. This is a critical global indicator and it shows the urgency for action everywhere."

Fig. 10.5 Pollution in a Chinese village

Case study

Biogas generation in Gujarat, India

Adapted from SATIS ASE unit

In Gujarat, Northwest India, are many small villages that are not connected to an electricity supply. Cooking is the main user of energy and this comes from firewood, dung or kerosene. Irrigation uses some power as well. A technologically simple biogas generator plant can provide enough methane gas for cooking for families in the village and reduce smoke pollution from burning fires within houses. Biogas generators are simple pits which are filled with animal dung mixed with water. By excluding air, anaerobic fermentation starts in the generator as bacteria use the dung as a food source and release methane as a waste product. The methane is piped into village houses to the kitchens. (Solar voltaic cells can provide electricity for lighting and solar cells heat up water for domestic use.) By using renewable energy resources locally, the villagers can provide an integrated solution to their power needs without the need for outside supplies of energy.

Questions

1 Why is wood considered a form of renewable energy?

2 List two environmental problems created by burning wood for heating and cooking.

3 What is the main ingredient in biogas?

4 What are the advantages and disadvantages of burning biogas (methane)?

5 Methane from landfills is often allowed to escape directly to the atmosphere. Why?

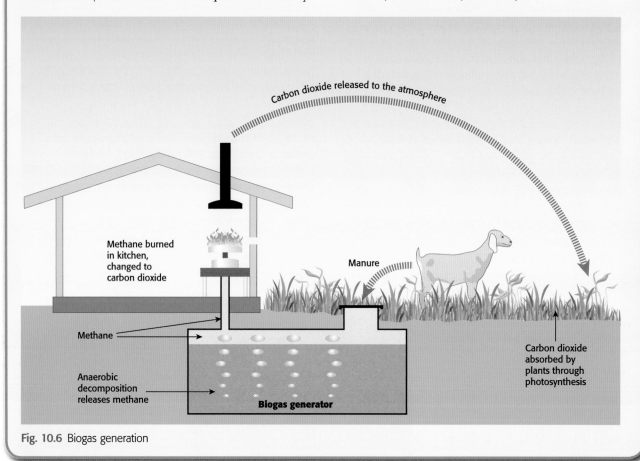

Fig. 10.6 Biogas generation

The Athabasca oil sands

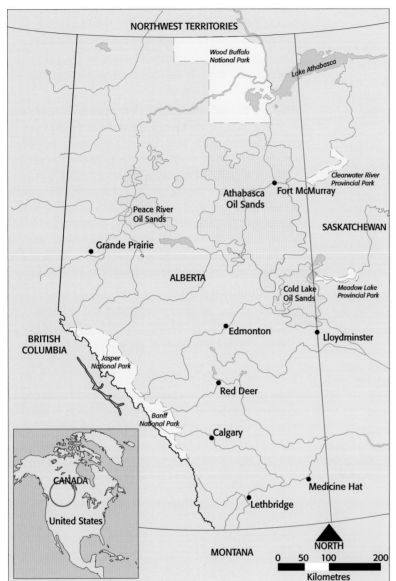

Fig. 10.7 Map of the Athabasca oil sands

In Alberta, Canada, there is a lot of oil in tar sands. It is a very heavy crude oil, rich in bitumen which is used to make roads. It takes so much energy to refine this oil that it is only economic to do so if oil prices are high. The cost of extracting a barrel of this oil is about US$30. Prices for crude oil are high so it is being extracted, by surface mining from the Athabasca Oil Sands in Alberta. With some smaller deposits, these oil sands cover 54 000 square miles of land. Growing on the land is boreal forest and muskeg ecosystems. Muskeg is a type of peat bog with acidic conditions so vegetation is semi-decomposed. It can be up to 30 m deep, is 1000 years old and is a habitat to beavers, pitcher plants and mosses. Caribou deer (reindeer) are found there. Few humans live there. The amount of oil in these tar sands is roughly equal to the rest of the world's reserves put together. It is mined by open-pit mining which destroys the vegetation and changes the shape of the landscape as huge volumes of tar sands are removed. Mining companies do restore the land to pasture or plant trees but not to the original boreal forest or muskeg.

Not only does it cost to extract the oil, it also takes a lot of energy – one barrel of natural gas to extract two of crude oil, and it takes a lot of water – two barrels of water for each barrel of oil and the waste water has to be kept in large tailing (mining waste) ponds.

Rather than reducing, the carbon emissions of Canada are increasing, partly due to this mining of tar sands.

To think about

Biofuels – the answer or the problem?

Biofuels are fuels made from crops.

Why? In theory, they are greener as they are carbon neutral. All the emissions of carbon dioxide made from burning them are fixed by growing the plants to replace them. As 25% of carbon emissions are from transport, using biofuels in vehicles is a neat idea but things are not straightforward in complex systems.

What crops are used? Sugar cane has been used for decades in Brazil to make ethanol by fermentation. The ethanol (an alcohol) is sold alongside or mixed with petrol (gas) and 80% of cars sold in Brazil now have hybrid engines – they run on petrol, alcohol or a mixture of the two called gasohol. From 2000 to 2005, the world's output of plant ethanol has doubled and biodiesel, made from oily plants like oil palm and soya bean, has increased.

In the USA, maize production for ethanol is heavily subsidized and has increased five-fold and is set to increase more. In the EU, a regulation was passed that all fuels must contain 2.5% biofuel, rising to 5% and 10% by 2020.

What's the problem? Amazingly, in calculating the carbon balance of biofuels, no one considered the additional carbon costs in extracting the ethanol, transporting the crop to the extraction plant, and transporting the processed fuel. Some of those costs also apply to conventional fossil fuel extraction of course but there is also the cost of fertilizer applied to the crop (oil-based). Fertilizers release nitrous oxide which is at least 200 times more powerful a greenhouse gas than carbon dioxide. Some calculate that maize ethanol requires 30% more energy to make than it contains.

The other problem is the land it takes to grow biofuels. It would take 40% of the EU cropland to meet the 2020 10% biofuel target.

What are the consequences? Deforestation is happening in order to plant crops for biofuels and this releases the carbon trapped in the trees. Wetlands are being drained and grassland being planted as well. Indonesia has planted so much oil palm on what was forest land that it is now the third largest emitter of carbon in the world. US farmers are selling 20% of their maize for biofuel instead of food, so soya farmers there are switching to the more profitable maize. Brazilian farmers therefore grow more soya bean to export, so they plant it on grazing land, and cattle ranchers cut more forest in the Amazon to turn into grassland.

What should be done?

The calculations need to be checked and all costs included for both fossil fuels and biofuels. Cellulosic ethanol, made from wood chips, switchgrass or straw, may be the answer as its production will reduce carbon emissions far more than growing maize or soya bean.

There is inertia in economic and government systems, so once subsidies and legislation are in place, it will take time to change them if we need to. If biofuels are leading to increased greenhouse gas emissions, not reduced ones, and taking up land needed to grow food, what do you think should be done?

Key words

fossil fuels	wind energy
coal	wave energy
oil	tidal energy
natural gas	solar energy
renewable	geothermal energy
non-renewable	energy crisis
nuclear energy	biogas
hydroelectric energy	biofuel

11 Water resources

Key points

- Most of the water on the Earth is salt water with only about 3% as fresh water.
- In the water cycle, water is transformed and transferred.
- Human use of water may be sustainable or unsustainable.

The Earth's water budget

The Earth is not referred to as the "blue planet" without due reason. From space the presence of water on Earth is very obvious. About 70% of the Earth's surface is covered by water. However, what is less apparent is the balance of water by type within the planet. Only about 3% of all water is fresh water and about 97% of the water on the planet is ocean water (salt water). Fresh water is in quite short supply. About 69% of this water is in the polar ice caps and glaciers and about 30% in groundwater. Water on the surface of the Earth in lakes, rivers and swamps is only about 0.3% of the total. You may think the atmosphere holds a lot of water but only about 0.001% of Earth's total water volume is water vapour in the atmosphere. According to the US Geological Survey, if all of the water in the atmosphere rained down at once, it would only cover the ground to a depth of 2.5 cm.

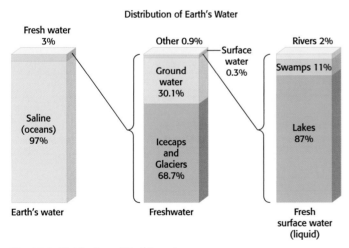

Fig. 11.1 Distribution of Earth's water

To do

Using the data given in Fig. 11.1, construct a table to list freshwater storages on Earth in decreasing order of size.

The **turnover time** is the time it takes for a molecule of water to enter and leave part of the system. Turnover times are very variable. In the oceans, it takes 37 000 years, icecaps 16 000 years, groundwater 300 years. Yet a molecule of water will stay in the atmosphere only 9 days and in rivers 12–20 days. So water can be seen as a renewable resource or a replenishable resource depending where it is.

Ocean currents and energy distribution

Ocean currents play an important role in the distribution of energy. Cold ocean currents run from the poles to the equator, whereas warm currents flow from the equator to the poles. The driving forces behind all these ocean currents are the wind and the Earth's rotation.

Examples of sea currents

Warm currents include the Gulf Stream (in the North Atlantic Ocean) and the Kuroshio current (Japanese for "black tide") which flows north on the eastern side of Japan.

Cold currents include the Humboldt current (off the coast of Peru) and the Benguela current (off the coast of Namibia in southwestern Africa). Some of the above currents are part of a global ocean current system, the so-called oceanic conveyer belt (Fig. 11.2).

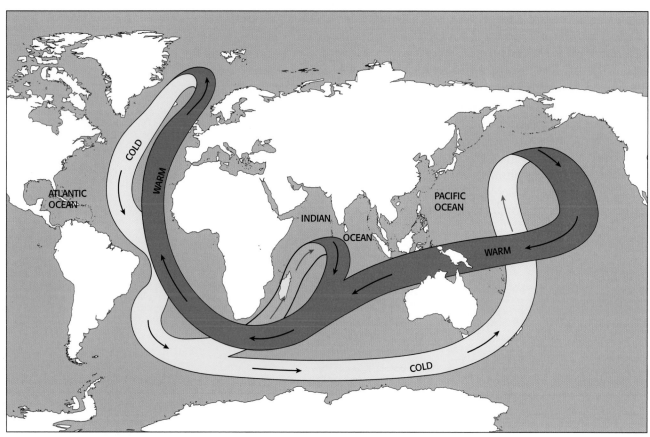

Fig. 11.2 The great oceanic conveyor belt

Ocean currents and climate

Water has a higher specific heat capacity than land. This means that water masses heat up and cool down more slowly than land masses. As a result, land close to seas and oceans has a mild climate with moderate winters and cool summers.

Ocean currents (Fig. 11.3) also affect local climate. One example is the warm Gulf Stream/North Atlantic Drift that moderates the climate of northwestern Europe, which otherwise would have a sub-arctic climate. Another example is the effect of the cold Benguela current which, under the influence of prevailing southwest winds, moderates the climate of the Namib desert. A third example is the effect of the Humboldt current on the climate in Peru (see below).

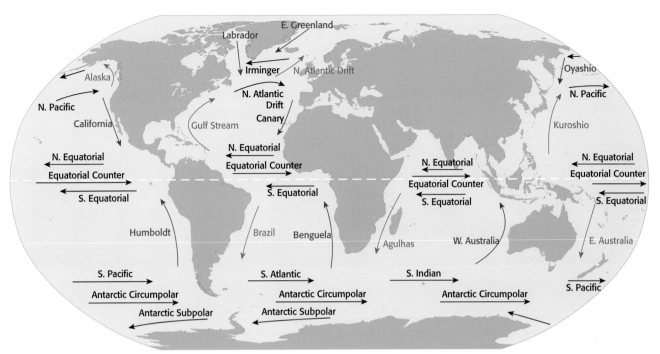

Fig. 11.3 The ocean currents

El Niño Southern Oscillation (ENSO)

El Niño Southern Oscillation (ENSO) is a phenomenon in the Pacific Ocean that has global consequences. Normally, air pressure in the eastern Pacific Ocean (South America) is higher than that in the western Pacific Ocean (Australia, Indonesia). This results in the trade winds that blow westward for most of the year (Fig. 11.4). The trade winds blow the warm surface water westward.

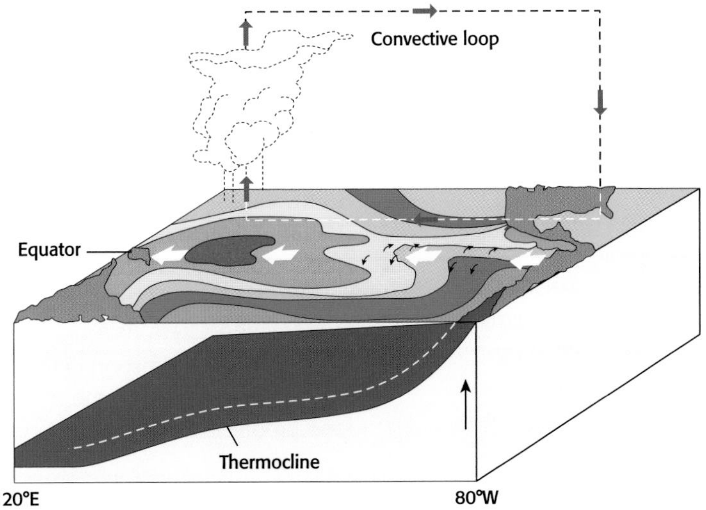

Fig. 11.4 Trade winds blowing warm surface waters of the Pacific westwards in a non-ENSO year

In some years however, the pressure difference across the Pacific Ocean is reversed (Fig. 11.5).

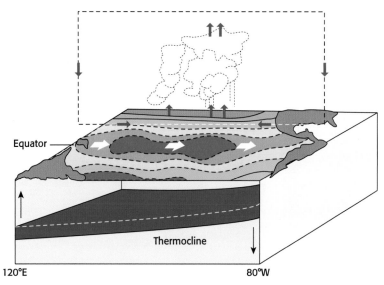

Fig. 11.5 An ENSO year when the winds blow eastwards across the Pacific

These pressure changes are the Southern Oscillation. The changed pressure difference alters the directions of both the wind and the warm surface current. This phenomenon was first discovered by Peruvian fishermen, who called it *El Niño* (the boy child), as it usually occurred just before Christmas. Because the reversal of sea and air currents is caused by the changed pressure difference, the whole phenomenon is often called ENSO: El Niño Southern Oscillation. El Niños occur about every two to eight years and last for about two years. They may be either weaker or stronger. Strong El Niño years are usually followed by several years of *La Niña* when the ocean temperature of the eastern Pacific is unusually cold after being unusually hot in the El Niño years.

El Niño has local and global effects. Local effects include collapse of anchovy fish stocks, massive death of sea birds, and storms and flooding in the coastal plain of Peru.

The anchovy fishery off the coast of Peru is extremely rich because of the occurrence of an upwelling when cold nutrient-rich waters come up from the ocean depths. Normally, productivity in oceans is quite low because of one of two limiting factors: light level and nutrient concentration. In the upper water levels light intensity is high, but nutrient levels are low and limit productivity. The available nutrients are taken in by phytoplankton and travel down the food chain. Dead organisms that are not eaten by some other organism sink and the nutrients stored within them end up on the ocean bottom. The lower water layers are therefore nutrient rich. But here, the absence of light makes photosynthesis impossible. West of the Peruvian coast, the prevailing eastern trade winds push the surface water westward. This water is replaced by cold, nutrient-rich water from the deep Humboldt Current, which originates in the Antarctic region and follows the South American coast to the north. The appearance of nutrient-rich water at the surface allows for high productivity, hence the high numbers of fish and their predators, the seabirds. During El Niño events, the upwelling disappears and the fish and seabirds starve.

During El Niño events the Peruvian coastal plain is subject to severe storms, accompanied by excessive rainfall. This is caused by the warm, extremely moist air being forced upward by the Andes Mountains.

The warm ocean water that moves east during El Niño events contains tremendous amounts of energy. The amount of energy is large enough to alter major air currents like the jet streams and, as a consequence, El Niño affects global weather. Examples of these changes are:

● Droughts in Australia, Indonesia, the Pacific Northwest of the United States and British Columbia (Canada). Forest fires are common in these areas.
● Heavy storms often resulting in flooding in California and the Midwest of the United States, Central Europe and eastern Asia.
● Absence of the monsoon in India. The Indian population depends on the monsoon rains for its food production.

While ENSO is a natural phenomenon, its effects are exacerbated by human pressures on the environment.

The Gulf Stream/North Atlantic Oscillation

The Gulf Stream is a current in the Atlantic Ocean that comes from the Gulf of Mexico across the Atlantic, where it is known as the North Atlantic Drift to western Europe. It carries warm waters in a current about 100 km wide and 1000 m deep on average. It makes Northern Europe warmer than it otherwise would be. As this water flows, some evaporates so its saltiness increases. By the time it reaches the north of Britain and Scandinavian coasts, it is so much saltier, and so much more dense, than the surrounding sea water, that it sinks and returns in a conveyor belt of the North Atlantic Deep Water current back to where it started.

There is some evidence that the Gulf Stream current is slowing down, possibly due to global warming. Some think it may stop completely in a few decades. Melting of the Greenland ice sheet may be adding so much fresh water to the North Atlantic that the saltiness of the North Atlantic Drift is reduced and the sinking and return of the water to the Gulf of Mexico is slower. If this were the case, climate in western Europe would be getting cooler, but it is warming instead.

The North Atlantic Oscillation (NAO) is a weather phenomenon in the Atlantic similar to ENSO in the Pacific. In high index NAO years, there is low pressure over Iceland and high over the Azores so westerly winds blow and winters are mild and summers cool and wetter. In low index NAO years, the pressure differences are lower, the westerlies reduced and winters are colder and summers have heatwaves. There is also an influence on eastern North America from the NAO.

Questions

1 Research the latest data on ENSO and NAO. Find out if there is an ENSO event now. What is the reduction in the Gulf Stream current now?

If human activity is altering climate, as most scientists believe, we are also exacerbating the effects of ENSO events as we put more pressure on the natural systems.

2 If the Gulf Stream were to stop suddenly (which we think has happened in the past), what would be the implications?

Water as a critical resource

These facts are from the World Water Council:

- 1.1 billion people live without clean drinking water
- 2.6 billion people lack adequate sanitation
- 1.8 million people die every year from diarrhoeal diseases
- 3900 children die every day from waterborne diseases
- Daily per capita use of water in residential areas:
  - 350 litres in North America and Japan
  - 200 litres in Europe
  - 20 litres in sub-Saharan Africa (or less)
- Over 260 river basins are shared by two or more countries, mostly without adequate legal or institutional arrangements
- Quantity of water needed to produce 1 kg of:
  - wheat: 1 000 L
  - rice: 1 400 L
  - beef: 13 000 L

To do

1 Assuming there are 6.6 billion people alive on Earth, what is the percentage of the world's population that have no clean drinking water?

2 How many children die per year from waterborne diseases?

3 What is the percentage of daily per capita use of water in sub-Saharan Africa compared with that in Japan and North America?

4 How many days of water use by one person in Europe would it take to produce one kilogram of beef?

5 How many kilograms of wheat would be produced by the equivalent of one year of water consumption by one person in sub-Saharan Africa?

Although there is a lot of water on Earth, most of it is saline. We can remove the salt from water in desalination plants but the costs in terms of energy are large and it is only currently possible in wealthy countries which are water-stressed and near the sea, e.g. Israel, Australia and Saudi Arabia. Unless we can find the technology to desalinate water cheaply, it is not a viable proposition at the moment.

The fresh water available to us is limited and the UN has applied the term "water crisis" to our management of water resources today. There is not enough usable water and it can be very polluted. Up to 40% of humans alive today live with some level of water scarcity. This figure will increase.

Humans use fresh water for:
- domestic water (water used at home for drinking, washing, cleaning)
- irrigation
- industry including manufacturing, mining, agriculture
- hydroelectric power generation
- transportation (ships on lakes and rivers)
- as a boundary marker between nation states (rivers and lakes).

The WHO (World Health Organization) states that each human should have access to a minimum of 20 litres of fresh water per day, but Agenda 21 says this should be 40 litres. But in much of the world this is not the case, while in other regions there is plenty (see Fig. 11.6).

Fig. 11.6 World map of water scarcity

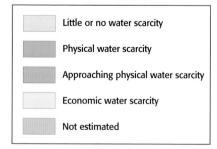

Little or no water scarcity

Physical water scarcity

Approaching physical water scarcity

Economic water scarcity

Not estimated

Water scarcity is not just a measure of how much water there is but of how we use it. There may be enough water in a region but it is diverted for non-domestic use. Agriculture uses tens of times more water than humans use, in irrigation and providing water for livestock. As human population expands, we need water to grow more food but, like food, it is not that there is not enough worldwide, it is the distribution of it that is uneven.

Egypt, for example, imports more than half its food as it does not have enough water to grow it and, in the Murray-Darling basin in Australia, there is water scarcity for humans as so much is used for agriculture.

Adding droughts and climate change, soil erosion and salinization to the story, you can see that water is and will become a major issue for nations and international organizations. Many major rivers run through several countries. It comes back to the question of who owns the water and the tragedy of the commons (see also page 188). The Danube River basin is shared by 19 countries and 81 million people. The Tigris-Euphrates rivers carry water that is extracted by Iran, Iraq and Syria. There have been and will be wars over water as it becomes increasingly needed and increasingly scarce.

Sustainability of freshwater resource usage

Sustainable use of a resource means that the resource is used at the same pace or slower than that at which it is reformed.

Sources of fresh water are **surface freshwater** (rivers, streams and lakes) and underground **aquifers**. The water can be extracted via wells. Aquifers are filled continuously by infiltration of precipitation. Often, however, this only happens in relatively small areas because of the presence of impermeable soil layers. Furthermore, water flow in aquifers is extremely slow (horizontal flows can be as slow as 1 to 10 metres per century). As a result, aquifers are often used unsustainably.

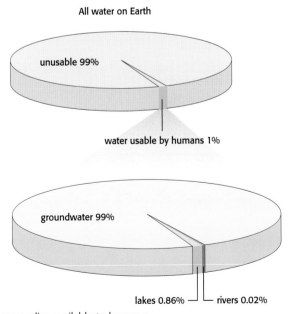

Fig. 11.7 Water supplies available to humans

Global freshwater consumption is increasing quickly because the human population is increasing and because the average quality of life is improving. This increased freshwater use leads to two types of problems: water scarcity and water degradation. Degradation means that water quality deteriorates, making it less suitable for use. Let's have a more detailed look at the problems related to freshwater use and some possible solutions.

Problems

- Low water levels in rivers and streams. The Colorado River in the USA, once a major river, now is not much more than a tiny stream when it enters the Gulf of Mexico, making navigation impossible.
- Slow water flow in the lower courses of rivers results in sedimentation, which makes the already shallow river even shallower.
- Underground aquifers are being exhausted. This simply means that the aquifer cannot be used anymore, which strongly affects agriculture. Buildings can be damaged when the soil is shrinking because the water has been taken away.
- Irrigation often results in soil degradation, especially in dry areas. Much of the water used in irrigation evaporates before it is absorbed by the crops. Dissolved minerals remain in the top layer of the soil, making it too saline (salty) for further agriculture. This process is called salinization.
- Fertilizers and pesticides often pollute streams and rivers.
- Industries release pollutants to surface water.
- Industries and electricity plants release warm water to rivers. Warm water can contain less dissolved oxygen than cold water, so aquatic organisms that take their oxygen from the water (fish, crayfish) may not get enough oxygen for respiration. Warm water outflow from power stations changes the species composition in the water.

Solutions

- Reduce domestic use of fresh water by using more water-efficient showers, dishwashers and toilets.
- Wash cars in car washers with a closed water system, or do not wash cars at all.
- Irrigation: Selecting drought-resistant crops can reduce the need for irrigation. (Some areas may simply be unsuitable for growing crops; cattle grazing may be better.) Using closed pipes instead of open canals can reduce evaporation, and using trickle systems instead of spraying water.
- Reduce the amount of pesticide and fertilizer used (using the smallest possible amount at the most appropriate time).
- Prevent overspray, e.g. spraying pesticide or fertilizer directly in a stream.
- Use highly selective pesticides instead of generic pesticides or use biological control measures.
- Industries can remove pollutants from their wastewater with water treatment plants. Often, they are forced by law to do so.
- Regulate maximum temperatures of released cooling water. Instead of releasing the warm cooling water, cooling towers that evaporate the water can be used.

To do

The water cycle

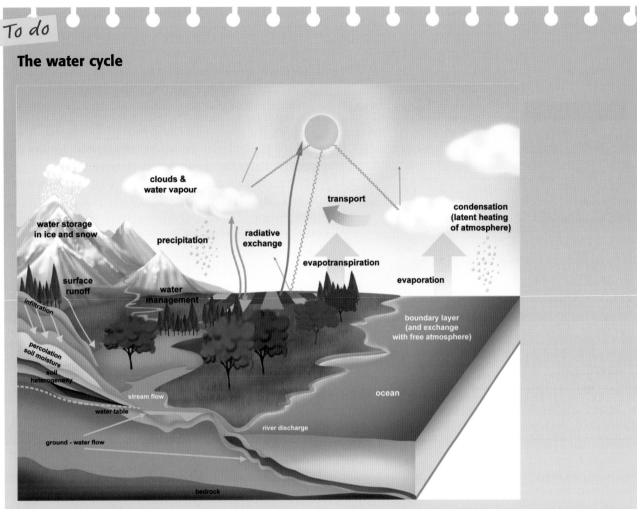

Fig. 11.8 The water cycle

The water or hydrological cycle (Fig. 11.8) consists of storages of water and the flows of water between the various storages. The **water budget** is a quantitative estimate of the amounts of water in these storages and flows. Water moves by transfer and transformation. It is transferred from river to ocean (same state) and transformed when it changes state from liquid water, evaporating to water vapour, condensing into rain and freezing into snow and ice. The storages include the oceans, icecaps and glaciers, lakes, groundwater and the atmosphere. The water cycle drives the world's weather systems.

1 Draw a systems diagram of the water budget and cycle showing the storages and flows given in Table 11.1. Make your storage boxes and width of flow arrows correspond to the proportions of these volumes. Label all storages and flows.

2 Six stores of water are shown in Fig. 11.8. List them in order of decreasing size (largest size first) and calculate the percentage of the total hydrosphere stored in each.

Storages	Water volume (km³ × 10³)
Snow and ice	27 000
Groundwater	9000
Lakes and rivers	250
Oceans	1350 000
Atmosphere	13
Soil	35 000
Flows	
Precipitation over oceans	385
Precipitation over land	110
Ice melt	2
Surface run-off	40
Evapotranspiration from land	70
Evaporation from sea	425

Table 11.1

3 Which of these stores can humans use and what percentage of all water is this?

Case study 1

The geopolitics of Israel's water shortage

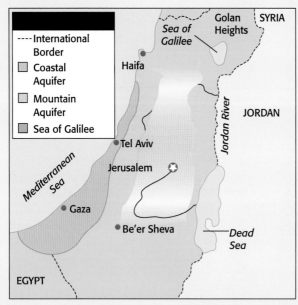

Fig. 11.9 Aquifers of the region

Water shortage looms in Israel after prolonged drought, but supply to Jordan continuing

International Herald Tribune 21 March 2008

Israel is suffering its greatest drought in the past decade and will have to stop pumping from one of its main drinking water sources by the end of the summer, an official said Wednesday. Water Authority spokesman Uri Schor said when Israel has to stop pumping from the Sea of Galilee—the source of about 40% of its drinking water—it will have to step up extraction from already-depleted aquifers, underground water-bearing seams of rock. "The situation is very, very bad," Schor said. "As we pump more from the aquifers, the quality of the water will go down."

Israel's water problem stems from population growth and an improvement in quality of life that brings a greater desire to water lawns and gardens, Schor said. This winter was the fourth that Israel got less than average rain, with only about 50–60% of the average in most areas, he said. Critics of government policy note that agriculture uses a large proportion of Israel's water, receiving heavily subsidized water rates. Since Israel in any event does not grow much of the food it needs, they say, irrigation for farming should be drastically curtailed. Israel's rainy winter season ends this month, though there can be occasional rainy days through June. The rainy season begins around October.

Despite the shortage, Israel will probably not reduce the amount of water supplied to Jordan according to the peace treaty between the countries, Schor said. Jordan's drought is much worse than Israel's, he said. Water is a contentious issue in the dry region (see Fig. 11.9), and one of the disputes Israeli and Palestinian negotiators hope to overcome in talks to work out a peace agreement.

In an effort to stem a serious shortage of water, Israel will launch a conservation campaign, targeting mostly household use. As part of the efforts, Israel has in recent weeks reduced by more than 50% the drinking water supplied to farmers, increasing their need for recycled water, Schor said. Water officials will this weekend debate raising the cost of drinking water in another attempt to cut household use, he said.

Israel has two desalination plants that supply about one-third of water needed by municipalities and households, Schor said. Three other plants scheduled to be completed by 2013 will double that amount. The next one is due to be operational next year.

Questions

1 List Israel's water sources (stores) and demands (flows) on this water.

2 What are the long-term environmental problems of overuse of water? Can you explain this in terms of your water cycle diagram?

3 What are the reasons for strained relations between neighbouring countries over water supplies?

4 How could Israel solve these problems in both the short term and the long term?

5 If some countries in the region can afford to desalinate sea water but others cannot, should they share the water?

6 In your opinion what would be a **sustainable** solution to water management in this part of the world?

Case study 2

China's Three Gorges Dam

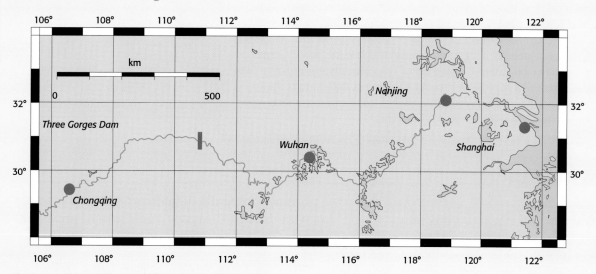

Fig. 11.10 Yangtze River and Three Gorges dam, China

The Three Gorges dam on the Yangtze River, China, is one of the largest engineering projects in the world. It will take about 17 years to build with 2011 as the completion date. The hydroelectric power produced from the dam could provide 9% of all China's output so the saving in carbon emissions would be huge. Up to a quarter of a million workers are building the dam and up to 1.2 million people are being resettled as the flood waters will cover their homes. The estimated cost is the equivalent of US$30 billion.

Apart from electricity generation, the dam is designed to stop the river flooding. These floods have drowned up to 1 million people in the past 100 years and flooded the major city of Wuhan for months.

Criticisms of the dam are that it will cause environmental degradation, create refugees, destroy archeological sites, threaten some wild species and silt up.

The Siberian crane is a critically endangered species and its winter habitat is wetlands that the dam will flood. The Yangtze River dolphin is probably extinct in the wild due to pollution and loss of water flow downstream. The Yangtze sturgeon will be affected as well.

Upstream of the dam, siltation will occur as the silt will get caught behind the dam. This not

only increases the danger of clogging the turbines but means that silt does not flow downstream. Lack of silt may make the river banks more liable to erosion. Shanghai, at the mouth of the Yangtze, is built on a silt bed and this may erode increasingly without replacement from upstream.

Behind the dam, water level will rise up to 100 metres and flood historical and cultural sites. The Ba people settled here 3000 years ago and buried their dead in coffins in caves on cliffs that will be submerged.

The resettlement of 1.2 million people and 100 townships is a major undertaking. The land to which they are going is less fertile than that which they leave.

Pollution by toxic wastes and materials used in building the dam may make the waters unsuitable for drinking and minimal sewage treatment may increase algal blooms and eutrophication in the dammed waters.

There is a risk to energy security as well. Having so much power from one dam means that, if it is destroyed, by terrorism or earthquake, all this power generation will be lost. It also has to be transmitted long distances to where it is used.

To think about

To have information is useful. To turn it into knowledge is difficult. Misinformation is all around us and a glut of information on every topic. As a critical thinker, always be aware of what is not written as much as what is written. Omitting essential facts that do not support the case you are arguing for is an easy thing to do. Consider either the Three Gorges Dam or any other project to manage water (e.g. the Aswan Dam, Aral Sea, Ogallala Aquifer). Put together a case *for* the project and *against* the project using the information you research. You can omit some facts and exaggerate others. Pseudoscience does just this.

Case study 3

The Aral Sea

The Aral Sea is in central Asia within boundaries in Kazakhstan and Uzbekistan. In the1940s the Soviet Union diverted the two rivers which used to feed the Aral Sea (the Amu Darya and the Syr Darya) away from it to provide water for irrigation to grow cotton. Uzbekistan is the world's second largest exporter of cotton as a result of this. Since then, the Aral Sea has no large rivers to replenish the waters so it has shrunk in size as the water evaporates. It has now lost 80% by volume and 60% by area of its former size. This causes the water that remains to become more saline and the increased salinity has killed most flora and fauna. It is also highly polluted from fertilizer runoff, weapons testing residues including anthrax from a secret island site, and industry. It used to be the world's fourth largest inland sea, there is now mostly desert there and it has been an ecological and human disaster. The former seabed is now mostly salt which the winds blow away. Former fishing fleets are now left kilometres away from the shoreline in ship cemeteries. Shrinkage only started about 20 years after the irrigation canals were built to divert the waters from the sea but then rates of shrinkage increased to 80–90 cm per year in the 1980s. The rate of shrinkage has been slowed by changing farming methods but the sea is still shrinking. With livelihoods gone, the people have a much increased disease rate and infant mortality is 30 times higher than it used to be.

The only drinking water is the salty Aral Sea water into which raw sewage is also pumped.

It is poisoning the inhabitants and food is scarce with the Red Cross and Red Crescent providing extra to prevent starvation.

The fishing industry, which used to employ 40 000 fishermen, died as did the industry of trapping of muskrats in the river deltas.

The Soviet Union experts expected the sea to shrink but some thought this was worth it for increased cotton production. There was consideration of diverting another river, the Ob, to refill this sea but this was abandoned due to costs and public resistance. The sea has split into a smaller North Aral Sea and larger South which is now also split into eastern and western seas.

The Uzbekistan cotton industry covers nearly 1.5 million hectares of land and brings in 60% of the state's hard currency in exports – about US$1 billion per year. Children harvest the cotton (whereas machines do this in other countries) and schools are suspended during the cotton harvest. Some restoration funding by the World Bank has improved the North Aral Sea to some extent. A dam was built to separate it from the South Aral Sea and water levels have risen, salt levels decreased and some fishing is possible. The sea has increased in depth again and restoration is improving it each year. But the eastern part of the South Aral Sea will probably dry out within 15 years. Exploration for gas and oil under the seabed by a consortium of Korean, Chinese and Uzbek companies is starting. Various possible solutions have been proposed including desalinating the waters in desalination plants; waterproofing

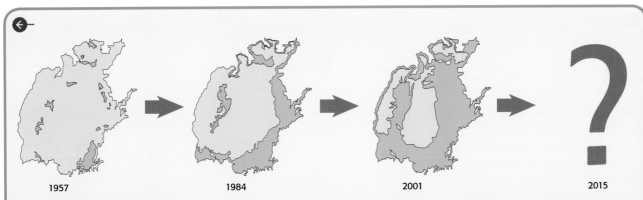

Fig. 11.11 The shrinking Aral Sea

the irrigation canals to reduce losses; using cotton varieties that require less water and less chemical input; redirecting other rivers (at a cost of up to US$50 billion) and melting glaciers and moving that water to refill the sea.

According to the Environmental Justice Foundation "The environmental consequences of Uzbekistan's water crisis are apparent over a space of approximately 400 000 km². Within this vast area, falling downstream water availability and increased salinity have led to the shrinkage of wetlands and lakes by up to 85%. These valuable ecosystems represent a prime habitat for a variety of wildfowl, and their loss is resulting in the widespread disappearance of native flora and fauna. As water availability declines, endemic plants are being replaced by invasive species more suited to the dry, saline environment.

Desiccation has also reduced and fragmented the Tugai forests. Russian experts estimate that in some parts of Uzbekistan as little as 15% to 20% of the Tugai forests remain. These unique riparian communities of poplar, willow, oleaster and reeds once stretched along the Amu Darya covering an estimated 100 000 hectares. Populated by 576 plant species, including 29 endemic to Central Asia, the Tugai provided habitat for amphibians, reptiles and birds, as well as reed cats, jackals, foxes, badgers, voles and wild boars. The Bukhara deer, once found throughout the riverine ecosystems of Central Asia, now numbers just 300 animals."

Questions

1. Why has the Aral Sea shrunk?
2. What is happening to the local ecosystem?
3. The outcome was known when the project started. Why did it go ahead? What were the perceived benefits?
4. Consider the solutions proposed and discuss the advantages and disadvantages of each.
5. Was the project of diverting the rivers a sustainable one?

Case study 4

The Colorado River story

The Colorado River is the longest river in the USA and supplies water to 17 million humans in seven states and Mexico. For over 100 years, the supply of this water and its quality have been contested and the river has been dammed and diverted by canals and aquifers to supply cities, industry and farming with fresh water. Southwest America is a desert and it is only because of the water from this river that large cities like Las Vegas and San Diego and agriculture that supplies some 15% of America's food can thrive there.

There are many demands on the supply of this water. It is used in hydroelectric power production, for irrigation, drinking by humans and livestock, watering golf courses, industrial production and so on. It is also needed by plants and animals in native ecosystems and there is tension between developers, ranchers, farmers, city dwellers and conservationists.

As the Colorado is such a big river, the agreements drawn up by the 1920s on which states take how much water, did, to some extent, work. But, as always, the pressure of an increasing human population with increasing demands on the finite water supply now means that the river literally dries up downstream in some years.

As much as 80% of the water goes to irrigation and cattle. But there are enormous wastages as the cost of the water is subsidized by the federal government. More efficient irrigation systems would reduce water evaporation but there is little incentive to put them in.

The quality of the water is also deteriorating. As water evaporates, the salinity increases and water running back into the river from irrigation has a high salt content. This is so high that downstream, in Mexico, the water cannot be used for irrigation as the salt content kills the crops. An expensive desalination plant has been put in to remove salt before the water reaches Mexico.

Seven major engineering projects to dam the river and to provide hydroelectric power have been constructed including the Hoover and Glen Canyon dams. In providing cheaper electricity, they have encouraged immigration to the region so increasing demand for more water and more power. The dams change water flow and cause erosion downstream as well as causing sediment to collect behind the dam and not be washed to the delta. The Colorado River delta used to be a rich habitat for fish, porpoises and plant life but now these have died out or are under threat through the lack of water and sediment.

One major problem stemming from managing the river was the formation of the Salton Sea. This formed in 1905 after an irrigation canal

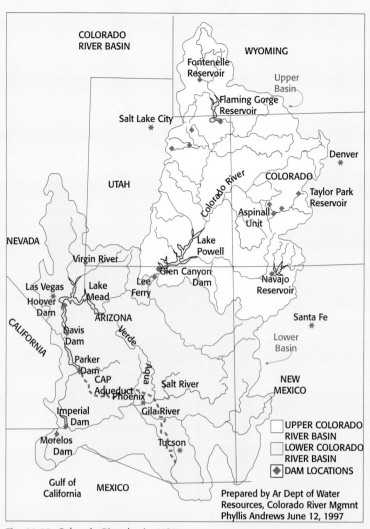

Fig. 11.12 Colorado River basin, USA

burst and the low-lying area flooded to form a large lake. It has been a habitat for many birds including pelicans but this is threatened by increasing levels of salinity and pollution. The sea is 200 feet below sea level. For over 50 years, discussions have been going on to address the salinity and pollution in the sea including cutting a sea-level canal to allow sea water to enter and so reduce salinity.

Questions

1 How would you determine the priorities for water use on the Colorado River?

2 List the demands and the groups involved and then consider who controls the water flow. Who are the winners and losers in this major manipulation of the environment?

3 Discuss whether you think water use in this region could be sustainable and justify your answer.

Sadly, these case studies are only a few examples of the many areas with water supply problems.

- The Yellow River, northern China. This ran dry for two-thirds of the year in 1997 and two-thirds of China's cropland is in the region.
- In the Middle East, water is perhaps more precious than oil.
- In India, the sacred Ganges is heavily depleted and the Sundarban wetlands of Bangladesh are drying up.
- In West Africa, the rivers Volta and Niger are polluted.
- Australia is the driest continent and the Murray-Darling basin is so heavily irrigated that salinization is a real problem.
- Turkey has dammed the Euphrates and Tigris for water and power and these rivers flow on through several countries that also need the water.
- The Zambesi in southern Africa is highly polluted.

1 Look up in an atlas where the rivers mentioned above are situated.
2 Prepare a short presentation for your classmates on one of these.
3 Research the availability of fresh water in the country in which you are living now.
   a Where does the water come from?
   b Is its use sustainable?
4 Do you think the River Amazon and the River Congo are being exploited unsustainably by humans? Explain your answer.

Test yourself

1 List strategies to improve water supplies.
2 Describe an ecocentric and a technocentric approach to improving water supplies.
3 What are the advantages and disadvantages of the two approaches?

Key words

evaporation	La Niña
condensation	upwelling
freezing	NAO
melting	Gulf Stream
transpiration	aquifer
evapotranspiration	salinization
global water budget	desalinization
hydrological cycle	ocean currents
El Niño	energy distribution
ENSO	heat capacity

Key points

- The soil system has storages, flows, transfers and transformations as do other systems.
- Sandy, clay and loam soils have different properties and these affect their productivity.
- Soils are degraded by human activities including irrigation, desertification and toxification.
- Soils can be conserved by measures that condition the soil or reduce wind or water erosion.

> "*Too many people have lost sight of the fact that productive soil is essential to the production of food.*"
>
> H.H. Bennett, 1943

What is soil?

We tend to take the soil around us for granted, but it is more than just mud or dirt. All the food that we consume depends ultimately on soil. Plants grow in soil and we eat either plants that grow directly in the soil or animals that have eaten plants. Soil is a habitat for many organisms. As well as holding water and mineral nutrients that plants depend on, soil acts as an enormous filter for any water that passes through it, often altering the chemistry of that water. Soils store and transfer heat, affecting atmospheric temperature, which in turn can affect the interactions between soil and atmospheric moisture.

Soils are the part of the lithosphere where life processes and soil-forming processes both take place.

We mentioned the four spheres of the earth in Chapter 4: the atmosphere, hydrosphere, lithosphere and biosphere. However there is also the pedosphere (soil sphere). This is a thin bridge between the biosphere and the lithosphere, acted on and influenced by the atmosphere and the hydrosphere.

To do

Make a systems diagram showing the relationship between soil, the biosphere, the hydrosphere and the atmosphere. Add flows and storages to your diagram.

What is soil made from?

Soils can be thought of as being made up of four main components:

- Mineral particles mainly from the underlying rock
- Organic remains that have come from dead plants and animals
- Water in the spaces between soil particles
- Air also in the spaces between soil particles.

Soil is a habitat for plants and animals. Soil is a highly porous medium typically with a 50:50 mix of solids and pore spaces. The pore spaces contain variable amounts of water and air.

It is the exact mix of mineral particles, organic matter, water and air (Table 12.1) that gives a soil its character but it is not just these that make a soil what it is. The soils within any environment are the result of a mix of complex soil-forming processes. Climate, parent rock material, the shape of the land, the organisms living on and within it, and time all contribute to and affect the finished soil.

Fraction	Constituents	Function
Rock particles	Insoluble, e.g. gravel, sand, silt, clay and chalk. Soluble, e.g. mineral salts, compounds of nitrogen, phosphorus, potassium, sulfur, magnesium etc.	Provides the skeleton of the soil and can be derived from the underlying rock or from rock particles transported to the environment, e.g. glacial till
Humus	Plant and animal matter in the process of decomposition	Gives the soil a dark colour. As it breaks down returns mineral nutrients back to the soil. Absorbs and holds on to a large amount of water
Water	Water either seeping down from precipitation or moving from underground sources by capillary action	Dissolved mineral salts move through the soil and are also available to plants. Rapid movement of water causes leaching of minerals. Large volumes of water in the soil can cause waterlogging leading to anoxic conditions and acidification
Air	Mainly oxygen and nitrogen	Well-aerated soils provide oxygen for the respiration of soil organisms and plant roots
Soil organisms	Soil invertebrates, micro-organisms and large animals	Soil invertebrates like worms help to break down dead organic matter into smaller particles. The small particles are then decomposed by soil micro-organisms recycling mineral nutrients. Larger burrowing animals help to mix and aerate the soil

Table 12.1 Constituents of soil

What does a soil look like?

If you dig a trench in the ground, the side of the trench creates a soil profile, a cross-section. This profile changes as it goes down from the surface towards the underlying base rock. It is a record of the processes that have created the soil, its mineral composition, organic content and chemical and physical characteristics such as pH and moisture.

Most soils have a more or less clear layered structure. The layers are called **horizons**.

When organisms die, they often end up on top of the soil. There, fungi, bacteria and many different kinds of animals will start to decompose the dead material. Often decomposition is incomplete and a layer of dark brown or black organic material is formed, the **humus layer**.

Under the humus layer, the soil consists of inorganic material. This inorganic material is formed by erosion of rocks. Within this, layers are formed by water moving either up or down. This movement is called **translocation**. In hot, dry climates, where P<E (precipitation is less than evaporation), water is evaporating at the soil surface and water from the lower soil layer moves upwards. When doing so, it dissolves minerals and takes them to the surface, where the minerals are left behind when the water evaporates. This also happens in irrigation and is called **salinization**. In colder and wetter climates, water flows down in the soil, dissolving minerals and transporting them downwards. This is **leaching** when P>E (precipitation is more than evaporation). The soil layer that loses minerals is called the eluvial layer. It is often rather pale in colour. The soil layer in which the minerals are deposited is called the illuvial layer. When this is intense, the soil becomes more acidic and **podsolization** occurs as the soil becomes a podsol with a nutrient-poor and bleached A horizon and a B horizon (see Fig. 12.1) full of iron oxide which is red.

In **gleying**, water cannot drain away, the soil is waterlogged and anaerobic conditions occur. Then iron oxide is reduced and becomes blue-grey forming a gley soil.

Many tropical soils are red due to the high temperature and rainfall that leaches the soil, speeds up chemical weathering and results in silica forming from clay breakdown. The silica is leached away leaving sequestered iron and aluminium oxides which are not translocated, so the soil looks red. This is **ferallitization**.

Horizons

In cross-section, soils have a profile (see Fig. 12.1) which is modified over time as organic material leaches (washes) downwards and mineral materials move upwards. Distinctive horizons (zones or levels) are often seen in the soil.

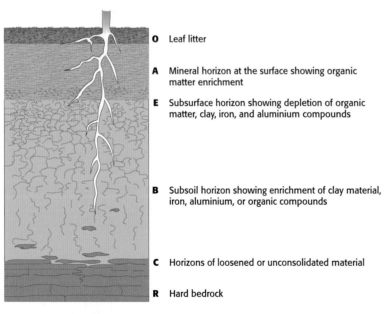

O Leaf litter

A Mineral horizon at the surface showing organic matter enrichment

E Subsurface horizon showing depletion of organic matter, clay, iron, and aluminium compounds

B Subsoil horizon showing enrichment of clay material, iron, aluminium, or organic compounds

C Horizons of loosened or unconsolidated material

R Hard bedrock

Fig. 12.1 A soil profile

- **O horizon:** Many soils also contain an uppermost layer of newly added organic material such as leaf litter. Where this is present it is called the O horizon.

- **A horizon:** Upper layer. This is where in many soils humus builds up. Humus forms from partially decomposed organic matter and is often mixed with fine mineral particles. In normal conditions, organic matter decomposes rapidly through the decomposer food web, releasing soluble minerals that are then taken up by plant roots. Humus that contains a high quantity of alkaline mineral is called *mull* humus whereas humus low in alkaline minerals forms acidic *mor* humus. Waterlogging reduces the number of soil organisms, which results in a buildup of organic matter and can lead to the formation of peat soils.
- **E horizon:** this is not always present but, in older soils, it is a layer where minerals and organic matter have been leached from the soil, leaving a pale layer which is mostly silica particles.
- **B horizon:** This is the layer where soluble minerals and organic matter tends to be deposited from the layer above. In particular, clay and iron salts can be deposited in this horizon.
- **C horizon:** Mainly weathered rock from which the soil forms.
- **R horizon:** Parent material (bedrock or other medium).

Not all soils contain all three A, B and C horizons; sometimes only two horizons can be distinguished while in other soils there may be no distinct layering.

Soil formation

This occurs in three stages:

1 Initial mechanical and chemical **weathering** processes, resulting in the inorganic component of the soil.
2 Introduction of living organisms, the biotic component.
3 Decomposition and the formation of an organic component.

Soils are formed by a combination of these processes. The abiotic component is formed from rock fragmentation by physical weathering and the alteration of these minerals by chemical weathering. Minerals are then redistributed both up and down in the soil by movement of water. The biotic component consists of plant roots and a vast variety of animals, which are responsible for mixing and opening up the soil. An even larger microbe population is responsible for the decomposition of plant and animal detritus leading to buildup of humus. Humus affects soil texture and contributes to chemical weathering and the waterholding capacity. These processes are further influenced by environmental factors, such as the underlying geology, the local vegetation, the climate and the topography. This means that in different places the processes combine in different ways to produce a variety of soils.

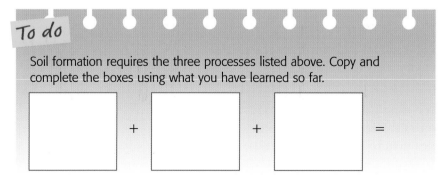

To do

Soil formation requires the three processes listed above. Copy and complete the boxes using what you have learned so far.

☐ + ☐ + ☐ =

Soil texture

The mineral portion of soil can be divided up into three particles based on size: sand, silt and clay. Most soils consist of a mixture of these soil particles and the soil texture therefore depends on the relative proportions of sand, silt and clay particles.

Particle diameter (mm)	Particle
< 0.002	Clay
0.002–0.05	Silt
0.05–2	Sand

Table 12.2

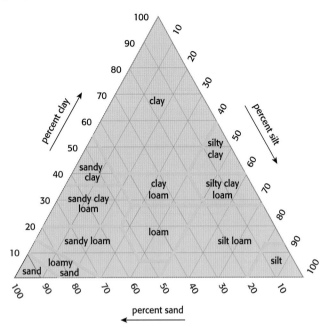

Fig. 12.2 Sandy, clay and loam soil types

The proportion of each of these particles gives soil its texture. Most soils also contain particles that are larger than 2 mm in diameter (pebbles and stones), but these are not considered in a description of soil texture.

It is possible to feel the texture of moist soil if you rub it between your fingers. Sandy soils are gritty and fall apart easily. Silty soils feel slippery like wet talcum powder and hold together better than sandy soils. Clay soils feel sticky and can be rolled up into a ball easily. Most soils contain a mixture of different soil particles and can be described as sandy clay or silty clays. If a fairly equal portion of each size is present then the soil is said to be a loam. Loam soils are fairly fertile, drain well and can be worked easily, making them the preferred soils for agricultural use.

It is possible to get a clearer picture of the proportions of soil particles by drying out a soil sample and passing it through a series of sieves of decreasing mesh size (first 2 mm, then 0.05 mm and finally 0.002 mm) separating the soil into sand, silt and clay particles.

Soil texture is an important property of a soil, as it determines the soil's fertility and the primary productivity. This is exemplified in Table 12.3.

Loam soils are ideal for agriculture. The sand particles ensure good drainage and a good air supply to the roots. The clay retains water and supplies nutrients. The silt particles help to hold the sand and clay particles together.

	Sandy soil	Clay soil	Loam soil
Composition (%)			
Sand	100	15	40
Silt		15	20
Clay		70	40
Mineral content	High	High	Intermediate
Potential to hold organic matter	Low	Low	Intermediate
Drainage	Very good	Poor	Good
Water-holding capacity	Low	Very high	Intermediate
Air spaces	Large	Small	Intermediate
Plants and animals	Low	Low	High
Primary productivity	Low	Quite low	High

Table 12.3 Comparison of three soil types

Nutrients

Plants make their own food by photosynthesis but they also need to take up nutrients from the soil. The main nutrients they need are nitrogen (N), phosphorus (P) and potassium (K) which are absorbed through the roots. Without these nutrients, photosynthesis and healthy plant growth cannot occur. Soil mineral particles may contain the elements that plants need as nutrients, but it is only when the elements are in a soluble form that they can be called nutrients and taken up by plants. Nutrients are not absorbed as elements but in the form of (compound) ions or compounds. Often nutrients are divided into **macronutrients** and **micronutrients**. Macronutrients are taken in in relatively large amounts, whereas micronutrients are only taken in in small amounts.

Macronutrients

Nitrogen (N)

Phosphorus (P)

Potassium (K)

Calcium (Ca)

Magnesium (Mg)

Sulfur (S)

Macronutrients are often present in short supply and limit productivity. This is why adding fertilizers usually results in higher productivity.

Fertilizers can be organic (farmyard manure, slurry, compost, seaweed, green manures) or inorganic (made industrially and containing N, P and K in known amounts). Organic fertilizers also improve soil structure by adding organic matter to the soil but the exact amount of nutrients added is not known. In applying inorganic fertilizers, farmers can know exactly how much they are adding, but soil structure is not improved. Crop yield can increase many times when fertilizer is added. Both organic and inorganic fertilizers can leach out of the soil and into waterways, causing eutrophication (see page 287).

Micronutrients

Iron (Fe; absorbed in much larger amounts than the other micronutrients)

Manganese (Mn)

Copper (Cu)

Nickel (Ni)

Zinc (Zn)

Molybdenum (Mo)

Cobalt (Co)

The micronutrients are in the mineral part of the soil. The supply is usually much larger than the demand.

Nutrient cycling and Gersmehl's nutrient flow model

Figure 12.3 shows the three major stores of nutrients in ecosystems represented by circles, and major flows of nutrients between them as arrows. In this model the size of the circles is proportional to the quantity of nutrients that are stored, and the thickness of the arrows is proportional to the amounts of nutrients transferred. This allows the nutrient cycles of different ecosystems to be easily compared.

The biomass circle represents nutrients stored in the forest vegetation and animal life.

The litter circle represents the nutrients trapped in fallen leaves and dead organisms.

The soil circle represents the nutrients present in soil humus, i.e. decomposing leaves and other dead organisms.

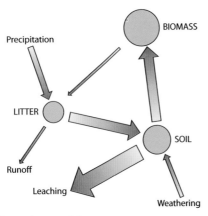

Fig. 12.3 Gersmehl's nutrient model

The nitrogen cycle

All living organisms need nitrogen as it is an essential element in proteins and DNA.

Nitrogen is the most abundant gas in the atmosphere but atmospheric nitrogen is unavailable to plants and animals, though some specialized micro-organisms can fix atmospheric nitrogen. For plants to take up nitrogen, it must be in the form of ammonium ions (NH_4^+) or nitrates (NO_3^-). Animals eat plants and so take in their nitrogen in the form of amino acids and nucleotides by feeding.

The nitrogen cycle can be thought of in three basic stages: nitrogen fixation, nitrification and denitrification (see Fig. 4.27, page 85).

Nitrogen fixation: atmospheric nitrogen (N_2) is made available to plants through the fixation of atmospheric nitrogen. This conversion from gaseous nitrogen to ammonium ions can be carried out in one of five ways:

1   By nitrogen-fixing bacteria free living in the soil (*Azotobacter*).
2   By nitrogen-fixing bacteria living symbiotically in root nodules of leguminous plants (*Rhizobium*). The plant provides the bacteria with sugars from photosynthesis, the bacteria provide the plant with nitrates.
3   By cyanobacteria (sometimes called blue-green algae) that live in soil or water. Cyanobacteria are the cause of the high productivity of Asian rice fields, many of which have been productive for hundreds or even thousands of years without nitrogen-containing fertilizers.
4   By lightning also causing the oxidation of nitrogen gas to nitrate which is washed into the soil.
5   The industrial **Haber** process is a nitrogen-fixing process used to make fertilizers. Nitrogen and hydrogen gases are combined under pressure in the presence of iron as a catalyst (speeds up the reaction) to form ammonia.

The last two processes are non-living nitrogen fixation.

Nitrification: Some bacteria in the soil are called nitrifying bacteria and are able to convert ammonium to nitrites (*Nitrosomonas*) while others convert the nitrites to nitrates (*Nitrobacter*) which are then available to be absorbed by plant roots.

Denitrification: Denitrifying bacteria (*Pseudomonas denitrificans*), in waterlogged and anaerobic (low oxygen level) conditions, reverse this process by converting ammonium, nitrate and nitrite ions to nitrogen gas which escapes to the atmosphere.

As well as nitrogen fixation, **decomposition** of dead organisms also provides nitrogen for uptake by plants. Decomposition of dead organisms supplies the soil with much more nitrogen than nitrogen fixation processes. Important organisms in decomposition are animals (insects, worms among others), fungi and bacteria. They break down proteins into different ions: ammonium ions, nitrite ions and finally nitrate ions. These ions can be taken up by plants which recycle the nitrogen.

Assimilation: Once living organisms have taken in nitrogen, they assimilate it or build it into more complex molecules. Protein synthesis in cells turns inorganic nitrogen compounds into more complex amino acids and then these join to form proteins. Nucleotides are the building blocks of DNA and these too contain nitrogen.

Humans and the nitrogen cycle

It is easy for humans to alter the cycle and upset the natural balance. When people remove animals and plants for food for humans, they extract nitrogen from the cycle. Much of this nitrogen is later lost to the sea in human sewage. But people can also add nitrogen to the cycle in the form of artificial fertilizers, made in the Haber process, or by planting leguminous crops with root nodules containing nitrogen-fixing bacteria. These plants enrich the soil with nitrogen when they decompose. The soil condition also affects the nitrogen cycle. If it becomes waterlogged near the surface, most bacteria are unable to break down detritus because of lack of oxygen, but certain bacteria can. Unfortunately they release the nitrogen as gas back into the air. This is called denitrification. Excessive flow of rainwater through a porous soil, such as sandy soil, will wash away the nitrates into rivers, lakes and then the sea. This is called leaching and can lead to eutrophication.

To do

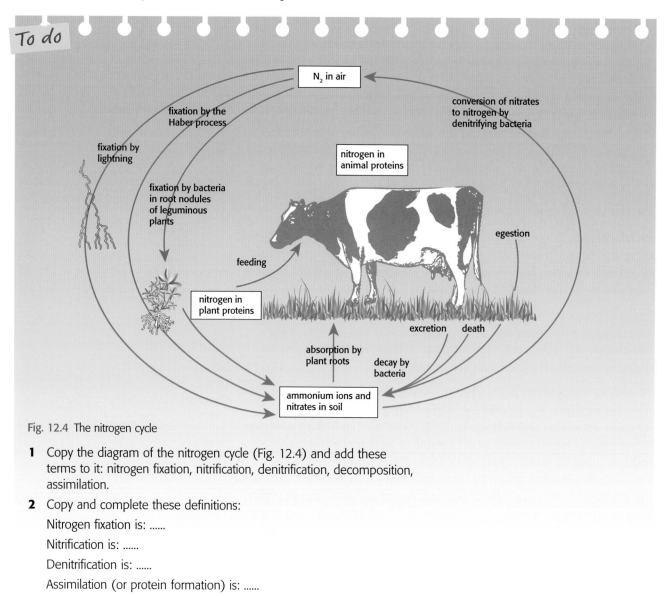

Fig. 12.4 The nitrogen cycle

1   Copy the diagram of the nitrogen cycle (Fig. 12.4) and add these terms to it: nitrogen fixation, nitrification, denitrification, decomposition, assimilation.

2   Copy and complete these definitions:

Nitrogen fixation is: ......

Nitrification is: ......

Denitrification is: ......

Assimilation (or protein formation) is: ......

Soil degradation

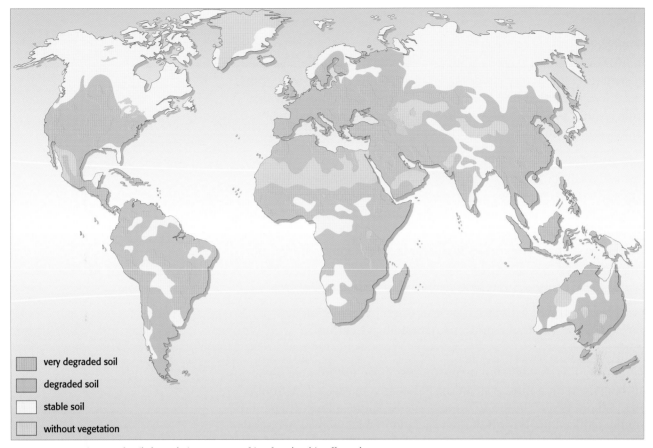

Fig. 12.5 An estimate of soil degradation – 15% of ice-free land is affected.
This is 11 million sq km affected by water erosion and 5.5 million sq km by wind erosion.

It could be argued that land degradation is the most pressing environmental and social problem facing society today. Climate change and loss of biodiversity have their place, but for the world's poor the most pressing issue is loss of soil and soil fertility and as a consequence loss of productive land. In addition there is depletion of fuel wood supplies due to unsustainable rates of use, lack of drinking water and poor sanitation. It is estimated that an area equal to the size of China and India combined is now classified as having impaired biotic function (damaged ecosystem structure) as a result of poor land management resulting in soil loss. As populations expand and as social and cultural changes occur, ever greater demands are being made on greater areas of landscape and soil.

In MEDCs where there has been a relatively long tradition of agriculture (agriculture on an industrial scale) there exists, within the agricultural culture, a knowledge of land management that aims for sustained soil fertility and strives to avoid soil erosion. However, even in MEDCs there are occasions when climate and intensive agriculture conspire to bring about unprecedented levels of soil erosion.

Two types of process can give rise to soil degradation.

- Processes that take away the soil (erosion). This mainly occurs when there is no vegetation on the soil. Wind and water can then simply take the soil away.

● Processes that make the soil less suitable for use. In these processes chemicals end up in the soil and make the soil useless in the long run.

Soil erosion

If natural vegetation covers a soil, processes that could damage the soil structure are largely eliminated. Leaves deflect heavy rain, roots hold the soil together and humus absorbs large quantities of water. If this natural cover is removed or even when it is replaced by agricultural vegetation, soils can become prone to erosion.

Three major processes of soil erosion exist:

1 Sheet wash: large areas of surface soil are washed away during heavy storms, moving as landslides in mountainous areas.

2 Gullying: channels develop following rainfall. Over time these channels become deeper.

3 Wind erosion: on drier soils high winds continually remove the surface layer.

Each of these processes can be triggered by a number of causes.

Examples of human activities that lead to soil degradation are overgrazing, deforestation and unsustainable agriculture.

Overgrazing and overcropping

Overgrazing occurs when too many animals graze in the same area. This happened on a huge scale in the Sahel area in Africa (just south of the Sahara desert) in the 1970s and 1980s. There, a long dry period strongly reduced the growth of the vegetation which was subsequently eaten by cattle. The soil particles were no longer kept in place by roots and were blown away by the wind. This resulted in the death of most of the cattle and, later on, in a terrible famine. As soil formation is a very slow process, it will take many years for the Sahel region to recover. In wet climates it is often rainwater that takes the soil particles away, especially when the rainwater is flowing down slopes.

Overcropping depletes soil nutrients. This reduces soil fertility as no nutrients are being returned to the soil. If the crop fails then the soil surface again becomes susceptible to erosion. This is especially true in dry regions where crop failure can lead to removal of topsoil by wind. During the 1930s, the American Midwest suffered a major period of wind erosion causing the "dust bowl" (Fig. 12.6). Through overuse of the land an area about twice the size of the United Kingdom, from Nebraska through to Texas, was affected by severe wind erosion. The wind moved soil and dust many thousands of kilometres.

Fig. 12.6 The Dust Bowl in North America, 1930s

Deforestation

Deforestation is the removal of forest. This can be done in different ways, ranging from careful removal of some of the trees to complete removal of all vegetation. Of course, the more vegetation is removed, the more the soil will be prone to erosion. As most forests are in relatively wet areas, the erosion will mainly be due to water.

Deforestation can have a massive effect on soil erosion, especially in tropical regions. The leaves of forest trees both deflect and slow down the progress of raindrops. This helps to stop them explosively removing soil particles. The root systems of forests help to bind the soil together and give it stability, while also absorbing large quantities of water from the soil directly. The absorbed water is eventually returned to the atmosphere via transpiration.

Unsustainable agriculture

Unsustainable agricultural techniques are techniques that cannot be applied over a long period of time without decrease in productivity or increased inputs of chemicals like fertilizers or energy. Several unsustainable agricultural techniques result in soil degradation:

- (Total) removal of the crops after harvest. This leaves the soil open to erosion.
- Growing crops in rows with uncovered soil in between. Again, erosion will occur, especially if the crops are grown on a slope and the rows are in the direction of the slope.
- Ploughing in the direction of the slope. This will leave ready-made channels for rainwater to flow down, taking the soil with it.
- Excessive use of pesticides. This will in the long run make the soil too toxic for further agricultural use. This process is called toxification.
- Irrigation. In many irrigation systems a major part of the water evaporates before reaching the crops. The minerals dissolved in the irrigation water remain in the top layer of the soil and form a hard, salty crust that will make the land unsuitable for growing crops. This process is called salinization (making the soil salty).

Improving the soil: conservation

A variety of measures can be taken to conserve soil and soil nutrients.

Addition of soil conditioners

Typical soil conditioners are lime and organic materials. Lime will increase the pH of soils that are affected by acid precipitation (see page 295). Water is not the only factor that affects soil acidity; some soil processes themselves also make the soil more acidic. The breakdown of organic matter releases carbon dioxide through respiration. This then dissolves into the soil water, creating carbonic acid. Nitrification of ammonium ions to nitrates increases acidity and the removal of basic ions through leaching all add to soil acidity. This in turn reduces soil fertility. For many centuries farmers have added lime to counter soil acidification. Limestone or chalk that has been crushed is scattered on the soil. This helps to increase pH and so counter acidity. Lime also helps clay particles to stick together so that they act more like sand particles. The larger particles created are more free draining than raw clay and they trap more air, helping to improve decomposition by soil micro-organisms.

Organic materials such as straw, or green manure crops that are ploughed back in, improve the texture of the soil and act as a supply of nutrients (after decomposition).

Wind reduction

The effect of the wind can be reduced by planting trees or bushes between fields (shelter belts) or by alternating low and high crops in adjacent fields (strip cultivation). An alternative is to build fences.

Soil-conserving cultivation techniques

Growing **cover crops** (fast-growing crops to cover the soil) between the rows of crops or between harvest and sowing can keep the soil particles in place.

Terracing is a method to reduce the steepness of slopes by replacing the slope with a series of horizontal terraces, separated by walls. Asian wet rice fields are constructed this way.

Ploughing helps to break up large lumps of soil and increases drainage. In northern countries ploughing occurs in autumn so that frost can attack the clods of earth and break them further. Ploughing also helps to mix humus through the top layers of the soil.

Contour farming is ploughing and cultivating along the contour lines, i.e. across the slope. By ploughing parallel to the slope (contour ploughing) instead of up and down hill, the water does not rush downhill and thereby erosion can be strongly reduced and the furrows and ridges act as small terraces trapping soil and water. This method has technical problems however: modern heavy machinery has a tendency to tip over when used parallel to the slope.

Improved irrigation techniques

By careful planning and construction of irrigation systems, evaporation and thereby salinization can be strongly reduced. Covering irrigation canals will prevent evaporation before the water reaches the land. (There are examples known where up to 50% of the water never even reaches the field.) Trickle flow irrigation systems consist of a network of pipes covering the field. The pipes have small openings next to the plants where water comes out drop-wise and can be taken in by the roots before it evaporates. It is this irrigation system that made it possible to grow roses in desert-like areas in Israel.

Fig. 12.7 Sprinkler irrigation of cotton in Israel

Stop ploughing marginal lands

Very poor land is simply not suitable for growing crops and could be more suitable for cattle grazing (but there is a risk of overgrazing). This may be the case in land at the boundaries of deserts.

Crop rotation

Some crops require more fertile soils than others. Legumes (see page 65) add nitrogen to the soil as the *Rhizobium* in their root nodules fix nitrogen from the atmosphere. Cereals take a lot of nutrients out of the soil. The earliest farmers recognized that growing the same crop on the same land year after year led to pest and disease buildup and impoverished the soil. Shifting cultivators in tropical rainforests move on after a few years to allow the soil to recover. Ancient civilizations practised crop rotation, sometimes by leaving ground fallow (with no crops), sometimes by growing several crops in a year to maximize yield. In the Islamic Golden Age between the 8th and 13th centuries, agricultural improvements such as irrigation, rotations, cash cropping (growing crops for the market) and exporting crops flourished and supported the growth of populations and civilizations. In Europe, a three-year rotation of winter wheat or rye, then spring oats or barley, then a fallow year, was practised until the four-year rotation of the British agricultural revolution introduced turnips and clover so that cropping could be continuous and soil fertility maintained.

Soils and agriculture

Saving soils

Human activity can have both positive and negative impacts on soils and soil fertility. Unmanaged and ignorant exploitation of soils will quickly lead to soil degradation, soil loss and lack of fertility. Soils, like any other part of the global system, exist in a dynamic state. Imbalances will lead to a change in the state of the system and thus the character of the soil both physically and chemically. This will also influence the ecosystem and biology associated with the soil.

Our modern global system depends heavily on soil and soil fertility to support a food production system that in turn supports a burgeoning population (90 million extra mouths per year). Soil is therefore an important resource that requires informed conservation. Approximately 24 billion tonnes of soil are eroded from the landscape annually. Croppable land is shrinking in volume annually (though production may be rising).

Soil is formed by a natural set of processes; soil erosion also occurs naturally. However, under natural conditions erosion is often offset by new soil formation, i.e. the rate of erosion is at or below the rate of accumulation within the soil profile. In natural systems, eroded soil will end up in water courses, estuaries and coastal waters. Here it provides essential minerals, organics and nutrients for ecosystems. Soil eroded from one geographical position may also form the basis of soil profiles within another. Eroded upland soils will wash into valleys, producing deep fertile floodplain soils.

Soil fertility

Soil fertility is a simple equation – as long as nutrients accumulate at a rate equal to or greater than the rate at which they are being used then soils remain fertile and sustainable.

In natural systems most organic material is cycled *in situ* (on site). In general, plants grow, die, decay and are assimilated into the soil profile. On occasions, herbivores will eat plant material and move to an adjacent ecosystem or habitat before producing waste. However, on balance the cycle only fails if the soil is washed away and enters rivers and seas or is blown away to stay in the atmosphere for a long time.

For example, in a deciduous woodland leaves are shed in autumn; they fall to the woodland floor and become an O horizon; as they decay over time they form part of the A horizon. As this humus layer continues to decay by biological and chemical processes, the nutrients released migrate downwards into the B horizon. These nutrients are now available once again for plants as food.

Agriculture operates under a different model. Plants are deliberately set as seed (planted) into a soil that has been cleared of the natural vegetation of that soil. The plants are often planted on to bare soil that has been previously conditioned (broken up) by ploughing (digging over). The plants are present in a monoculture. The plants are grown in relatively large numbers per unit area. Their ability to grow is often encouraged by making available additional water via irrigation and nutrients in the form of inorganic or organic fertilizers. The plant varieties are most often "nutrient hungry" – they extract from the soil quantities of nutrients often magnitudes greater than indigenous species. They are plants that have over time been genetically selected to produce large volumes of biomass, whether in tubers, seed heads or the general body of the plant.

At the end of the season the crop is harvested as a food supply for humans or to feed their livestock. Harvesting requires the removal of the biomass from the field, the soil and the ecosystem. There has been a net loss of biomass, humus, minerals and water from the system. The soil is more impoverished than it was. If another crop is extracted the soil fertility is reduced again. Gradually soil fertility is lost.

To think about

Ecocentric or technocentric view of soil fertility management?

As agricultural systems, even those on the relatively small scale, become more and more intensive and more and more commercial, greater and greater demands are made on soil per unit area. The soil is required to "work harder", produce greater volumes per unit area, cope with greater demands from genetically modified, nutrient-hungry plants. The soil in its natural state and operating under ambient environmental conditions can no longer fulfill the demands being made upon it, therefore the soil environment must be modified artificially with the addition of fertilizers and additional irrigation water. This is the technocentrist's response. But very quickly the farmers can find themselves on a treadmill with spiralling costs as they attempt to maintain or increase productivity per unit area while at the same time maintaining soil health and thus guaranteeing long-term sustainability of the industry. It costs relatively large amounts to buy synthetic fertilizers and the hardware for irrigation. Therefore the farmer must produce more to get the revenue to pay for this investment in chemicals and hardware. In producing more, more strain is put on the health of the soil system, thus more external inputs are required. Potentially a vicious circle is set in motion.

What would the ecocentric farmer do?

Case study 1

Danum Valley, Sabah, Malaysia

Fig. 12.8 Danum Valley rainforest

The Danum Valley area of Sabah, East Malaysia, is an area of lowland tropical forest. Annual rainfall is over 2500 mm with little seasonality in the rain pattern. This means that forestry operations occur throughout the year. Large parts of the forest are selectively felled (only certain trees are taken out of the forest rather than cutting down the entire forest area) but even where this occurs there is an increase in stream flow compared to areas with full canopy cover. This is because of the loss of transpiration returning water back to the atmosphere. However, this increase is much less than in clear felled areas.

Because of impermeable geology in the Danum catchment, water tends to flow in the top surfaces of the soil rather than percolating down into bedrock. This means that heavy rainfall can create large flash floods. This makes the soils more sensitive to disturbance, especially from the large vehicles used by foresters.

The Danum catchment is covered by a network of channels, most of which continually flow with water. This makes it very difficult for forestry vehicles to cross the terrain without damaging the banks of stream channels and adding to soil erosion. Also where forestry companies have tried to drain areas, or keep logging roads clear of water, some of these drainage channels have developed into wide gullies that carry eroded soil into the river systems.

Another major factor affecting soil erosion in Danum are the logging roads that have been put in place. Many of these have been built at very steep angles up slopes. The result of this is often mass movement of soil in large landslides. These eroded soils end up in the local streams and rivers and are lost from the forest.

Soil as a non-renewable resource

Soil formation takes a very long time. Under the best conditions (wet, temperate climate) only a few millimetres of soil are formed per year. And this is only after the initial chemical and physical weathering has occurred and fine material and soil organisms are present. As a consequence, soil use often exceeds soil formation and therefore soil should be considered a **non-renewable resource/ natural capital**.

Review

1 Copy, fill in the gaps and delete incorrect options in the paragraph below.

All living organisms need various nutrients for healthy development and growth, and there is a finite amount of these nutrients. The plants take up the nutrients from the soils, and once they have been used are passed on to the carnivores/herbivores/photosynthesizers/producers and then the _____ which feed upon them. As organisms die, they _____ and nutrients are returned to the system. As for all systems, there are inputs, _____, storages and _____. Nutrients are stored in _____ main compartments: the biomass (total mass of living organisms), the soil and the _____ (the surface layer of vegetation which may eventually become humus).

2 The nutrient cycles vary according to the climate and type of vegetation. The size of each of the storages and size of the transfer (flow) can be different. Figure 12.9 shows a hypothetical Gersmehl's nutrient flow model (see also page 230) where the sizes of the storages and flows are the same. In real ecosystems, these will vary depending on the climate, soils and vegetation.

a Redraw this model and fill in the names of the processes in the green boxes using some of the words from this list: decomposition, death, excretion, defecation, assimilation, growth.

b What is the main nutrient flow *from* the soil? Why does this happen?

c Is transport of minerals from one soil layer to another a transfer or a transformation process?

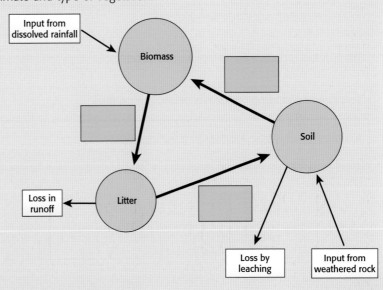

Fig. 12.9 Gersmehl's model

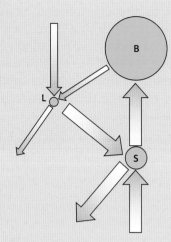

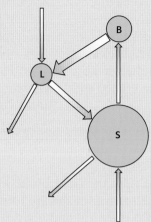

Fig. 12.10 Gersmehl's models for tropical rainforest and continental grassland ecosystems

3 Look at the two nutrient models in Fig. 12.10, which represent a tropical rainforest and a continental grassland (prairie) ecosystem.

    a Label each with its respective ecosystem name.

    b Copy and complete Table 12.4 of comparisons between the two ecosystems.

Comparison	Ecosystem	Explanation
Which ecosystem stores most nutrients in biomass?		
Which ecosystem has most undecomposed detritus?		
Which ecosystem has least humus?		
In which ecosystem is plant uptake of nutrients greatest?		
In which ecosystem is decomposition slowest?		
Which ecosystem loses nutrients from biomass quickest?		
In which ecosystem are most nutrients lost due to heavy rain?		
In which ecosystem does rainfall supply many nutrients?		

Table 12.4

Fig. 12.11 Ed Miller, a farmer, summerfallowing in 1932

The North American Prairies and commercial farming

The prairies cover 1.4 million square miles in central North America from Alberta, Manitoba and Saskatchewan in Canada to Texas and Oklahoma in the USA. The Rocky Mountains lie to the west of the prairies. The prairies have low rainfall, hot dry summers and cold winters and the vegetation is grasses and herbaceous (non-woody) plants. Droughts are common. There are very few trees. Herds of buffalo and elk used to graze the prairies. In natural prairies, fires sweep through and plants survive by having underground storage structures. Animals returned nitrogen and organic matter to the soil and there was a sustainable ecosystem. Nearly all prairie land, except for about 2%, is now under cultivation for cereals, grazed by cattle or built on. Winter and spring wheat, barley and maize are grown on prairie land. Cattle ranching is on poorer soils that cannot support cereal growing.

Increasing salinity, soil erosion and loss of soil fertility are all issues facing prairie farmers and ranchers. In the natural ecosystem, a mat of decomposing grasses protects the soil from blowing or washing away and, when this decomposes, the soil is enriched with organic matter. When the grasses are removed in grazing or crops taken in the harvest, the soil is bare and rains can wash it away. Erosion by the wind is more likely though as the soils dry out and there are no roots to hold it in place. In the 1930s, the infamous Dust Bowl of the "dirty 30s" was caused by poor farming practices and a drought. The dry soil blew away and much of it ended up in the oceans. Eventually 2.5 million people were displaced, many going to California, as the land, without topsoil, could not support them.

Soil salinity is increased by the practice of summerfallow when soil is left bare with no crops for a year. Any weeds are killed by herbicides or by tilling and ploughing. Leaving the soil bare conserves moisture for the next crop as water is limiting in the prairies. It also increases organic matter in the soil so less fertilizer is required when the crop is planted and breaks any pest life cycles. But we now know that summerfallow is not sustainable. Leaving bare soil means it is likely to be washed or blown away. Organic matter does not increase over the long term as crops are removed. Salinity increases as the water table rises to the level of crop roots and carries salts upward with it. Some 40% of prairie soils are affected by high salinity levels and, if these are too high, the crops will not grow at all.

What can be done?

To reduce salinity – This is difficult but not impossible. Summerfallowing or leaving bare soil for long periods should be stopped or reduced. Leaving crop residues on the surface, instead of ploughing them in (zero tillage), allows them to trap snowfall and conserve moisture. Soil does not blow or wash away with crop residues on the surface. Deep-rooted (to lower the water table) and salt-tolerant crops can be planted. Snow fences or barriers enable snowdrifts to pile up which provide water when they melt in.

To reduce erosion – Leaving stubble (the roots and cut straw stalks of cereals) in the fields protects the soil. This is called conservation tillage. Cover cropping (growing any quick crop to keep the soil covered) both slow erosion and add organic matter if the crop is ploughed back into the soil. Contour ploughing – along the contour lines instead of up and down slopes – traps soil and water. Strip cropping – growing alternate strips of cereals and other crops such as flax and tall wheatgrass at right angles to the wind – reduces the effect of the wind. Shelter belts of trees reduce windspeed.

To increase soil fertility and improve structure – Loss of organic matter from soil reduces the nutrient level and decreases the quality of soil structure. Plants need nitrogen for healthy growth. This can be added by inorganic fertilizers but they do not improve soil structure. The better way is to practise crop rotation and include a leguminous crop that adds nitrogen to the soil. Peas, beans, soya and clover are all legumes. By keeping soil covered with crops, crop residue or cover crops, organic matter increases as well.

To conserve moisture – summerfallowing only conserves about 20% of the precipitation for the next year's crop. A better method is to trap snowfall which melts and the water percolates into the soil. This can be improved by leaving strips of taller straw in fields, shelterbelts, strips of tall wheatgrass and stubble mulching (not ploughing but leaving the cut stubble on the surface) which all trap snow.

Case study 3

Burkina Faso stone lines and subsistence farming

Fig. 12.12 Planting trees along a stone line in Burkina Faso

Burkina Faso is in West Africa and a remarkably simple solution has been adopted to conserve water in the north of the country. As the Sahara advances southwards (desertification) and the human population increases, there is more pressure on the land to produce enough food. With increasing droughts, the herds of cattle and the goats which graze the savanna south of the Sahara have been forced southwards to find enough grazing and this is where the farmers are growing their crops. Low rainfall with occasional flash floods make farming difficult enough and there is too much pressure on the land to give the soil time to lie fallow. In the past, farmers would crop for three years and then leave the soil to recover organic matter for at least twice as long. With increasing demand, cropping is now continuous and crop yields falling. There was a need to conserve moisture and organic matter in the soil and prevent overgrazing by stock.

In 1979, the charity Oxfam worked with local farmers to extend a practice that had been carried out for some time. This is simply putting stones in lines – stone lines – along the contours of the land. The difficulty was in placing the stones accurately along a contour line when the slope was shallow. The solution was to have a flexible transparent hosepipe filled with water. As water finds its own level, all that was needed was to align the water levels at each end of the pipe which tells you they are at the same level, and then place the stones accordingly. By doing this, any rainfall that would previously run away on the soil surface was stopped by the stone line. Any organic matter, dead leaves and debris washed away by the rain is also stopped by the stone line. This makes such a difference that seeds germinate in this moister, rich organic soil and trees can be planted here. Once trees are established, the soil moisture and organic matter increase further so a virtuous cycle is created. Crop yield may be half as much again with stone lines in place. Now the north of Burkina Faso is covered in the stone lines.

To do

Consider the three case studies in this chapter. Copy and complete Table 12.5.

Case study	1	2	3
Causes of soil erosion or degradation			
Effects on vegetation			
Effects on humans			
Solutions			

Table 12.5

Key words

soil profile
horizon
parent material
decomposition
soil formation
weathering
soil texture
clay
loam
sandy soil
humus
leaching
aeration
nitrification
denitrification

nitrogen fixation
leguminous plants
Rhizobium
nutrients
macronutrients
micronutrients
soil degradation
erosion
overgrazing
deforestation
unsustainable agriculture
toxification
irrigation
salinization
soil conservation

Food resources

Key points

- There is enough food in the world but an imbalance in its distribution.
- Terrestrial and aquatic food production systems have different efficiencies.
- Different food production systems have different impacts and make different demands on the environment.
- Food production is closely linked with culture, tradition and politics.
- Subsistence and commercial farming systems manage soil in different ways.

Some basic food facts

- Wheat is a staple food for over one-third of the world's population. (Staple foods provide a large part of the diet, are generally plants high in starch, easy to store and relatively cheap.)
- Wheat, rice, maize (coarse grains), potato, barley, sweet potato, cassava, sorghum and millet are the staple carbohydrate foods for most humans.
- Grain production provides half the human population's calories.
- World food production is concentrated in the northern hemisphere temperate zones.
- There are 23 billion livestock animals on Earth, including over 3 billion cows and other ruminants and 1 billion pigs.
- There are nearly three times as many chickens on Earth as people.
- Livestock need 3.5 billion hectares of land.
- Crops take up 1.5 billion hectares of land – about 11% of the total land area.
- There are 13.5 billion hectares of land on Earth but 90% of this is too dry, wet, hot, cold, steep or poor in nutrients for crops.
- In Africa, only 7% of the total land area is cultivated.
- LEDCs have 80% of the world's human population and eat 56% of the world's meat.

"Abundance does not spread, famine does."

Zulu proverb

"This is a sad hoax, for industrial man no longer eats potatoes made from solar energy; now he eats potatoes partly made of oil."

Howard T. Odum, an American ecologist, referring to the common perception that modern agriculture has freed society from limits imposed by nature, when in fact it is highly dependent on non-renewable fossil fuels (1971)

Fig. 13.1 Rice, a staple crop for human populations

Setting the scene

Using data from 2007, the Food and Agriculture Organization of the United Nations (FAO) estimates that 854 million people in the world do not get enough energy from their food; they suffer from **undernourishment**. This has risen from 853 million a year before. Of these, about 200 million are children and infants. Chronic undernourishment during childhood years leads to permanent damage: stunted growth, mental retardation and social and developmental disorders. Many people are also suffering from **malnourishment**. Their food contains enough energy, but lacks essential nutrients like proteins, vitamins and certain minerals.

According to the Hunger Site (www.thehungersite.com) about 10% of undernourished people die each year from starvation or malnutrition and three-quarters of these are children under the age of 5. There are incomprehensible figures to most of us but put it into context: there are about 6.6 billion humans alive on Earth now. Why do about a sixth of the world (approximately 13%) not have enough food when there are, apparently, large surpluses stored in some MEDCs?

Food is potentially one of the most important resource issues facing global society today, alongside potable water (drinking water). As populations increase, as global trade expands and as market choice develops, greater and greater demands are being made on food supplies and food production systems. These are primarily agricultural systems operated on an industrial scale – agribusiness. In the last 50 years technology and science have made huge advances in agricultural practice and agricultural production but then human population has increased too.

In many MEDCs, the cost of food is relatively cheap. Most people purchase foods out of choice and preference rather than basic nutritional need. Seasonality of produce has disappeared in most MEDCs. Exotic foods are freely available all the year round. Modern technology and transport systems mean that New Zealand lamb, beans from Kenya, dates from Morocco and bananas from the Windward Islands can be bought in almost any MEDC supermarket anywhere in the world.

In LEDCs, many populations struggle to produce enough food to sustain them. There may be political and economic agendas as well as simple environmental limitations on food production. Cereals can be grown for export and revenue generation rather than to feed indigenous populations – cash cropping. Crops other than food crops can be grown as cash crops: coffee, hemp, flax and biofuels. However, they occupy land that could be used for food production and arable land is in finite supply.

The rising cost of food

There will be billions more human mouths to feed in the next 50 years so an increase in the demand for food is not going to go away. Human population was 6.1 billion in 2000 and is predicted by the UN to be 8.0 billion in 2025 and 9.2 in 2050. Because it is profitable at the moment, there is a massive increase in growing crops for biofuel (living plants converted to fuel to replace fossil fuels) and this uses land that would otherwise be used to grow crops for food. We cannot have it both ways as crop land is limited.

In 2008, there were food riots (e.g. in Burkina Faso, Egypt and Bangladesh) when people protested about the rising price of food. In Haiti, the poorest country, the government was overthrown due to the high cost of food. According to the FAO, 37 countries are facing food crises. Food costs a higher proportion of income for the poor than for the rich and its price has rocketed to all-time highs. For 100 million people or more, this could mean starvation and many are going without meals because they cannot afford them.

The World Food Programme (the UN food aid agency) could only buy half as much food in 2008 as in 2007 with the same amount of money and is requesting more. Wheat, soya, rapeseed oil and palm oil prices are all very high. Biofuels are one of the reasons for this as farmers get subsidies for growing them so do not put their crops into the food chain. The poor harvest in Australia hit wheat supplies and the futures trading in commodities, including food, has also driven up prices. The increasing wealth of India and China means that more people can afford and want more meat to eat so demand for meat has not just increased because there are more of us but because many of us eat more meat.

The United Nations projects that biofuels will be "one of the main drivers" of projected food price hikes of 20% to 50% by 2016.

While there is still enough food, we need to review the strategies on global food production.

Questions

1 What has happened to food prices recently? Do your own research on this.
2 Do your own research on biofuels and read the section on them in Chapter 10. Evaluate the costs and benefits of growing biofuels on agricultural land.
3 Is the country in which you live a net importer or exporter of food?

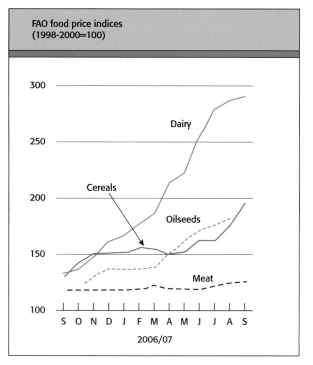

Fig. 13.2 Rise in food prices 2006–7

Food production and distribution around the world

Let's look first at agricultural production (Fig. 13.3).

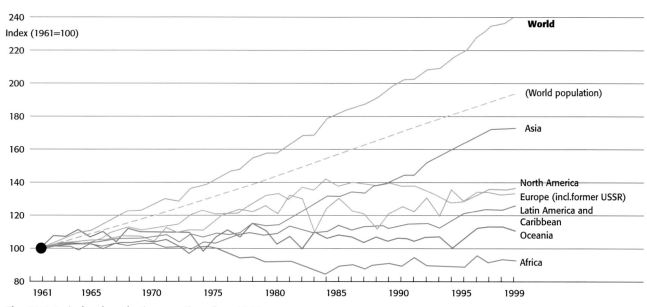

WORLD AGRICULTURAL PRODUCTION

Fig. 13.3 Agricultural production over time since 1961

So why is there such a difference in world agricultural production? If you search the Internet or textbooks, many will say that there is plenty of food to go round. The American Association for the Advancement of Science suggests that there is an average of 2790 calories of food available each day for every human on the planet – 23% more than in 1961 and enough to feed everyone. Food production has actually kept up with world population growth, so why are there still so many problems with famine, hunger and malnutrition?

There are many factors that may be considered, but is the main one distribution? It is well known that countries like the USA, Canada and Australia have more food than their population needs. Consider who should pay to ship the excess foods from where they are stored to countries like Bangladesh, Ethiopia and Sudan, and is it the type of food they want?

There is also the political angle, one that probably affects far more than we ever realize. If excess food is not paid for, does this put the receiving country forever under the power of the exporting country? What happens when corrupt governments do not distribute to those who need the food, and who decides who needs the food anyway? These are some of the questions that need to be considered where the issues involved in the imbalance in global food supply are concerned.

Huge differences exist between the diets of people in MEDCs and LEDCs. In MEDCs people's food contains on average 3314 calories per capita per day of energy. The average value for LEDCs is only 2666 calories per capita per day. As these values are averages, the differences between individual countries are far larger (for example, USA 3774 calories per capita per day and only 1512 calories per capita per day in Eritrea).

Not only the total amount of food is different in MEDCs and LEDCs; also the composition of the food is rather different. The menu in MEDCs contains more meat and fish than the menu in LEDCs. In contrast, cereals play a larger role in LEDCs. To start answering these issues, it is necessary to have an overview of the global pattern of world food supply. Table 13.1 compares food intake in LEDCs and MEDCs. The rest of the food will be in the form of non-cereal vegetables, fruits and fats. Consider the differences in these figures.

Recent doubts

So far in the history of the human population, food supply has kept pace with population, confounding the Malthusians among us (see page 171). But very recently, some are doubting that technology, efficiency and innovation will allow us to feed a world of 9 billion.

As we adapt more and more of the NPP on Earth to human needs, use and degrade more land, demand more meat, we must be reaching the limits of growth.

The 1.1 billion living in poverty appear to be increasing, not decreasing, and getting hungrier.

Annual grain yields per hectare have slowed their rate of increase since the Green Revolution (see page 252) as the benefits have been realized and we may be near the limit of productivity.

To do

Compare and contrast the agricultural production trends shown in Fig. 13.3 (about 100 words).

Hint: Be sure to compare various areas, do not just describe each line; pick contrasting trends as well as similar ones. Do not attempt to explain these trends; this is not what the question asks.

Food composition, in %*		
	MEDCs	LEDCs
Meat	12.9	7.3
Fish, seafood	1.4	0.9
Cereals	37.3	56.1

Table 13.1 Comparison of food composition in MEDCs and LEDCs in the late 1990s

To do

Look at the FAO website to find a map showing per capita calorie availability and low-income food deficit countries (LIFDCs).

In about 200 words, comment on the distribution of LIFDCs in the world. (*Hint*: Distribution means spread – are they all grouped together in one geographical area? Are there any common factors?)

Annual Increase in Grain Yields by decade

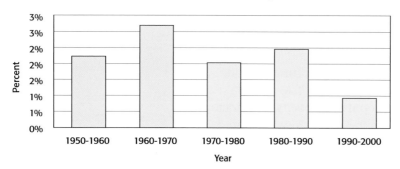

Fig. 13.4 Annual increase in grain yields by decade

Agriculture and society

Agriculture (farming) is general accepted as "the growing of food plants and the husbandry of animals for food". Farming began approximately 10 000 years ago, with the first selective sowing of cereal grasses in the Middle East. By 6000 years ago agriculture had advanced sufficiently in terms of productivity to allow urban populations to develop. Populations were then not dependent on producing their own food and the market of food supply and demand was born. Food had become an economic resource that could be bought, sold and exchanged for other goods. Cultivation encouraged settled communities.

Bush meat is any wild animal killed for food. In some countries, this is called game, in others, hunting wild animals. The term bush meat is highly politicized as many see it as illegal hunting, particularly in Central and West Africa and tropical rainforests. Although bush meat can be many species, the emotive element focuses on the killing of the great apes for meat, often orphaning their young.

Trade in bush meat is increasing. The reasons are that logging roads built into the forests make access easier for hunters, and the high price of meat in markets.

The cane rat (Fig. 13.5) is a large rodent pest of crops in West Africa and is hunted for bush meat. Now, some farmers are successfully starting to farm cane rats for meat in Benin and Togo.

Fig. 13.5 A cane rat

Farming – agriculture

Types of farming systems

Subsistence farming is the provision of food by farmers for their own families or the local community; there is no surplus. Usually mixed crops are planted and human labour is used a great deal. There are relatively low inputs of energy in the form of fossil fuels or chemicals. With low capital input and low levels of technology, subsistence farmers are unlikely to produce much more than they need. They are vulnerable to food shortages as little is stored. **Cash cropping** is growing crops for the market, not to eat yourself.

Commercial farming takes place on a large, profit-making scale, maximizing yields per hectare. This is often by a monoculture of one crop or one type of animal. Often this means planting one crop or

Case study 1

Palm Oil – rainforest in your shopping

Fig. 13.6 Oil palm

The oil palm is a tropical palm tree indigenous to West Africa and Central America but imported to Southeast Asia in the early 1900s. Here it is grown for its oil in large plantations with about half being in Malaysia and the rest in Indonesia and other Southeast Asian countries.

The benefits of oil palm plantations are in providing employment and exports. Growing a few oil palms can bring an income for a subsistence farmer, and large oil plantations and processing plants provide much-needed employment.

Palm oil is used in making cooking oil and margarine, put into processed foods, cosmetics and soaps (e.g. Sunlight soap, Palmolive), as a lubricant, and more recently as a biofuel. It is high in saturated fats and semi-solid at room temperature.

However, oil palm plantations often replace tropical rainforest and, in Malaysia and Indonesia, primary rainforest has been cleared for oil palm. Often this forest is on peat bogs which are then drained and habitats lost.

To maintain the monoculture of oil palms, herbicides and pesticides are used on the plantations and these poison other animal species. Animals that were in the rainforest, such as elephants, move into the plantations seeking food and are killed as pests.

According to Friends of the Earth, a UK environmental charity, demand for palm oil is the most significant cause of rainforest loss in Malaysia and Indonesia.

Palm oil is found in one in ten food products, including chocolate, bread and crisps, and in detergents and lipsticks. In Indonesia, the area of land occupied by palm oil plantations has doubled in the last 10 years and is still increasing.

Look at the products that you use and list which contain palm oil. (It may also be called: *Elaeis guineensis*, isopropyl palmitate, palmitic acid, palm olein, cooking oil or sodium palm kernelate.) If you could not use these products, with what would you replace them?

rearing one type of animal. High levels of technology, energy and chemical input are usually used with corresponding high outputs.

Farming may also be described as **extensive** or **intensive**. Extensive farming uses more land with a lower density of stocking or planting and lower inputs and corresponding outputs. Intensive farming uses land more intensively with high levels of input and output per unit area. Animal feedlots are intensive.

Pastoral farming is raising animals, usually on grass and on land that is not suitable for crops. **Arable** farming is sowing crops on good soils to eat directly or to feed to animals. **Mixed** farming has both crops and animals and is a system in itself where animal waste is used to fertilize the crops and improve soil structure, and some crops are fed to the animals.

Farming system	Shifting cultivation	Cereal growing	Rice growing	Horticulture and dairying
Where	Amazon rainforest	Canadian prairies	Ganges Valley	Western Netherlands
Type	Extensive subsistence	Extensive commercial	Intensive subsistence	Intensive commercial
Inputs	Low – labour and hand tools	High use of technology and fertilizers	High labour, low technology	High labour and technology
Outputs	Low – enough to feed the family	Low per hectare but high per farmer	High per hectare, low per farmer	High per hectare and per farmer
Efficiency	High	Medium	High	High
Environmental impact	Low – only if enough land to move to and time for forest to regrow	High – loss of natural ecosystems, soil erosion, loss of biodiversity	Low – padi rice has a polyculture, stocked with fish. Also grow other crops	High – greenhouses for salads and flowers are heated and lit. In dairying, grass is fertilized, cows produce waste

Table 13.2 Types of farming system

Farming's energy budget

The efficiency of a farming system can be measured in a number of ways. One is the energy contained within the crop of harvested product per unit area. Therefore we can compare the relative energy returns from cereal and root crops, from wheat and corn, from beef or lamb. However this calculation is problematic. Does the calculation consider biomass harvested or does it consider only the marketable or edible portion of the harvest? In livestock terms do we consider liveweight or dressed deadweight of animal (just the meat)?

More scientifically we could look at the efficiency of agricultural systems: a system with inputs, outputs, storages and flows. At the end of the system there is a marketable product that is usually sold by weight, e.g. cereal (maize, wheat, rice, etc.). It is possible to calculate the energy contained in a food per unit volume (joules per gram). In order to calculate the energy balance of the farming system all you need to do is calculate the energy it took to produce that food and deliver it to the market. You would need to consider the fuel, labour and any other energy that was used to prepare the soil, sow the seed, harvest the crop, prepare it and appropriately package it for the market, then transport it to market. You would also have to include the energy cost of dealing with waste products associated with the farming system.

Some statistics do exist on the efficiency of agricultural systems, and for all but cereal growing the energy output to energy input ratio is less than one. So more energy is used to get the foodstuff to our tables than is produced in the process of growing or raising it.

Dairying	0.38
Cattle rearing (beef)	0.59
Sheep (lamb)	0.25
Pigs (pork)	0.32
Cereal growing	1.9

There are even statistics available for food products

White bread	0.525
Chicken	0.10
Battery eggs	0.14
Lettuce produced under glass	0.0017

While total energy out is less than energy in when all factors are considered, remember that the quality of the energy is different. Fats and proteins contain more energy content per gram than carbohydrates. You need to eat less meat and fish than cereals to get the same amount of energy. A higher energy content food costs less to transport as it has a lower volume. To equate measures, the **grain equivalent** in kilograms (the quantity of wheat grain that would have to be used to produce one kilogram of that product) is sometimes used.

A brief history of farming

Livestock farming – why farm animals?

Humans have been associated with animals for longer than we have been farming cereals. Animal domestication shadowed the advent of crop farming; first the dog, then sheep, goats, pigs and cattle were domesticated and used to fulfill a variety of needs. Dogs were used as hunting companions and ultimately as herding animals. In some cultures the dog was also a source of food. Sheep, goats and pigs probably were hunted as wild prey and then corralled and domesticated to provide a more convenient and more reliable source of food. Cattle, reindeer, horses, donkeys, yaks, camels and llamas were (and are) used as food and as beasts of burden. In addition they provided wool and hides for clothing and even their bones, antlers and teeth had a value as tools and decorations.

Livestock are a useful means of converting plant material unsuitable for human digestive systems – mostly grass – into high-value protein. Sheep and goats living in Mediterranean scrub forests browse on woody scrubs and trees, digesting vegetation that is unpalatable to humans; pigs were often kept on farms as a valuable way of processing waste products and producing valuable protein.

To think about

The European Union's (EU) Common Agricultural Policy (CAP)

About half of the EU budget goes towards the CAP, a system of subsidies for agriculture. In this farmers are guaranteed a price for their produce and a tariff and quota are placed on imports of foodstuffs. The initial aims of the CAP were laudable: to ensure productivity, give farmers a reasonable standard of living, allow food stocks to be secured and provide food in shops at affordable prices. However, it has cost a great deal to implement and it is said that a cow in the EU gets more money in subsidies than a poor person in Africa.

The CAP is being reformed with subsidies for individual animals phasing out and a single farm payment being given per hectare of cultivatable land.

One big issue with the CAP is that it is accused of being protectionist – keeping products from other countries out of the EU to the benefit of the 5% of Europeans who work in agriculture – "Fortress Europe". It has also been accused of keeping food prices artificially high, causing massive food stocks to build up as producers keep producing even if no consumers want to buy as the EU will buy the products (the "wine lake", the "butter mountain") and allowing farmers to pollute with high levels of chemicals to increase production. All these issues are being addressed now.

Questions

1 What do you think about protectionism in the trade of food (stopping or adding tariffs to imports from other countries so your own farmers can sell theirs)?

2 Should small EU farmers have been allowed to go bankrupt and give up farming?

The Green Revolution of the 20th century

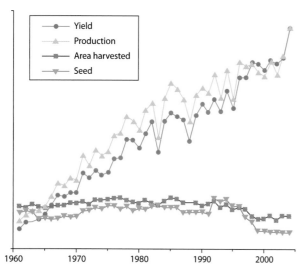

Fig. 13.7 Coarse grain production 1961–2004 (coarse grain is wheat, maize, rice, barley)

From the 1940s to the 1960s, plant breeding of wheat and rice and then other cereals was undertaken to produce varieties that were less prone to disease, had shorter stalks (so they did not lodge or fall over in rain) and gave higher yields. This was done by artificial selection of the varieties and individual plants that had the traits wanted and the results were spectacular in terms of crop yields. (It was before we knew about the possibility of or could carry out genetic engineering which is switching genes from one species to another.)

In Mexico, wheat yields increased such that the country became self-sufficient in wheat and then exported the surplus. In India, the IR8 variety of rice, an HYV or high yielding variety, gave five times the yield of older varieties with no added fertilizer and ten times with fertilizer. This led to a fall in the cost of rice, and India exporting the surplus. However, in Africa, there was little difference in crop yields.

Some have been critical of the Green Revolution varieties as the result has been that more fertilizer, irrigation and pesticides have been used on crops (if farmers can afford them) and these cause eutrophication, salinization and chemicals to accumulate in food chains. They have also reduced genetic diversity in the crops as most farmers use them. While some say that the poor have become poorer because of this, there is little doubt that the HYVs have allowed us to produce enough food for the population.

Questions

1 Explain the relationships between the four graphs in Fig 13.9.

2 Evaluate the impacts of the Green Revolution on world food supply and the environment.

Rice production in Borneo and California

In this section, two food production systems are discussed:

- traditional, extensive rice production in Indonesian Borneo (Kalimantan)
- intensive rice production in California.

Traditional extensive rice production is characterized by low inputs of energy and chemicals, high labour intensity and a low productivity. Nearly all energy added is in the form of labour and seeds. The rice yield is rather low. However, the energy efficiency (defined as energy output over energy input) is high. As no fertilizer or pesticides are used, the rice yield is the only output (no pollution).

Intensive rice production is characterized by high inputs of energy and chemicals, low labour intensity and a high productivity. The energy inputs are in the form of diesel or petrol, not in the form of labour. Large amounts of fertilizer (N, P) and pesticides (insecticides and herbicides) are used. As a result, high rice yields are obtained. However, because of the large energy inputs, the energy efficiency is much lower than in the traditional extensive rice production example. In some intensive agricultural production systems, the energy inputs are larger than the energy outputs. The intensive rice production system also has extra outputs compared to the extensive system: pollution in the form of excess fertilizer and pesticides. Note that these are not shown in Table 13.3: they were not measured.

		Energy (1000 kcal ha^{-1})	
		Borneo	**California**
Inputs			
Direct energy	labour	0.626	0.008
	axe and hoe	0.016	–
	machinery	–	0.360
	diesel	–	3.264
	petrol	–	0.910
	gas	–	0.354
Indirect energy	nitrogen	–	4.116
	phosphorus	–	0.201
	seeds	0.392	1.140
	irrigation	–	1.299
	insecticides	–	0.191
	herbicides	–	1.119
	drying	–	1.217
	electricity	–	0.380
	transport	0.051	0.121
Outputs	rice yield	7.318	22.3698
	(protein yield)	(141 kg)	(462 kg)
Energy efficiency		7.08	1.55

Table 13.3 Rice production systems in Borneo and California

Fisheries – industrial hunting

According to the FAO more than 70% of the world's fisheries are fully exploited, in decline, seriously depleted or under drastic limits to allow a recovery. The global fish catch is in decline even though technology has improved. Demand is high and rising but fishermen cannot find or catch enough fish. They are no longer there.

Other Fish in the Sea, But For How Long?

Copyright © 2003 Earth Policy Institute

A recent review of marine fisheries concluded that a startling 90% of the world's large predatory fish, including tuna, swordfish, cod, halibut and flounder, have disappeared in the past 50 years. This 10-year study by Ransom Myers and Boris Worm at Canada's Dalhousie University attributes the decline to a growing demand for seafood, coupled with an expanding global fleet of technologically efficient boats.

Once thought to be inexhaustible, the world's fisheries are now showing their vulnerability. The United Nations Food and Agriculture Organization (FAO) estimates that three-quarters of the world's oceanic fisheries are being fished at or beyond their sustainable yields. The innovations that have allowed us to pull more fish out of the oceans—larger and more-powerful boats (some with on-deck processing facilities), improved fishing gears, and navigational and fish-finding technologies—may undermine the oceans' presumed resilience.

Data show that once large boats target a fishery, they can deplete populations in a matter of years. Within 15 years, some 80% of the large fish are lost. Smaller species may initially flourish, but often their populations soon crash too, either because of a limited food supply, overcrowding and disease, or because they become

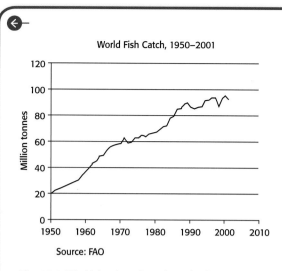

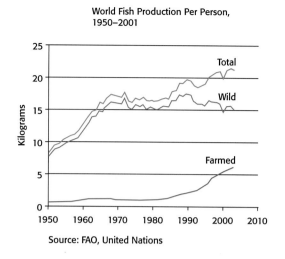

Fig. 13.8 World food catch and production per person

targets for those who are "fishing down the food web". The average size of top predatory fish is now only one-fifth to one-half that in the past, in part because the fish left to breed are the ones small enough to escape from nets. Another problem is that slow-maturing fish are often caught before they are old enough to reproduce.

Fishing gears are frequently indiscriminate. Trawlers drag enormous nets over a vast area, virtually clearcutting the seabed, destroying marine habitat and taking up untargeted species. Worldwide, almost one-fourth of the fish catch is discarded dead at sea, either because the fish are not marketable or because fishers have exceeded their catch allotment. Whales, dolphins and porpoises also become part of this collateral damage. In certain fisheries, like the Gulf of Mexico shrimp fishery, the weight of this "bycatch", as it is known, may exceed that of the profitable take by 10 to 1.

After growing steadily for decades, the world fish catch has stalled at between 85 million and 95 million tonnes since 1986 (see Fig. 13.8). In 2001, the world's fishing fleets landed some 92 million tonnes of fish, according to FAO data. An analysis by Reg Watson and Daniel Pauly at the University of British Columbia reveals that overreporting by China, the world's largest fisher, and climate-related fluctuations in Peruvian anchoveta populations, may have masked an actual decline in the global catch of some 660 000 tonnes per year since 1988.

From 1950 to 1988, the world fish catch climbed from 19 million to 89 million tonnes. This fivefold expansion dwarfed the growth in global beef production during that period (from 19 million to 54 million tonnes). In per capita terms, the annual fish catch per person peaked at 17 kilograms in 1988, up from 8 kilograms in 1950. For most of the last half-century, we could count on a steadily growing oceanic catch to help meet the growing demand for animal protein. That era is over.

Prospects for the 1 billion people throughout the world who rely on fish as their primary source of protein and the 200 million involved with fishing and fish-related industries rest on careful management of wild fish stocks and farms. Ecologists liken fish stocks to a bank account. With a certain balance preserved in the bank, we can live off the interest. But if we continue to dip into the principal, eventually we are left with an empty account.

The collapse of the Newfoundland cod fishery is a case in point. For centuries it was one of the world's most productive fisheries, yielding 800 000 tonnes of fish and employing 40 000 people at its peak in 1968. Then its stocks plummeted as a result of overharvesting and habitat damage. In 1992, the fishery was closed in an effort to save it. But it may have been too late: a decade has passed, but stocks have not recovered.

This collapse was local in scale, but the issue is much larger. Fisheries operating over the entire North Atlantic Ocean now catch half as many of the popular species—such as cod, tuna, flounder and hake—as 50 years ago, despite tripling their efforts. Cod stocks in the North Sea and to Scotland's west are on the verge of collapse. In a 2001 report entitled *Now or Never: The cost of Canada's cod collapse and disturbing parallels with the UK*, Malcolm MacGarvin urges Europe to avoid the same fate as Newfoundland's fishers.

The deterioration of oceanic fisheries can be reversed. Granting fishers an ownership stake in fish stocks is one way to help them understand that the more productive their fishery is, the more valuable their share. For example, fishers in Iceland and New Zealand have used marketable quotas, allowing them to sell catch rights, since the late 1980s. The upshot is smaller but more profitable catches and rebounding fish populations. The classic "tragedy of the commons" problem is averted.

Because of the complexity of marine ecosystems, some scientists are pushing for management of whole ecosystems rather than single species. In addition, studies have shown that well-positioned and fully protected marine reserves, known as fish parks, can help replenish an overfished area. By giving fish a refuge to breed and mature in, reserves can increase the size and total number of fish both in the reserve and in surrounding waters. For example, a network of reserves established off St Lucia in 1995 has raised the catch by adjacent small-scale fishers by up to 90%. Preservation of nursery habitats like coral reefs, kelp forests and coastal wetlands is integral to keeping fish in the sea for generations to come.

Consumers can promote healthy fishery production by eating less fish and buying seafood from well-managed, abundantly stocked fisheries. The *Seafood Lover's Guide* from Audubon's Living Oceans program is one valuable reference. Chilean seabass, for example, makes the list of fish to avoid because stocks are on the verge of collapse and illegal fishing abounds. The list also distinguishes between wild Alaska salmon, which comes from a healthy fishery, and farmed salmon, which is fed meal made from wild fish and thus does not relieve pressure on marine stocks. Proper labels are needed to allow consumers to make wise purchasing decisions. The Marine Stewardship Council, a new independent international accreditation organization, has thus far certified seven fisheries as being sustainably managed with minimal environmental impact.

The capacity of the world's fishing fleet is now double the sustainable yield of fisheries. Myers and Worm from Dalhousie University believe that the global fish catch may need to be cut in half to prevent additional collapses. Reducing bycatch, creating no-take fish reserves, and managing marine ecosystems for long-term sustainability instead of short-term economic gain are all policy tools that can help preserve the world's fish stocks. If these are coupled with a redirection of annual fishing industry subsidies of at least $15 billion to alternatives such as the retraining of fishers, there could be a big payoff. It is difficult to overestimate the urgency of saving the world's fish stocks. Once fisheries collapse, there is no guarantee they will recover.

Questions

1 Describe the "tragedy of the commons".
2 State two reasons why we are overfishing the oceans.
3 Why has the world fish catch stalled?
4 What actions can be taken to reverse overfishing?
5 What actions can you take?

Although stocks are shrinking, governments around the world continue to invest an estimated US$50 billion of subsidies into a grossly overcapitalized industry supporting a fishing industry which could not otherwise stay afloat. That is not only a problem of the industrialized nations subsidizing their modern fleets which have grown with demand but outstripped supplies. The Chinese Xinhua news agency reports that the Indonesian government plans to procure 81 069 fishing vessels between 2008 and 2013.

The tragedy of the commons

This metaphor illustrates the tension between the common good and the needs of the individual and how they can be in conflict. If a resource is seen as belonging to all, we all tend to exploit it and overexploit it if we can. This is because the advantage to the individual of taking the resource (be it fish, timber, minerals, apples) is greater than the cost to the individual as the cost is spread among the whole population. In the short term it is worth taking all the fish you can because, if you do not, someone else will. This assumes that humans are selfish and not altruistic and it has caused much debate among economists and philosophers. The solution is often by regulation and legislation by authorities which limits the amount of common good available to any individual. This may be by permit, set limits or by co-operation to conserve the resource.

Exploitation of the oceans is a good example of the tragedy of the commons. The Grand Banks off the coast of Newfoundland were once among the richest fishing grounds on Earth. Since the 1400s, they have been fished by fleets from Spain, Portugal, England, France, and later Newfoundland, Canada, Russia and the United States, and others. In the early 1990s, fish stocks crashed and by 1995, cod and flounder fishing there had been closed in an attempt to conserve the remaining stocks. So far, there has been little, if any, recovery of fish numbers.

Fishing resources are a resource under pressure, being exploited by overfishing – fishing at an unsustainable level. We are just too good at finding and catching fish on an industrial scale. Commercial fishing is informed by the latest satellite technology, GPS navigation and fish-finding scanning technology of military quality. Fishing fleets have become larger and, with modern refrigeration techniques, including blast freezing, they can stay at sea for weeks or an entire season. Within a fleet there will be a suite of vessels including fishing vessels, supply vessels and factory ships that process the catch at sea.

To do

The **United Nations Convention on Law of the Sea (UNCLOS)** is an international agreement written over decades that attempts to define the rights and responsibilities of nations with respect to the seas and marine resources. Most, but not all, countries have signed and ratified the convention. It defines these categories of waters:

- Internal waters – next to a country's coastline where foreign ships may not travel and the country is free to set its own laws and regulate use.
- Territorial waters – up to 12 nautical miles from the coastline where foreign ships can transit on "innocent passage" but not spy, fish or pollute. In island states, e.g. the Maldives, a boundary line is drawn around the whole archipelago.
- Beyond the territorial waters is another 12 nautical miles of contiguous zone where a state can patrol smuggling or illegal immigration activities.
- In the exclusive economic zones (EEZs), the state has sole rights to exploit all natural resources. Foreign nations may overfly or navigate through this zone. If a country's continental shelf is greater than 200 miles from its coast, it also has exclusive rights to exploit this.

The International Seabed Authority, created by the UN, controls and monitors seabed exploitation in international waters.

Landlocked states are given a right of free access to and from the sea under UNCLOS rules.

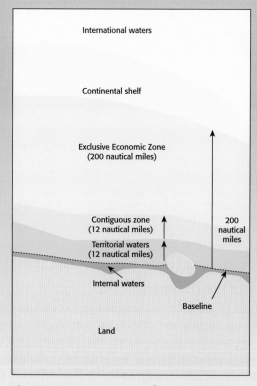

Fig. 13.9 Ocean zones according to UNCLOS

1 Do you think that UNCLOS was a good idea?
2 Why would some nations not sign up to it?
3 What would have happened without UNCLOS?

All of this is a far cry from the hunter-gatherer ethos of fishermen going to sea in primitive craft with simple fishing equipment to land a fish catch for their own family or community.

When we talk of wars over resources, we are usually thinking about wars over fuel and energy supplies, however a significant number of serious international crises and near war events have taken place over fishing and fishing rights in the last 60 years. In the 1970s, Iceland banned all foreign vessels from fishing in Icelandic waters. This led to three "cod wars" between Britain and Iceland. In 1994, British and French fishermen competed with Spanish fishermen for tuna in the Bay of Biscay. In 1995, there was the turbot (also called halibut) war between Canada and Spain when Canada fired on and captured the crew of a Spanish fishing boat, the *Estai*, having chased it from national into international waters. The *Estai* cut its trawl net when it was being pursued but the Canadians recovered it from the seabed and found it had a net with a mesh size smaller than that permitted and this caught the smaller turbot. Eventually, Canada and Spain agreed on a solution to the turbot war and the result was an increased regulation of fisheries.

What happens to the fish?

The world fish catch is just over 90 million tonnes per year of which about 20 million tonnes are not wild fish but from aquaculture (fish farming). About a quarter of the global catch goes into fish meal and fish oil products to feed animals … to feed us. About 20 million tonnes are bycatch – unwanted fish that are thrown back into the oceans from the fishing boats. This bycatch may be the wrong species (non-target), fish that are legally too small to catch, over quota fish. Most of these are dead or dying when they are thrown back into the seas.

Globally, 15% of animal protein eaten by humans comes from fish. In Japan nearly half of the animal protein diet is fish.

To do

Draw a pie chart to illustrate the quantities in the paragraph alongside.

Species	Reason	Alternatives
Atlantic cod (from overfished stocks)	Species listed by World Conservation Union, IUCN. Some stocks close to collapse, e.g. North Sea	Line-caught fish from Icelandic waters
Atlantic salmon	Wild stocks reduced by 50% in last 20 years	Wild Pacific salmon. Responsibly and/or organically farmed salmon
Chilean seabass (Patagonian toothfish)	Species threatened with extinction by illegal fishing, also high levels of seabird bycatch	None
Dogfish/spurdog	Species listed by IUCN	None
European hake	Species heavily overfished and now scarce	South African hake (*M. capensis*)
European seabass	Trawl fisheries target pre-spawning and spawning fish also high levels of cetacean by catch	Line-caught or farmed seabass
Grouper	Many species are listed by IUCN	None
Haddock (from overfished stocks)	Species listed by IUCN	Line-caught fish from Icelandic and Faroese waters
Ling (*Molva spp*)	Deep-water species and habitat vulnerable to impacts of exploitation and trawling	None
Marlin	Many species are listed by IUCN	None
Monkfish	Long-lived species vulnerable to exploitation. Mature females extremely rare	None

North Atlantic halibut	Species listed by IUCN	Line-caught Pacific species. Also farmed N Atlantic halibut
Orange roughy	Very long-lived species vulnerable to exploitation	None
Shark	Long-lived species vulnerable to exploitation	None
Skates and rays	Long-lived species vulnerable to exploitation	None
Snapper	Some species listed by IUCN**, others overexploited locally	None
Sturgeon	Long-lived species vulnerable to exploitation. 5 out of 6 Caspian Sea species listed by IUCN	None although this species is now farmed
Swordfish	Species listed by IUCN	None
Tuna	All commercially fished species listed by IUCN except skipjack and yellowfin which is overfished	"Dolphin-friendly" (EII monitored) skipjack or yellowfin. Preferably pole & line caught
Warm-water or tropical prawns	High bycatch levels and habitat destruction	Responsibly farmed prawns only

Table 13.4 What not to eat. Source: Marine Conservation Society

If you eat fish, research whether the species that you eat are (a) farmed and (b) fished sustainably.

To think about

In a world committed to a free market – in a global context – have we the right to criticize or control those who use technology and labour efficiencies to produce more food at less cost? Without intensive agriculture, we could not feed the human population.

1 Why does food harvesting/production need to be ethical?
2 What do we actually mean by ethical food?
3 What is Fair Trade? Does it help farmers out of poverty or does it make consumers in MEDCs feel better, or both?

Atlantic salmon and fish farming

Wild Atlantic salmon live in the North Atlantic and Baltic seas. Due to overfishing at sea and in rivers where they go to spawn, the commercial market for wild salmon crashed. Now they are farmed in fish farms (a form of aquaculture). Although commercially successful, these farms create pollution as the fish are kept in high densities and their uneaten food, faeces and chemicals used to treat them enter the oceans. Sea lice are a problem in the farms and escape of some farmed salmon which then interbreed with wild salmon reduces the genetic diversity of the wild stock as farmed salmon are bred for fish farm life, not the wild.

As wild stocks of many fish species decline, fish farming has increased. Although there are environmental problems created by keeping fish of one species in high densities, the benefits of not chasing fish stocks across the oceans are large. Some have likened fishing in the seas to hunting on land. We now get most of our terrestrial meat from farmed livestock and little from hunting. It makes economic sense to do the same with our aquatic meat supply.

To research

Research the history and current state of one of these fisheries:

- Pacific wild salmon
- North Sea herring
- Grand Banks fisheries
- Peruvian anchovy

or any other that is near to where you live.

Maximum sustainable yield (MSY)

A sustainable yield (SY) is the increase in natural capital, i.e. natural income, that can be exploited each year without depleting the original stock or its potential for replenishment. Sustainable yield of an aquifer is the amount that can be taken each year without permanently decreasing the amount of water stored. For commercial ventures, it is **maximum sustainable yield** that is of interest: the highest amount that can be taken without permanently depleting the stock. In fisheries, this is a crucial amount – how many fish can be taken of what size in one year and not impair the harvest in subsequent years.

The carrying capacity for each species depends on the reproductive strategy of the species, its longevity and the indigenous resources of the habitat or ecosystem. Each breeding season or year, new individuals enter the population (either new offspring or immigrants). If the number recruited to the population is larger than the number leaving (dying or emigrating), then there is a net increase in population. If the difference in population from initial size to new population size is harvested, the population will remain the same. This number is the maximum sustainable yield (MSY) for this population.

SY = (total biomass or energy at time t+1) − (total biomass or energy at time t)

or

SY = annual growth and recruitment − annual death and emigration

In practice, harvesting the maximum sustainable yield normally leads to population decline and thus loss of resource base and an unsustainable industry or fishery. There are several reasons for this. The population dynamics of the target species are normally predicted (modelled) rather than the species numbers being quantitatively measured (counted). It is often impossible to be precise about the size of a population. Estimates are based on previous experience. Therefore if you extract or harvest at the MSY, you will deplete a population in poor recruitment years. Also the model does not allow for monitoring of the dynamic nature of the harvest in terms of age and sex ratio. If the harvest primarily targets reproductive females this will have a much greater impact on future recruitment than the targeting of mature or old males. Targeting young immature fish will also equally impact on future recruitment rates and thus overall population size.

A much safer approach is to adopt the harvesting of an **optimal sustainable yield** (OSY). This is calculated as half the carrying capacity. It has a much greater safety margin than MSY but still may have an impact on population size if there are other environmental pressures within a system.

Fishing **quotas** are often set as a percentage or proportion of the OSY per fleet per year. The quota is set as a weight of catch, not number of fish.

Terrestrial and aquatic food production systems compared

In terrestrial food production systems food is usually harvested at the first (crops) or second trophic level (meat usually originates from primary consumers like cows, pigs and chickens). This ensures that these production systems are making fairly efficient use of solar energy.

In aquatic food production systems, most food comes from higher trophic levels. Typical food fish tend to be carnivorous and are quite often at trophic level 4 or higher. Because of the energy "losses" at each trophic level, the energy efficiency of aquatic food production systems is lower than that of terrestrial systems. Note that, although the energy conversions along the food chain tend to be more efficient in aquatic ecosystems, the initial intake of solar energy is less efficient than in terrestrial ecosystems because of the absorption and reflection of sunlight by water.

Table 13.5 shows other typical differences between traditional extensive and intensive agricultural production systems:

Traditional extensive agriculture	Intensive agriculture
Limited selective breeding	Strong selective breeding
No genetically engineered organisms	Genetically engineered organisms
Polyculture	Monoculture

Table 13.5 Traditional extensive and intensive systems compared

For the case studies that follow, you will need to follow the advice given by Rudyard Kipling in his poem, The Elephant's Child:

"I keep six honest serving-men
(They taught me all I knew);
Their names are What and Why and When
And How and Where and Who."

So let's break down the information that we can find into those six categories.

Case study 2

Sudan

Sudan is Africa's largest country and agriculture is the main industry. In the south, there used to be enough rainfall for grazing herds belonging to nomadic tribes and for cropping. In the north, where the Nile and other rivers flow, irrigation is possible and cotton, peanuts, sugarcane, citrus fruits, dates, coffee and tobacco are grown. Cotton is the main cash crop and sorghum, millet, barley and wheat the staple cereals. Gold and oil are also present.

What has happened? Sudan is a country that has been the subject of much political dissension in recent history. Since the 1980s, civil war has raged and political agencies have interfered with humanitarian aid; consequently, the people of Sudan have suffered immensely. Food has been scarce and the economy is in ruins. The ongoing

civil war has led to 2 million people in refugee camps. Drought has led to frequent food crises, lack of drinking water, desertification and soil erosion.

Why has it happened? Civil war has been one of the main reasons and this has been politically motivated. Sudan's government and the pro-government Arab militias are fighting the region's black African population: an estimated 200 000 have died.

When? The war started in 1986 and supposedly ended in 2005. In 1988 there was a huge appeal because of famine. In recent years, war has flared again.

How did it all start? It was between the mainly Muslim north and the Animist and Christian south. Southern rebels said they were battling "oppression and marginalization". The official language is Arabic with Islam the main religion; however, because the country has a large non-Arabic speaking and non-Muslim population there has been rejection of the attempts by the government in Khartoum to impose Islamic "Sharia" law on the country as a whole.

Where did all this happen? Recently, one of the main regions targeted was Darfur which is in Western Sudan.

Who is involved? The military are in conflict with Sudanese rebels who mainly consist of farmers. The Sudanese support the military – is this where all the humanitarian financial aid is going? Is the Sudanese government corrupt? What is the truth behind the claims that ethnic cleansing is happening?

How is the presence of a refugee camp likely to affect the environment?

Construct a food chain for the area. Where do the energy inputs and outputs exist? What are the processes involved?

Case study 3

Australia

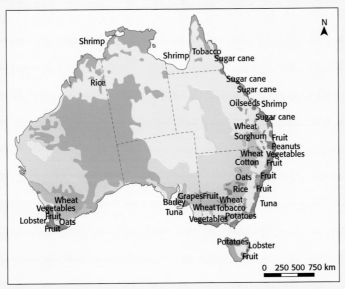

Fig. 13.10 Agriculture in Australia

What? Australians farm livestock and grow crops.

Where? Sheep are found mainly in Queensland, West and South Australia, Victoria and Tasmania. Half a million tonnes of wool is produced. The finest beef is produced in Tasmania, Victoria, New South Wales and Southeast Queensland. Wheat is a dominant crop with 45% of the available land under cultivation (New South Wales and Western Australia are the main areas). At least half is exported (representing 15% of the world's trade).

Who gains? Farms are huge with a team of stock men (often aboriginal). Annual meat production in Australia is in excess of 1 million tonnes, of which over half is exported and about 10% is canned. As much as 75% goes to the US followed by UK and Canada.

What has happened? Dairy products have expanded since the beginning of the last century but milk yield is still low. It is more restricted than beef due to the need for lush pasture, therefore it is found near coasts of Victoria, New South Wales and South Queensland. Over 50% of milk is used for butter (85% of this goes to the UK), 20% fresh milk and 10% for cheese.

Why has this been so successful? Let's look at north-western Australia; the Ord and the Fitzroy are the two largest rivers. Rainfall is 500–700 mm per annum and there is a summer monsoon.

90% of the rain falls between November and March so at the end of the wet season valleys dry up from floods, leaving vast fertile plains.

How have they dealt with adverse environmental problems? The growing season is only 14 weeks so farming is intensive. Although it is a fertile area, the soil requires heavy applications of nitrogen and phosphates and aerial spraying is necessary because of the tendency of the area to have insect pests.

As in all farming, water is very important in Australia. So, for this environmental system to survive in balance with human enterprise, irrigation and fertilization are essential.

Consider the energy flows, inputs and outputs of this system. Draw a systems diagram and label the storages and flows for an Australian farm.

To do

The Food and Agriculture Organization (FAO) is a UN agency. Find its website (www.fao.org/worldfoodsituation) from where Fig. 13.11 is taken, and answer the questions below.

1 Describe the changes in Fig. 13.11.
2 Read the 10 FAQs on the world food situation.
3 Answer these questions in your own words:
    a Why are world food prices rising?
    b What is the effect of biofuels on food prices?
    c Who are the winners and losers?
4 Describe the current food situation in Zimbabwe.

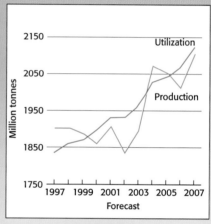

Fig. 13.11 World food supply and use, 1997–2007

Future global food supply

Feeding the world's population in the future is a challenge, as the population is growing and the area available to agriculture is decreasing. Not only many new mouths need to be fed, also today's undernourished and malnourished people need more and better quality food. Factors that contribute to this decrease in agricultural land are erosion, salinization and desertification. In order to ensure a sufficient food supply, changes in the management of food production systems are needed.

We will need to maximize the yield of food production systems. A problem is, of course, how to do this without unsustainable practices.

Another way to increase the amount of food available is to improve storage and distribution methods. This will lead to smaller losses and thereby to a more efficient use of the food.

Changing people's diet can also improve the efficiency of food production. If we obtain more food from lower trophic levels, we will greatly increase the amount of food available. People in MEDCs eat more meat than they actually need, so they could simply replace some of their meat with food taken from the first trophic level. However, the trend is the other way. More people in LEDCs are eating more meat. On average, in 2000, each human consumed 38 kg of meat per year. In the USA, it was 122 kg, UK and Brazil 77 kg, China 50 kg, India 5 kg. We cannot all eat more meat. Even though animals graze on some land that would not support crops, it is energetically inefficient to feed our grain to animals which we then eat.

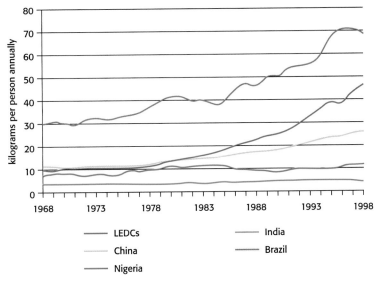

Fig. 13.12 Meat consumption in selected countries, 1968–98

In MEDCs, we mostly eat more than we need to sustain ourselves. Obesity has become a problem in some countries. We just do not seem able to get the balance right.

The FAO has reported its predictions on food supply for 2030 as follows:

1 Human population will grow to about 8 billion by 2030, a slowdown in growth rate to 1.1% per year from 1.7% in 2000. Demand for agricultural products will also slow but not in LEDCs where consumption will increase.

2 The human population will be increasingly well-fed with per capita calorie consumption increasing to 3000 kcal per day from 2800 now.

3 Numbers of hungry people will decrease to about 440 million but in some areas (Saharan Africa) this will not fall as much.

4 More people will eat more meat.

5 An extra billion tonnes of cereals will need to be grown.

6 LEDCs will have to import cereals.

7 Most increased production will be from higher yields and more irrigation, not from more land.

8 GM (genetically modified) crops, no tillage planting, soil conservation measures, and improved pest control will all increase productivity.

9 Aquaculture will increase.

Perhaps we are at the start of a new agricultural revolution. Land area for crops and animal rearing is finite so efficiency has to be improved to increase yields. Here are some examples of developments that may increase production:

- Genetic engineering of crops and livestock. Inserting into cereals the gene from legumes that allows them to fix nitrogen would save the need for nitrogenous fertilizers. (See box on GM crops, below.)
- Mixed cropping and interplanting conserves water and the soil.
- No-plough tillage, drilling seeds into the stubble of the previous crop, also conserves water and soil.
- Biological control of pests and integrated pest management reduce losses.
- New crops are being developed.
- Trickle irrigation is less wasteful of water.
- Fish farming and hydroponics.
- Improved crop storage to reduce spoilage.
- Less food processing so less energy is used in food processing, packaging and transport.
- Education about food so we eat neither too much nor too little and eat the right sorts of food.

Can you think of others?

GM crops

GM or genetically modified crops have DNA of a different species inserted into the crop species to form a transgenic plant. GM crops of soya bean, cotton and maize are the most common but the issue of GM crops is surrounded by controversy over ethics, food security and environmental conservation.

Proponents of GM crops say they are part of the solution to increasing food production. By making a crop disease- or pest-resistant, fewer chemicals need to be used and less is lost in spoilage. We are only doing what selective breeding has done since farming began but on a molecular scale. "Golden rice" has been made to synthesize beta-carotene, the precursor to vitamin A so, as a humanitarian tool, it could stop vitamin A deficiency suffered by 124 million people. (It has not yet been grown commercially.)

Opponents of GM crops say we do not know what we are releasing into the environment. GM opponents ask the following questions: Could the GM plants cross-pollinate with other varieties so the introduced DNA escapes into the wild populations? Could the DNA cross the species barrier? If a GM crop can kill a pest species that feeds on it, will that species die out? What will the effect on food chains be? Labelling of GM foods is now demanded and some countries have rejected them entirely.

Key words

agriculture	grain equivalents
commercial	biofuel
subsistence	bush meat
intensive	aquaculture
extensive	bycatch
shifting cultivation	tragedy of the commons
arable	energy budget
pastoral	energy subsidy
staple food	GM crop
coarse grains	

14 Succession

"An understanding of succession provides a basis for resolving man's conflict with nature."

Eugene P. Odum

Key points

- Succession is the change in species composition in an ecosystem over time.

- It may occur on bare ground (primary succession) where soil formation starts the process or where no soil has already formed, or where the vegetation has been removed (secondary succession).

- Early in succession, gross primary productivity (GPP) and respiration are low and so net primary productivity (NPP) is high as biomass accumulates.

- In later stages, while GPP may remain high, respiration increases so NPP may approach zero and the production:respiration ratio is near 1.

- A climax community is reached at the end of a succession when species composition stops changing.

- In agricultural systems, succession is often deliberately stopped when NPP is high and crops are harvested.

- Species biodiversity is low in early stages and increases as succession continues, falling a little in a climax community.

Review

Review these terms: GPP, NPP, respiration, species biodiversity. (See chapters 3 and 5.)

Ecological succession

Bare land almost anywhere on the planet does not stay bare for very long. Plants very quickly start to colonize the bare land and over time an entire plant community develops.

This change is directional as one community is replaced by another. Ecologists call this process **succession**. Succession is a natural increase in the complexity of the structure and species composition of a community over time.

Primary succession or prisere

During a succession an ecosystem develops by the gradual colonization of a lifeless abiotic substrate. A succession passes through several stages called **seral** stages. A **sere** is a set of communities that succeed one another over the course of succession at a particular location. For example, a lithosere is the colonization of bare rocky ground leading to a woodland community. During a hydrosere, vegetation fills a lake and develops into forest. In a **plagiosere**, human activity arrests a natural succession so that the climax or subclimax community is not reached.

Stages in primary succession

Table 14.1 illustrates the stages of a sere.

Bare, inorganic surface	A lifeless abiotic environment becomes available for colonization by pioneer plant and animal species. Pioneers are typically *r*-selected species (see Chapter 8), showing small size, short life cycles, rapid growth and production of many offspring or seeds. Soil is little more than mineral particles, nutrient poor and with poor water-holding capacity.
Seral stage 1 Colonization	First species to colonize an area are pioneers, adapted to extreme conditions. Simple soil starts from windblown dust and mineral particles.
Seral stage 2 Establishment	Species diversity increases. Invertebrate species begin to visit and live in the soil, increasing humus (organic material) content and water-holding capacity. Weathering enriches soil with nutrients.
Seral stage 3 Competition	Microclimate continues to change as new species colonize. Larger plants increase cover and provide shelter, enabling *K*-selected species to become established. Temperatures, sun and wind are less extreme. Earlier pioneer *r*-species are unable to compete with *K*-species for space, nutrients or light and are lost from the community.
Seral stage 4 Stabilization	Fewer new species colonize as late colonizers become established, shading out early colonizers. Complex food webs develop. *K*-selected species are specialists with narrower niches. They are generally larger and less productive (slower growing) with longer life cycles and delayed reproduction.
Seral climax	The final seral stage or climax community is stable and self-perpetuating. It exists in a state of dynamic equilibrium (see below). The seral climax represents the maximum possible development that a community can reach under the prevailing environmental conditions of temperature, light and rainfall and is called a **climatic climax community**.

Table 14.1 The stages of a sere

Primary succession involves the colonization of newly created land by organisms. It occurs as new land is either created or uncovered such as at river deltas, volcanic lava fields, sand dunes or glacial deposits.

Simple mineral soils evolved from erosion are slowly invaded by mosses and lichen. These and early plants are adapted to survive periods of drought as water drains quickly away from the mineral soils. These contribute organic matter to the soil as they spread and die, creating conditions that allow larger mosses to invade. These help add more organic matter to the soil, which improves its water-holding capacity, and provide a habitat for soil organisms that help speed up the breakdown of organic matter and release of nutrients.

Conditions then become favourable for grasses and other higher plants to establish on the primitive soils, and humus forms as more organic matter is added. Eventually shrubs and trees invade, first from wind-dispersed seeds and then by animal dispersal. Eventually over time a stable woodland community develops.

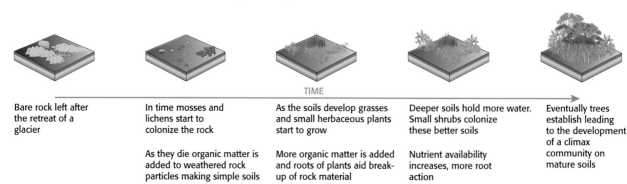

| Bare rock left after the retreat of a glacier | In time mosses and lichens start to colonize the rock | As the soils develop grasses and small herbaceous plants start to grow | Deeper soils hold more water. Small shrubs colonize these better soils | Eventually trees establish leading to the development of a climax community on mature soils |
| | As they die organic matter is added to weathered rock particles making simple soils | More organic matter is added and roots of plants aid break-up of rock material | Nutrient availability increases, more root action | |

Fig. 14.1 Primary succession after glacial retreat

Succession progresses in stages from **pioneer species**, that are adapted to live in limiting environments, to a stable developed community. This final community is termed a **climax community**.

An example of primary succession can be seen in the development of the natural broadleaf forest that covered much of Northern Europe following the end of the last Ice Age. We know that following the retreat of ice around 10 000 years ago until around 7500 years ago, a boreal community formed: first juniper, then birch and later pine trees. As the climate warmed so the community changed from a dominance of birch to oak with abundant wych elm, alder and lime, marking a change to the warm, moist Atlantic period until about 5000 years ago. Much of Northern Europe would still be covered in this mixed broadleaf forest if Neolithic humans had not started changing the plant community around them as agriculture developed.

Ecological period	Years ago	Community
Pre-Boreal	10 000 – 9500	Tundra with patches of willow, birch and pine
Boreal	9500 – 7500	Hazel, pine
Atlantic	7500 – 5000	Hazel, oak, elm, lime , ash, alder
Sub-Boreal	5000 – 2500	Mixed oak with many cleared areas either being farmed or abandoned and returned to woodland

Table 14.2 Community types in Northern Europe in the last 10 000 years

If primary succession starts on dry land it is called a **xerosere**. A succession in water is a **hydrosere** (Fig. 14.2).

Deep freshwater, no rooted plants because of lack of light in deep water

Community only micro-organisms and phytoplankton

Sediments get carried into the pond allowing rooted submerged and floating to start to grow

Sediments continue to build up

Reeds and grasses develop around pond margin, trapping more sediment

A marsh community builds up around the pond margins

Reeds take over more of the pond as more silt buildup

As the soils around the edge dry from waterlogged to damp, tree species such as willow and alder become established

Fig. 14.2 A hydrosere succession

Ponds and lakes get continuous inputs of sediment from streams and rivers that open into them. Some of this sediment passes through but a lot sinks to the pond bottom. As plant communities develop they add dead organic material to these sediments. Over time these sediments build up, allowing rooted plants to invade the pond margins as the pond slowly fills in. This eventually leads to the establishment of climax communities around the pond margins and in smaller ponds the eventual disappearance of the pond.

In regions where rainfall is high, the xerosere climax community may not establish after a hydrosere. The wet conditions create the development of raised bogs as the climax following hydrosere succession.

Secondary succession

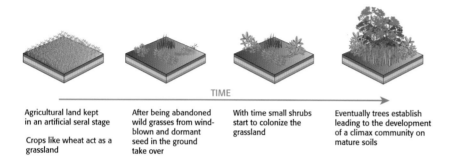

TIME

Agricultural land kept in an artificial seral stage

Crops like wheat act as a grassland

After being abandoned wild grasses from wind-blown and dormant seed in the ground take over

With time small shrubs start to colonize the grassland

Eventually trees establish leading to the development of a climax community on mature soils

Fig. 14.3 Stages of secondary succession in abandoned agricultural land

Where an already established community is suddenly destroyed, such as following fire or flood or even human activity (ploughing), an abridged version of succession occurs.

This **secondary succession** occurs on soils that are already developed and ready to accept seeds carried in by the wind. Also there are often dormant seeds left in the soil from the previous community. This shortens the number of seral stages the community goes through.

Changes occurring during a succession

During a succession the following changes occur:

- The size of organisms increases, with trees creating a more hospitable environment.
- Energy flow becomes more complex as simple food chains become complex food webs.
- Soil depth, humus, water-holding capacity, mineral content and cycling all increase.
- Biodiversity increases because more niches (lifestyle opportunities) appear, and then falls as the climax community is reached.
- NPP and GPP rise and then fall.
- Production : respiration ratio falls.

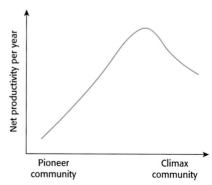

Fig. 14.4 Productivity changes in a succession

In early stages, gross primary productivity is low due to the initial conditions and low density of producers. The proportion of energy lost through community respiration is relatively low too, so net productivity is high, i.e. the system is growing and biomass is accumulating.

In later stages, with an increased producer, consumer and decomposer community, gross productivity continues to rise to a maximum in the climax community. However, this is balanced by equally high rates of respiration particularly by decomposers, so net productivity approaches zero and the production:respiration (P:R) ratio approaches 1.

Studies have shown that standing crop (biomass) in succession to deciduous woodland reaches a peak within the first few centuries. Following the establishment of mature climax forest, rate of increase of biomass tends to fall as trees age, growth slows and an extended canopy crowds out ground cover. Also older trees become less photosynthetically efficient and more NPP is allocated to non-photosynthetic structural biomass such as root systems.

Early stage	Middle stage	Late stage
Low GPP but high percentage NPP	GPP high	Trees reach their maximum size
Little increase in biomass	Increased photosynthesis	Ratio of NPP to R is roughly 1
	Increases in biomass as plant forms become bigger	

Table 14.3 Biomass accumulation and successional stage

Species diversity in successions

Early stages of succession tend to be marked by few species within the community. As the community passes through subsequent seral stages so the number of species found increases. Very few pioneer species are ever totally replaced as succession continues. The result is increasing diversity: more species. This increase tends to continue until a balance is reached between possibilities for new species to establish, existing species to expand their range and local extinction. Evidence following the eruption of the Mount St Helens volcano in 1980 has provided ecologists with a natural laboratory to study succession. In the first 10 years after the eruption, species diversity increased dramatically but after 20 years very little additional increase in diversity occurred.

Disturbance

Early ideas about succession suggested that the climax community of any area was almost self-perpetuating. This is unrealistic as communities are affected by periods of disturbance to a greater or lesser extent. Even in large forests trees eventually age, die and fall over, leaving a gap. Other communities are affected by flood, fire, landslides, earthquakes, hurricanes and other natural hazards. All of these have an effect of making gaps available that can be colonized by pioneer species within the surrounding community. This adds to both the productivity and diversity of the community.

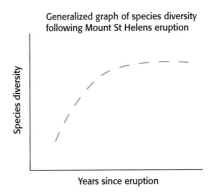

Fig. 14.5 Species diversity changes in a succession

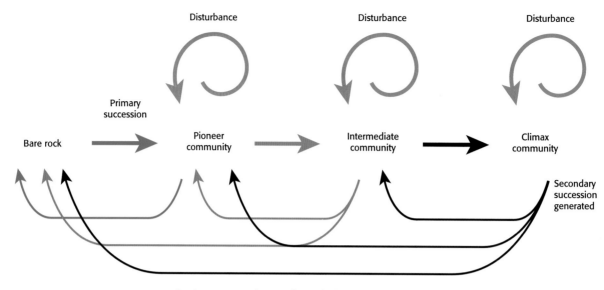

Disturbance can send any seral stage back to an earlier seral stage or create gaps in a later community that then regenerate increasing both productivity and diversity of the whole community

Fig. 14.6 Effects of disturbance in a succession

Case study 1

Succession on sand dunes

On the southern coast of England in Dorset is Studland Bay where, since the 16th century, the dunes have developed parallel with the shore. So the newest dune is nearest to the bare sand on the beach. The conditions are windy and there is little fresh water for plants to take up.

Fig. 14.7 Sand dunes at Studland Bay, Dorset, UK

This begins with a bare surface of sand. Vegetation colonizes the sand. The pioneer plants tend to be low growing: why? They have fat fleshy leaves with a waxy coating and are able to survive being submersed, temporarily.

Later, the predominant plant species is marram grass on the seaward side due to its ability to cope with the environmental conditions. It, like the pioneer grasses, has leaves which are able to fold to reduce their surface area. Leaves are waxy to reduce transpiration and can be aligned to the wind direction. Marram grass incorporates silica into its cell structure to give the leaves extra strength and flexibility.

As a result of the humus from the previous stages, a sandy soil has now developed. This is now able to support "pasture" grasses and bushes. Species such as hawthorn, elder, brambles and sea buckthorn (which has nitrogen-fixing root nodules so can thrive in nutrient-poor soil) are present. As the scrub develops, shorter species will be shaded out.

The oldest dunes will have forest, first pine and finally oak and ash woodland growing on them; the **climatic climax vegetation** (CCV) for the area. Here the species diversity declines due to competition.

In every case, vegetation colonizes in a series of seral stages.

The final sere is in dynamic equilibrium with its climatic environment and hence is known as climatic climax vegetation. In the UK this is temperate deciduous forest. The type of CCV is dictated by local conditions.

From east to west, there is increase in vegetation cover, soil depth and humus content, soil acidity, moisture content and sand stability.

Kite diagrams are often used to illustrate changes in species in a succession or in zonation (change over distance of an environmental factor). In a kite diagram, the width of the "kites" represents percentage cover or abundance of the species so, at a glance, you can see what grows where and how much of it there is. Figure 14.8 is a kite

diagram of species on a different sand dune in England. (A slack is a region in sand dunes that is below the water table.)

Questions

1. What is the name of a succession that develops on sand dunes? (*Hint*: see Key words on page 275.)

2. What is the seral climax community that develops on the sand dunes?

3. Describe the formation of the seral climax from the marram grass stage outlined above.

4. In the climax vegetation of the oldest dunes, for what would the plants be in competition?

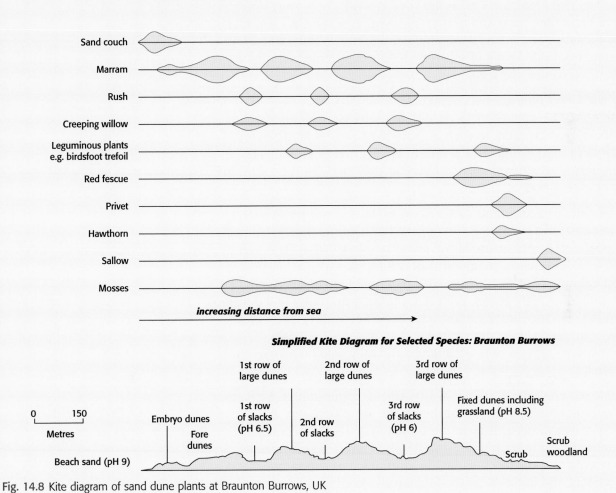

Fig. 14.8 Kite diagram of sand dune plants at Braunton Burrows, UK

Succession and zonation

Some communities such as a rocky seashore show a zonation of organisms caused by an abiotic gradient, which in this case is the length of time the sea covers different parts of the shore. Zonation is spatial and static but succession is dynamic and the process takes place over long periods of time. Sand dune colonization is unusual in that the succession is dynamic but one also observes vegetation zones of the various seral stages of the process.

Succession should not be confused with zonation. Succession is how an ecosystem changes in time. Zonation is how an ecosystem is changing along an environmental gradient like water content in the soil, altitude or salt percentage. Easy-to-recognize examples of zonation are vegetation changes on mountain slopes and on coasts. On many mountains clear vegetation zones can be seen. In the Swedish fjäll area, lower areas are covered with coniferous forest. At higher altitudes, grass and herbs replace the trees and at even higher altitudes rock and lichen dominate. On sea shores, often a clear change from salt-loving or salt-resistant species to more common inland species can be seen.

Arrested and deflected successions

A sere may be stopped or "arrested" at a seral stage by an abiotic factor, e.g. soil conditions such as waterlogging, or a biotic factor such as heavy grazing. This results in an arrested or **subclimax** community which will only continue its development if the limiting factor is removed.

Under other circumstances a climax community may be affected by either a natural event, e.g. fire or landslide, or human activity such as agriculture, regular use of fire or habitat destruction. This will lead to a deflected or **plagioclimax** community such as pasture, arable farmland or plantations with reduced biodiversity. Again, if the human activity ceases, the plagioclimax community will develop into the climatic climax community. These **secondary** successions reach the seral climax more quickly than priseres (**primary** successions) because soils are mature and may contain a large bank of seeds.

Stable old ecosystems are not always climax communities. Ecologists have found out that the established lodgepole pine forest in Yellowstone National Park in the United States is *not* a climax community. In this national park, frequent forest fires interrupt succession and the forest should be considered as a *disturbed* ecosystem and not as a climax community. Such ecosystems are often referred to as equilibrium communities or subclimax communities.

An example of primary succession: Krakatau

On 27 August 1883, an explosion was heard from Australia to the islands of the coast of Africa. The volcanic island of Krakatau, Indonesia, had exploded. Most of the island disappeared and islands in the vicinity were stripped of their wildlife. When scientists arrived at the islands a few months later they discovered that the islands were virtually sterilized: not even one species of plant or animal was found.

Year	Number of species		
	1897	1924	1989
Sea-dispersed flowering plants	23	53	59
Animal-dispersed flowering plants	2	48	110
Wind-dispersed flowering plants	14	48	75
Ferns (all wind dispersed)	13	51	81

Table 14.4 Number of plant species found on Krakatau

This is an example of primary succession, a prisere, in the absence of soil, on bare rock. Because our planet is several billion years old, primary succession is quite rare. Secondary succession is much more common. Examples of secondary succession include regrowth of forest after clear cutting, and forest taking over unused farmland.

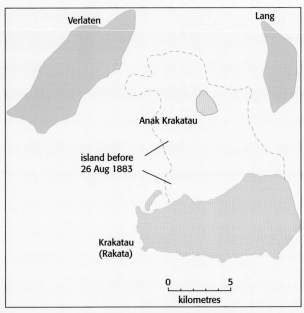

From Simkin and Fiske, 1983

Fig. 14.9 Map showing the Krakatau Islands before and after the 1883 explosion

Between 1883 and 1935 scientists paid annual visits to Krakatau in order to study how nature would recover. In May 1884 the only life form found was a small spider. By September 1884 a few blades of grass were growing. After about 40 years the forest canopy closed and by around 1930 hardly any open terrain remained.

More recent volcanic eruptions have been disturbing the succession on the Krakatau islands. In the 1930s new volcanic activity resulted in the formation of the small island volcano Anak Krakatau ("child of Krakatau"). Also on this island succession occurred; however, frequent eruptions have cleared this island of vegetation several times. Anak Krakatau's eruptions have also disturbed the succession on the islands Sertung and Panjang (these islands' old Dutch colonial names were Verlaten and Lang respectively; see Fig. 14.9). Because of the wind, the southernmost island, Rakata, has not been affected by these volcanic eruptions and therefore the recovery from the 1883 explosion can be studied there.

On the Krakatau Islands, the forest has returned and looks dense and mature. However, in 1983, 100 years after the eruption, only about 300 species of plants had returned to the islands. Similar areas on the nearby mainland contain between 1200 and 1500 plant species. It is estimated that it will take more than 1000 years before the forest of the Krakatau Islands reaches a similar biodiversity.

Significance of changes during succession

During succession energy flow, gross and net productivity, diversity and mineral cycling change. In early stages gross productivity is low due to the initial conditions and low density of producers. The proportion of energy lost through community respiration is relatively low too, so net productivity is high, i.e. the system is growing and biomass is accumulating. In later stages, with an increased consumer community, gross productivity may be high in a climax community. However, this is balanced by respiration, so net productivity approaches zero and the production : respiration (P : R) ratio approaches 1. **Biodiversity** increases during succession as more species arrive and then decreases slightly if a stable climax community is reached. **Mineral cycling** tends to be slow at the early stages of succession but increases quickly during the succession process. (See Fig. 14.10.)

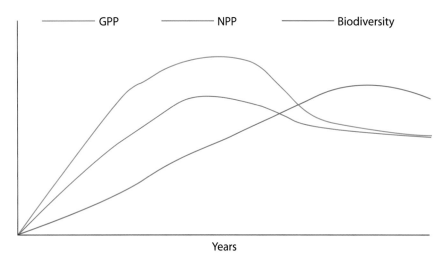

Fig. 14.10 Changes during succession

There is a basic conflict between natural succession and human requirements in agriculture. We want to achieve high rates of productivity with no standing crop left. The natural system leads to increasing complexity, longer food chains, higher biodiversity, more biomass and a well-organized stratified ecosystem. Food production aims for a simple system where weed plants are controlled and there is maximum yield of a monoculture that does not reach a climax community. But we have not placed a value on other services – natural income that natural systems provide – a balance in the carbon cycle, nutrient cycling, climate buffer of forests and oceans, clean water provision, esthetic services that natural systems provide. Less productive areas are as valuable as productive fields.

To research

In 1980, Mt St Helens volcano erupted in Washington State, USA. Research what has happened to the vegetation in the area since then. Start here http://vulcan.wr.usgs.gov/Volcanoes/MSH/Recovery/framework.html

Sketch curves for gross productivity, net productivity and respiration as a function of time in one graph for Mt St Helens since 1980. Indicate the different seral stages in the graph.

Fig. 14.11 Mount St Helens erupting in 1980

To do

1 Define succession and define zonation.
2 Describe the main differences between succession and zonation.
3 Describe the difference between primary and secondary succession.
4 For a named habitat, describe the process of succession, giving named examples of species in the different seral stages.

5 What is a plagioclimax?
6 Why do farmers want to maintain their crops at a plagioclimax?
7 Evaluate the costs and benefits of replacing natural ecosystems with crops or livestock grazing.

Key words

succession	climax community
primary succession	mineral cycling
prisere	hydrosere
secondary succession	plagiosere
seral stages	plagioclimax
seres	psammosere
pioneer community	production : respiration
subclimax community	zonation

15 Pollution management

Key points

- Pollution occurs when human activity adds a substance to the environment that affects organisms and at a rate greater than that at which it can be rendered harmless.

- Pollution from point sources is often easier to manage than that from non-point sources.

- Types of pollutant include gases from burning fossil fuels and industry, solid and liquid waste from industry, agriculture and homes.

- Pollution can be measured directly and indirectly in the atmosphere, water and soil.

- Examples of pollution are eutrophication; solid domestic waste; depletion of stratospheric ozone; urban air pollution; and acid deposition.

- Pollution management of these issues can be considered in the "replace (with alternatives), regulate (the release) and restore (the environment)" model. But factors such as culture, politics and economics influence strategies used to manage pollution.

"Pollution is nothing but the resources we are not harvesting. We allow them to disperse because we've been ignorant of their value."

Richard Buckminster Fuller
(US engineer and architect, 1895–1983)

"Unfortunately, our affluent society has also been an effluent society."

Hubert H. Humphrey

The nature of pollution

Pollution is the addition to the biosphere of a substance or an agent (such as heat) by human activity, at a rate greater than that at which it can be rendered harmless by the environment, and which has an appreciable effect on the health of organisms within it. It can be matter (gases, liquids or solids) or energy (noise, light, heat). Since humans have been on Earth, we have polluted to a greater or lesser extent. Sewage and waste are products of human communities and burning wood and then coal has caused air pollution for 1000 years. Pollution may be an inevitable side-effect of the economic development that has given most humans a far higher standard of living than we would otherwise have had. Since the industrial revolution pollution has increased but how we deal with it has also improved as we monitor industries and legislate against excessive pollution.

Pollutants and pollution

It is sometimes said that a pollutant is a substance in the wrong place, in the wrong amount or at the wrong time. Could this be true of carbon dioxide, ozone or nitrate?

Major sources of pollutants

Table 15.1 lists some major **sources** of pollutants. We shall be considering some of these in this chapter.

Fig. 15.1 Pre-1950 poster from the USSR encouraging production, saying that the smoke from chimneys is the breath of Soviet Russia

Major source	Pollutant	Effect
Combustion of fossil fuels	Carbon dioxide	Greenhouse gas – climate change
	Sulfur dioxide	Acid deposition – tree and fish death, respiratory disease in humans
	Nitrogen oxides	Respiratory infections, eye irritation, smog
	Photochemical smog including tropospheric ozone, PANs (peroxyacyl nitrates), VOCs (volatile organic compounds) Secondary pollutants (formed from others in the atmosphere)	Damage plants, eye irritation, respiratory problems in humans
	Carbon monoxide	Binds with haemoglobin in red blood cells instead of oxygen – can lead to death by suffocation
Domestic waste	Organic waste (food and sewage)	Eutrophication, water-borne diseases
	Waste paper	Volume fills up landfill sites, forests cut to produce it
	Plastics – containers, packaging	Volume fills up landfill sites, derived from oil
	Glass	Energy required to manufacture it (as with all products), can be recycled but often goes into landfill sites
	Tins	Can be recycled but also goes into landfill sites
Industrial waste	Heavy metals	Poisoning, e.g. mercury, lead, cadmium
	Fluorides	Poisoning
	Heat	Reduces solubility of gases in water, so less dissolved oxygen so aquatic organisms may die
	Lead	Disabilities in children
	Acids	Corrosive
Agricultural waste	Nitrates	Eutrophication
	Organic waste	Eutrophication, spread disease
	Pesticides	Accumulate up food chains

Table 15.1 Major sources of pollutants and their effects

Point source and non-point source pollutants

Non-point source (NPS) **pollution** is the release of pollutants from numerous, widely dispersed origins, e.g. gases from the exhaust systems of vehicles, chemicals spread on fields.

Point source pollution is the release of pollutants from a single, clearly identifiable site, e.g. a factory chimney or the waste disposal pipe of a sewage works into a river.

Clearly, point source pollution is easier to manage as it can be found more easily. It is also easier to see who is polluting – a factory, car or house. NPS pollution may have many sources and it may be virtually impossible to detect exactly where it is coming from. Rainwater can collect nitrates and phosphates which are spread as fertilizer as it infiltrates the ground or as runoff on the surface. It may travel many kilometres before draining into a lake or river and increasing the concentration of nitrates and phosphates so much that eutrophication occurs. It would not be possible to say which farmer spread the excess fertilizer. Air pollution can be blown hundreds of kilometres and chemicals released from open chimneys mix with those from others. So one solution is to set limits for all farmers and all industries to reduce emissions and then monitor what they actually do.

To think about

The Prisoner's Dilemma

A big question about us is whether we are, by nature, loving or aggressive, noble or selfish, nice or nasty. Do we refrain from stealing or cheating because we may be found out or because we know it is wrong? Is it our default position to be kind and helpful to each other, or to come out on top even if, or particularly if, it hurts someone else?

Scientists, sociologists, philosophers, politicians and all thinking people want to know about our innate nature and why we react as we do.

There is a type of game that you can play as an example of game theory and it is called the prisoner's dilemma. Here is a version of it.

Two people A and B are suspected of a crime and arrested. There is not enough evidence to convict them unless they confess. The police separate them and offer each one the same deal. If one admits that they both did the crime and betrays the other, that one goes free and the other goes to prison for 10 years. If both stay silent, they both go to prison for a year. If both confess, they both go to prison for 5 years. What should they do? The best scenario for one is to confess and the other stays silent. But they don't know what the other will do. What has this to do with pollution? Quite a lot.

The best economic scenario for a polluter is to keep polluting as long as he or she is not found out. Not to confess. The cost of the pollution is then shared between everyone and the polluter does not have to spend money reducing their own personal or business pollution. If the polluter confesses, they may be punished by a fine, imprisonment or having to spend money in reducing the pollution.

But, just as in the prisoner's dilemma, while keeping silent and polluting is fine in the short term, in the longer term, the best scenario is "tit for tat": if I cooperate with you – stop polluting, you will cooperate with me – stop polluting too, and the world will be a cleaner place – we both gain. If we keep betraying each other, we will both be losers at the end. And that is where we are with pollution. If we pollute with NPS pollutants, we are unlikely to be found out and everyone pays for the cleanup. An individual, company or country can gain from non-compliance in the short term if the others comply.

But what will happen in the long term?

Think of two particular types of pollution (one in the atmosphere and one in water) that could be examples of NPS pollution.

What do these examples mean for international agreements on pollution?

Detection and monitoring of pollution

Pollution can be measured directly or indirectly. Direct measurements record the amount of a pollutant in water, the air or soil. Indirect measurements record changes in an abiotic or biotic factor which are the result of the pollutants. Direct measurements of air pollution include measuring the acidity of rainwater; amount of a gas, e.g. carbon dioxide, carbon monoxide, nitrogen oxides in the atmosphere; amount of particles emitted by a diesel engine or amount of lead in the atmosphere. Direct measurements of water or soil pollution include testing for nitrates and phosphates, amount of organic matter or bacteria and heavy metal concentrations. Indirect measurements of pollution include measuring abiotic factors that change as a result of the pollutant (e.g. oxygen content of water) and recording the presence or absence of **indicator species** – species that are only found if the conditions are either polluted (e.g. rat-tailed maggot in water) or unpolluted (e.g. leafy lichens on trees).

Biochemical oxygen demand

The **biochemical** (or biological) **oxygen demand** or **BOD** is a measure of the amount of dissolved oxygen required to break down the organic material in a given volume of water through aerobic biological activity (by micro-organisms).

Use of BOD to estimate pollution levels in water

Measurement of the rate of oxygen uptake is used as a standard *indirect* test to detect the polluting capacity of effluent (sewage, farm wastes). The greater the quantity of organic pollutant present, the more oxygen-consuming microbes are required to break it down, and therefore the greater the BOD. The test measures the mass (in milligrams) of dissolved oxygen consumed per litre of water when a water sample is incubated in a dark chamber at 20°C for five days.

Table 15.2 shows examples of BOD values.

Source of pollutant	BOD (mg dissolved oxygen l⁻¹)
Unpolluted river	<5
Treated sewage	20–60
Raw domestic sewage	350
Cattle slurry	10 000
Paper pulp mill	25 000

Table 15.2

Test yourself

Four factories discharge effluent containing organic matter into rivers. Table 15.3 shows the volume of discharge into the river and the resulting biological oxygen demand of the river water.

Factory	Volume of effluent/ 1000 l day⁻¹	BOD / mg l⁻¹
A	14.0	27
B	1.0	53
C	3.0	124
D	0.8	33

Table 15.3

1 Explain whether these pollution data are for point source or non-point sources.
2 Which pollution source, point source or non-point source, is easier to regulate? Explain your choice.
3 Which factory is adding most to the BOD of the river into which it discharges?
4 If factory C discharges water at a temperature of 50°C, give three possible effects on the organisms in the river.

Biotic indices and indicator species

Indicator species are plants and animals that show something about the environment by their presence, absence, abundance or scarcity. They are the early warning signs that something may have changed in an ecosystem and are the most sensitive to this change. For example, canaries were once taken down coal mines in Britain because they are more sensitive than humans to poisonous gases (carbon monoxide, methane). If gases were present in the mine the bird would die and so warn the miners in time for them to escape.

A **biotic index** is a scale (1–10) that gives a measure of the quality of an ecosystem by the presence and abundance of the species living in it.

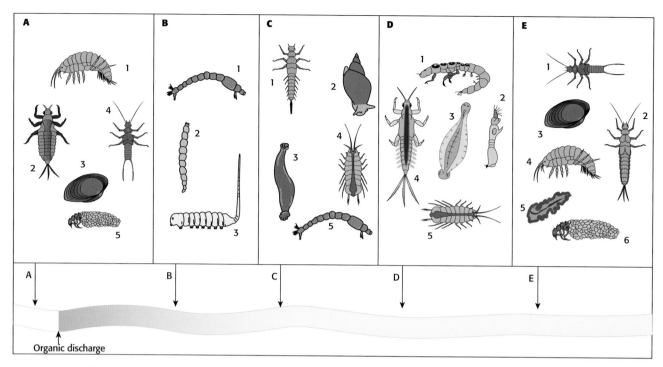

Fig. 15.2 Invertebrates found in fresh water

Using biotic indices and indicator species is another indirect method of measuring pollution. The pollutants are not measured directly but their effect on biodiversity is measured.

Figure 15.2 shows some typical invertebrates that might be found above and at various points below a point source sewage outfall pipe.

Invertebrates are used to estimate levels of pollution, as they are sensitive to decreases in oxygen concentration in water, caused by the action of aerobic bacteria as they decompose organic matter. The presence of certain indicator species that can tolerate various levels of oxygen is used to calculate a biotic index, a semiquantitative estimate of pollution levels.

Biotic indices based on indicator species are usually used at the same time as BOD. BOD gives a measure of pollution at the *instant* a water sample is collected whereas indicator species give a summary of *recent history*.

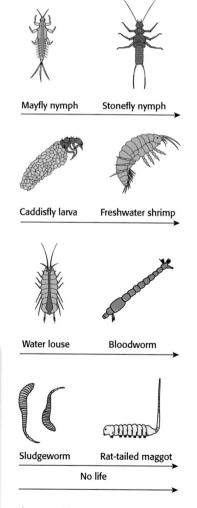

Fig. 15.2A Key

Case study 1

DDT and malarial mosquitoes
The manufacture and use of DDT was banned in the US in 1972, on the advice of the US Environmental Protection Agency. The use of DDT has since been banned in most other MEDCs, but it is not banned for public health use in most areas of the world where malaria is endemic. DDT was recently exempted from a proposed worldwide ban on organophosphate chemicals. DDT for malarial control involves spraying the walls and backs of furniture, so as to kill and repel adult *Anopheles*

mosquitoes that carry the malarial parasite. Although other chemicals could be used, DDT is cheap and persistent and good at the job. DDT is not used outside buildings because of its persistence and toxicity. Also, its persistence means that mosquitoes become resistant to DDT (the ones that survive go on to breed and develop a population of DDT-resistant mosquitoes).

The question is whether banning DDT did more harm than good.

It is believed that malaria kills 2.7 million people a year, mostly children under the age of 5, and infects 300–500 million a year. It is also thought that DDT prevented millions of deaths due to malaria. So why the ban? In her book, *Silent Spring*, Rachel Carson discusses the effect of DDT on birds of prey in thinning their eggshells and reducing their population numbers. But some say the evidence for bird eggshell thinning was slight; DDT is an effective insecticide against the malarial mosquito and should not have been banned.

Malaria incidence is increasing, partly due to resistance, partly to changes in land use and migration of people to areas where malaria is endemic. In treating the cause, DDT use is just one tool with other chemicals, mosquito nets and removal of stagnant water where mosquitoes breed.

There is much controversy in the debate on DDT but malaria probably does not receive enough funding for research as it is mostly a disease of the poor.

Questions

1   Do your own research on DDT. What evidence can you find for both sides of the argument? Be careful in looking at sources. Are they biased? Can they substantiate their claims?

2   Do you now think that the DDT ban was the right thing to do at the time?

3   Justify the arguments for continued use of DDT to combat malaria in humans.

Three-level model of pollution management

Figure 15.3 shows a model for reducing the impact of pollutants, taken from the *Environmental Systems and Societies Guide*.

We shall call it the "replace (with alternatives), regulate (the release) and restore (the environment)" model. Pollution can cause long-term effects in the environment and it is usually better not to cause it in the first place than to treat the symptoms or clean it up. Often a comprehensive approach is used with strategies from all three boxes. We shall return to this model in the pollution examples that follow.

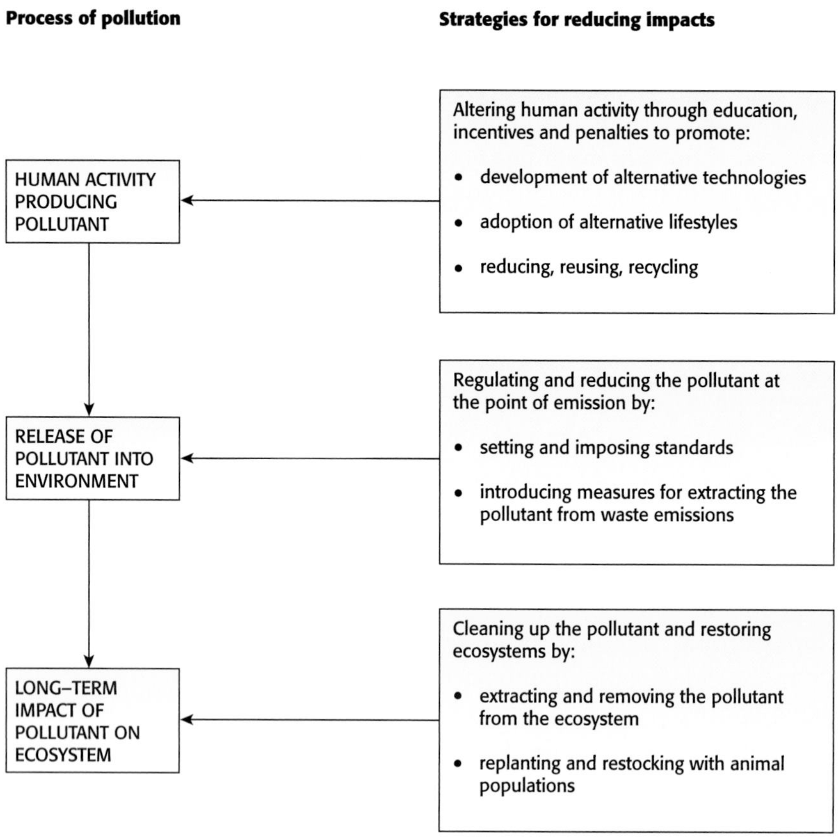

Process of pollution   Strategies for reducing impacts

```
┌──────────────────┐        ┌─────────────────────────────────────────┐
│ HUMAN ACTIVITY   │ ◄──────│ Altering human activity through education,│
│ PRODUCING        │        │ incentives and penalties to promote:      │
│ POLLUTANT        │        │                                           │
└────────┬─────────┘        │ • development of alternative technologies │
         │                  │                                           │
         │                  │ • adoption of alternative lifestyles      │
         │                  │                                           │
         │                  │ • reducing, reusing, recycling            │
         │                  └───────────────────────────────────────────┘
         ▼
┌──────────────────┐        ┌─────────────────────────────────────────┐
│ RELEASE OF       │ ◄──────│ Regulating and reducing the pollutant at  │
│ POLLUTANT INTO   │        │ the point of emission by:                 │
│ ENVIRONMENT      │        │                                           │
└────────┬─────────┘        │ • setting and imposing standards          │
         │                  │                                           │
         │                  │ • introducing measures for extracting the │
         │                  │   pollutant from waste emissions          │
         │                  └───────────────────────────────────────────┘
         ▼
┌──────────────────┐        ┌─────────────────────────────────────────┐
│ LONG–TERM        │ ◄──────│ Cleaning up the pollutant and restoring   │
│ IMPACT OF        │        │ ecosystems by:                            │
│ POLLUTANT ON     │        │                                           │
│ ECOSYSTEM        │        │ • extracting and removing the pollutant   │
└──────────────────┘        │   from the ecosystem                      │
                            │                                           │
                            │ • replanting and restocking with animal   │
                            │   populations                             │
                            └───────────────────────────────────────────┘
```

Fig. 15.3 "Replace, regulate, restore" model of pollution management

Pollution management strategies

It is not straightforward to select the best management strategy for a pollution problem. While not polluting may seem the obvious choice, in being alive we all cause the emission of carbon dioxide and greenhouse gases and produce waste. It would be impossible not to. Economies depend on production of goods and these need raw materials. Politicians have to make difficult choices which sometimes come down to the choice between jobs or the environment. Culturally, we may not be willing to change what we do or we may not know how to change.

Case study 2

Read the four mini case studies below. Consider the three-level model and discuss at what level pollution management strategies are implemented, if at all, in these cases. What factors are influencing the choice of strategy?

Copper

Chuquicamata is a state-owned copper mine in Chile. It is one of the largest open pit copper mines in the world and produces about 5% of the world's copper. Mining copper is a dirty job and pollution is so bad that, under new laws on pollution, the town of Chuquicamata is too polluted for people to live there. The solution is to move all the inhabitants of the town to Calama, 17 km away. The deserted town will be covered by copper waste. The mining company has cleaned up to some extent but says it would cost too much to do more. The more economic option is to move the inhabitants. As the mine brings in hundreds of millions of US$ to Chile, the economic benefit of continuing it is huge.

Arsenic

In **Bangladesh,** in the 1970s, many people were suffering from the effects of drinking unclean water polluted with sewage from surface ponds. The solution was for the World Bank and UNICEF to fund a project to sink 900 000 tube wells to tap cleaner water. This was done. But there was no standard test for arsenic in water and the water from the wells contained arsenic from natural sources, probably the Himalayas. Many people reported symptoms of arsenic poisoning (blistered hands and feet, internal organ damage, cancerous ulcers) but it took until 1998 for the authorities to accept that the water was to blame. Up to 77 million people are still drinking this water. It may take 10 years for symptoms to show but some wells have higher arsenic levels than others. Solutions suggested are to sink deeper wells (to get clean water), paint wells red to show high arsenic levels, use collection devices for rainwater but, as there are so many wells, most have not yet even been tested. This is pollution from a natural source but altruistic organizations in thinking they were doing the best thing have created a huge problem for which there is no simple solution.

Nuclear waste

Sellafield nuclear power complex is on the coast of Cumbria in England. It is on the edge of a national park, the Lake District. Since 1947, nuclear power has been generated there and it employs 10 000 people in an area with few jobs. On the site are a number of old reactors: two plutonium-producing reactors, the world's first four commercial nuclear reactors, a defunct advanced gas-cooled reactor and two nuclear fuel reprocessing plants, reprocessing Magnox fuel. Britain's main plutonium stockpile and nuclear waste stores are also at the Sellafield site. In 60 years of activity, there have been 20 serious incidents involving radioactive materials. In some, radioactive waste was released onto beaches or into the sea. There are more cases of childhood leukemia and non-Hodgkin's lymphoma around Sellafield than in comparable populations elsewhere. Electricity is no longer generated at Sellafield but reprocessing of nuclear waste continues.

Pesticides

Cotton is the world's most important non-food agricultural commodity, yet this industry is responsible for the release of US$2 billion worth of chemical pesticides each year, of which at least US$819 million worth are considered toxic enough to be classified as hazardous by the WHO. Cotton accounts for 16% of global insecticide releases: almost 1 kg of pesticide is applied for every hectare under cotton.

Between 1% and 3% of agricultural workers worldwide suffer from acute pesticide poisoning. A single drop of the pesticide aldicarb, absorbed through the skin, can kill an adult. Aldicarb is commonly used in cotton production and in 2003 almost 1 million kg were applied to cotton grown in the USA. It is also applied to cotton in 25 other countries worldwide.

In Uzbekistan, the world's second largest cotton exporter, toxic agrochemicals first applied to cotton 50 years ago now pollute the country's land, air, food and drinking water. Despite the substantial damage that these chemicals cause to human health and the environment, the government still sanctions the use of cotton pesticides so toxic that they were banned under the Soviets. In India and Uzbekistan children are directly involved in applying pesticide, while in Pakistan, Egypt and Central Asia children work in fields either during or following the spraying season.

Hazardous cotton pesticides are now known to contaminate rivers in India, Pakistan, Uzbekistan, Brazil, USA, Australia, Greece and West Africa.

Since the 1980s the global consumption of cotton has risen dramatically, almost doubling in the last 30 years. With demand now in excess of 25 million tonnes annually, the world's consumers buy more cotton today than ever before. While the bulk of global cotton production occurs in LEDCs, the majority of cotton products are sold to consumers in MEDCs, with North America alone responsible for 25% of global household cotton product consumption, and Europe accounting for a further 20%.

Organic cotton production offers a strong alternative to current production methods.

Domestic waste

Solid domestic waste (SDW) or municipal solid waste (MSW) is our trash, garbage, rubbish. It is a mixture of paper, packaging, organic materials (waste food), glass, dust, metals, plastic, textiles, paint, old batteries. It is collected from homes and shops and, although it only makes up about 5% of total waste, which includes agricultural and industrial waste, it is waste that we can control.

Waste is material which has no value to its producer. If it is not recycled it becomes a problem to be disposed of. We create waste in most of the processes we carry out – energy production, transport, industrial processes, construction, selling of goods and services, domestic activities. Just think about (or collect) the waste that you produce in 24 hours.

We produce nearly 3 kg per capita of waste per day in the USA, according to the US Environmental Protection Agency (EPA), and 500 kg per capita per year in the EU, according to the European Environment Agency. This has risen from 300 kg per capita per year in 1985. Consider where this waste goes, how much could be recycled and whether it is recycled.

Strategies to minimize waste

These can be summarized in three words: **reduce**, **reuse**, **recycle**.

The best action we can take is producing less waste in the first place – reducing. We can reuse it instead of it becoming waste and we can recycle it. Sorting waste into separate containers for recycling before it leaves the home helps. In Germany, for example, each household has four bins for this. In the UK, there is discussion about charging households more if they produce more than the standard amount of waste. In India and China, very little waste is food waste as this is either not thrown away or is fed to animals. In MEDCs, up to 50% of waste is food waste.

Recycling

Recycling involves collecting and separating waste materials and processing them for reuse. (If materials are separated from the waste stream and washed and reused without processing in some way, this is reuse.) The economics of recycling determine whether it is commercial or not and this can vary with the market cost of the raw materials or cost of recycling. Some materials have a high cost of production from the raw material and so recycling of these is particularly worthwhile commercially. Aluminium cans are probably the best example of this.

Aluminium is found in bauxite and extraction from the ore is very energy intensive. It is far, far cheaper to recycle aluminium than to mine the bauxite. It uses only 5% of the energy to recycle aluminium as it takes to extract it from bauxite. Aluminium can be recycled indefinitely with no loss of quality and it is the most cost-effective material to recycle. In the USA alone, over 100 billion cans are produced per year. About half of these are not recycled. A recycled aluminium can saves enough energy to run a television for three hours.

Fig. 15.4 Recycling symbol

Steel cans can also be recycled and recycling a steel can uses only 25% of the energy required to make a new steel can from iron ore.

Glass bottles and jars can also be recycled indefinitely, saving energy which would be used in processing the raw material – sand – to make new glass. The broken glass produced is called cullet and is used to make more bottles and jars.

Fig. 15.5 Don't drop litter symbol

The success of the plastic bag

Plastic bags are everywhere and are clean, cheap to produce, waterproof, convenient. It costs one US cent to produce a plastic bag and about four to make a paper bag. They are so cheap that stores give them away. We are so keen to use them that an estimated 500 billion to one trillion are made each year. Most are used once and then thrown away as they are so thin. A few degrade in sunlight if they are made of biodegradable starch polymer materials but most are made from oil. When we have finished with them, they may end up in landfill sites, or in trees, oceans, turtle stomachs, on deserted islands. Everywhere. Although they take up less room in a landfill site than a paper bag, they take 200–1000 years to break down. Burning them releases toxins. We only started using them in quantity in the late 1980s. Before that we carried reusable shopping bags. We can do this again.

The plastax – plastic bag tax – may be the answer. In South Africa, Ireland, Australia, Taiwan and Bangladesh, governments have acted to ban or tax plastic bags. In Ireland a tax on the bags resulted in a decrease in their use of 95%. In South Africa, thin bags were banned and the thicker ones can be reused, have to be paid for and do not float around the country. China, said to use 3 billion bags a day, has banned the use of the free and flimsy bags and from June 2008, bags must be charged for.

Disposal of waste

If waste, materials are not recycled or reused, the options are to put them in landfill sites or incinerate them, dump them in the seas or to compost organic waste.

Landfill is the main method of disposal. Waste is taken to a suitable site and buried there. Hazardous waste can be buried along with everything else and the initial cost is relatively cheap. Landfill sites are not just holes in the ground. They are carefully selected to be not too close to areas of high population density, water courses and aquifers. They are lined with a special plastic liner to prevent **leachate** (liquid waste) seeping out. The leachate is collected in pipes. Methane produced as a result of fermenting organic material in the waste is either collected and used to generate electricity or vented to the atmosphere. Soil is pushed over the waste each day to reduce smells and pests. New landfill sites are getting harder to find as we fill up the ones we have at a faster and faster rate.

Fig. 15.6 A waste truck unloading in a landfill site, Wales

Incinerators burn the waste at high temperatures of up to 2000°C. In some, the waste is pre-sorted to remove incombustible or recyclable materials. Then the heat produced is often used to generate steam to drive a turbine or heat buildings directly. This is called waste-to-energy incineration. In others, all the waste is burned but this practice can cause air pollution, particularly release of dioxins from burning plastics, heavy metals (lead and cadmium) from burning batteries and nitrogen oxides. But the ash from incinerators can be used in road building and the space taken up by incinerated waste is far smaller than that in landfills. Plants are expensive to build though and need a constant stream of waste to burn so do not necessarily encourage people to reduce their waste output.

Organic waste can be composted or put into anaerobic digesters. The methane produced can be used as fuel and the waste later used as fertilizer or soil conditioner.

> **To do**
>
> Make a table listing the advantages and disadvantages of landfill, incineration and recycling as waste disposal methods.

The issue of plastic cups

from http://www.eia.doe.gov/kids/energyfacts/saving/recycling/solidwaste/plastics.html

A paper or a plastic cup (styrofoam) or a ceramic mug that you wash and reuse? Plastic cups are made from non-renewable oil; paper from renewable wood; ceramic mugs from non-renewable clay. Which should you choose?

A study by Canadian scientist Martin Hocking shows that making a paper cup uses as much petroleum or natural gas as a polystyrene cup. Plus, the paper cup uses wood pulp. The Canadian study said, "The paper cup consumes 12 times as much steam, 36 times as much electricity, and twice as much cooling water as the plastic cup." And because the paper cup uses more raw materials and energy, it also costs 2.5 times more

than the plastic cup. But the paper cup will degrade, right? Probably not. Modern landfills are designed to inhibit degradation so that toxic wastes do not seep into the surrounding soil and groundwater. The paper cup will still be a paper cup 20 years from now.

Surely it is kinder on the environment to use a ceramic mug? Well, it depends how you clean it. If you consider the energy cost of making it, the use of hot water and detergent in a dishwasher in cleaning it, you would need to use the mug 1000 times to get down to the environmental impact that the plastic cup has.

What do you use?

How many years?

How long do you think these objects take to break down in a landfill?

1	Disposable nappy/diaper	**6**	Plastic bottle
2	Cotton T-shirt	**7**	Paper bag
3	Leather belt	**8**	Banana peel
4	Styrofoam cup	**9**	Aluminium can
5	Glass bottle	**10**	Block of wood

Answers:
1 500–600 years, **2** 6 months, **3** 50 years, **4** 1 million years, **5** 1 million years, **6** 1 million years, **7** 2 months, **8** 1 month, **9** 500 years, **10** 20 years.

Eutrophication

Eutrophication is the addition of excess nutrients to a freshwater ecosystem. It can be a natural process but anthropogenic eutrophication has accelerated it. The nutrients are usually nitrates and phosphates and they come from:

● detergents
● fertilizers
● drainage from intensive livestock rearing units
● sewage
● increased erosion of topsoil into the water.

These sources may be point sources, e.g. wastewater from cities and industry or animal feedlots, overflows from storm drains or sewers, and runoff from construction sites. Harder to identify are the sources of pollution from non-point sources. These may be runoff from crops and grassland, urban runoff, leaching from septic tanks or leaching from landfill sites or old mines.

The process of eutrophication

- Fertilizers wash into the river or lake.
- High levels of phosphate in particular allow algae to grow faster (as phosphate is often limiting).
- Algal blooms form (mats of algae) that block out light to plants beneath them, which die.
- More algae mean more food for the zooplankton and small animals that feed on them. They are food to fish which multiply as there is more food so there are then fewer zooplankton to eat the algae.
- Algae die and are decomposed by aerobic bacteria.
- But there is not enough oxygen in the water so, soon, everything dies as food chains collapse.
- Oxygen levels fall lower. Dead organic material forms sediments on the lake or river bed and turbidity increases.
- Eventually, all life is gone and the sediment settles to leave a clear blue lake.

In lakes or slow-moving water bodies, eutrophication leads to a series of damaging changes, which severely reduces biodiversity. In fast-moving water, eutrophication leads to a temporary reduction in biodiversity downstream which can be followed by recovery and restoration of clean water.

Eutrophication management strategies

Table 15.4 uses the "replace, regulate and restore" model of pollution management strategy to suggest some actions for reducing eutrophication:

Draw an annotated diagram to illustrate eutrophication. Indicate on your flow diagram an instance of positive feedback.

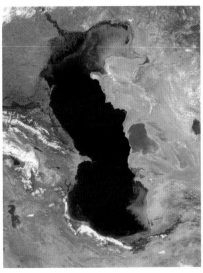

Fig. 15.7 Eutrophication in the Caspian Sea (lighter areas of algal bloom)

Strategy for reducing pollution	Example of action
Altering the human activity producing pollution	Ban or limit detergents containing phosphate (it is there to improve the performance of the detergent in hard water areas)
	Plant buffer zones between fields and watercourses to absorb the excess nutrients
	Stop leaching of slurry (animal waste) or sewage from their sources
Regulating and reducing the pollutants at the point of emission	Pump air through the lakes
	Divert or treat sewage waste
	Minimize fertilizer dosage on agricultural lands or use organic matter instead
Clean-up and restoration	Dredge sediments with high nutrient levels from the river and lake beds
	Remove excess weeds physically or by herbicides and algicides

Table 15.4 "Replace, regulate and restore" model of pollution management for eutrophication

Impacts of eutrophication

Anthropogenic eutrophication leads to unsightly rivers, ponds and lakes covered by green algal scum and duckweed. They also give off foul-smelling gases like hydrogen sulfide. Other changes include:

- oxygen-deficient (anaerobic) water
- loss of biodiversity and shortened food chains
- death of higher plants (flowering plants, reeds, etc.)
- death of aerobic organisms – invertebrates, fish and amphibians
- increased turbidity (cloudiness) of water.

To do

1 Copy Table 15.4 of strategies and actions on eutrophication and add a third column, entitled Evaluation. Complete this column.

2 Figure 15.8 shows eutrophication by point source river pollution due to nutrient input from an outfall pipe discharging sewage.

   a Copy the figure, and using coloured pencils draw curves to show changes in detritus (sewage), turbidity (suspended solids), bacterial growth (BOD), oxygen concentration, invertebrate and fish biodiversity, clean water species. Make a key to your curves.

   b Copy the figure again and complete the diagram to show the effects of high nutrient levels, e.g. from over-fertilized fields being washed into a stream. Using coloured pencils draw curves to show changes in nutrient (nitrate and phosphate), algal growth (bloom), detritus increase (due to higher plant death), bacterial growth (BOD), oxygen concentration, invertebrate biodiversity, clean water species. Make a key to your curves.

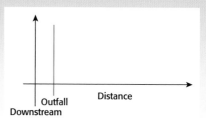

Fig. 15.8

Introduction to ozone

Ozone is found in two layers of the atmosphere – the troposphere where it is considered to be "bad" and the stratosphere where it is "good". In the troposphere, at ground level, it is a pollutant in urban smog and is a danger to human and other life. In the stratosphere it blocks incoming ultraviolet (UV) radiation from the sun and protects life from damaging ultraviolet radiation. It is the same chemical in both, O_3, and is made up of three oxygen atoms. Oxygen gas is diatomic O_2.

Confusion may arise as well because ozone is also a greenhouse gas (GHG).

Depletion of stratospheric ozone

The ozone layer

Ozone is a reactive gas of which the majority is found in the so-called ozone layer in the lower stratosphere. The highest ozone concentrations are usually seen at altitudes between 20 and 40 km (at the poles, between 15 and 20 km). But it is a very thin layer of about 1–10 parts ozone per 1 million parts of air and the ozone is constantly breaking down and reforming in a chain reaction.

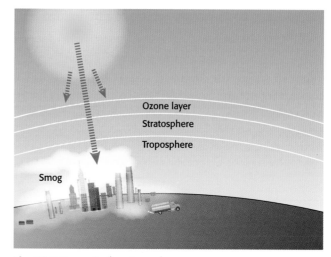

Fig. 15.9 Ozone in the atmosphere

In the ozone layer, ozone is continuously made from oxygen (O_2) and also continuously reacts back to form oxygen. So, the ozone layer is an example of a dynamic equilibrium. In both the formation and the destruction of ozone, UV radiation is absorbed. Under the influence of UV radiation, oxygen molecules are split into oxygen atoms. Oxygen atoms are extremely reactive, and can combine with an oxygen molecule to ozone. Ozone can also absorb UV radiation and then splits into an oxygen molecule and an oxygen atom. The oxygen atom can react with another ozone molecule, making two oxygen molecules.

The absorption of UV radiation by the ozone layer is crucial, for without it, life on land would be impossible. UV radiation is usually divided into UVA, UVB and UVC radiation. UVC radiation has the highest energy (shortest wavelength) and is therefore the most harmful type of UV radiation, while the longer wavelength (low energy) UVA radiation is relatively harmless. The ozone layer absorbs more than 99% of the UVC radiation and about half of the UVB radiation.

Damaging effects of ultraviolet radiation

Increased exposure to UV radiation will have a variety of damaging effects:

- mutation and subsequent effects on health
- damage to photosynthetic organisms, especially phytoplankton
- damage to consumers of photosynthetic organisms, especially zooplankton.

Ultraviolet radiation can cause mutations, changes in a species' DNA. Especially in Australia, the number of cases of skin cancer in humans has risen sharply. People are advised to wear clothes on the beach and to use sun blocks to protect their skin. It also causes cataracts in the lens of the eye when the protein of the lens denatures and turns cloudy instead of clear.

Photosynthetic organisms are sensitive to UV radiation. This can have disruptive effects on food pyramids.

Air pollution and the ozone hole

Since the 1950s, British scientists have been measuring the amount of ozone in the stratosphere above Antarctica. They discovered what later would be called the ozone hole: the amount of ozone decreased markedly during the spring (September and October) and increased again in November. Apart from this annual ozone cycle, the British scientists discovered that the ozone hole is growing. During the last 30 years, the minimum thickness of the ozone layer has reduced drastically and recovery has been taking longer. These results were later confirmed by NASA satellite data. Reductions in the amount of stratospheric ozone have been observed in other areas as well, including the Arctic region.

This ozone depletion is the result of air pollution. The most important ozone-depleting gases are chlorofluorocarbons or CFCs, but there are others too.

Substance	Use/source	Remarks
Chlorofluorocarbons (CFCs or freons)	Propellants in spray cans Plastic foam expanders Refrigerants	Release chlorine atoms
Hydrochlorofluorocarbons (HCFCs)	As replacements for CFCs	Release chlorine atoms, but have a shorter lifetime in the atmosphere
Halons	Fire extinguishers	Release bromine atoms
Methyl bromide	Pesticide	Releases bromine atoms
Nitrogen oxides (NO, NO_2, N_2O, often summarized as NO_x)	Bacterial breakdown of nitrites and nitrates in the soil (intensive farming) High flying supersonic aircraft	The nitrogen oxides are converted to NO, which reacts with ozone

Table 15.5 The most important ozone-depleting substances (ODS)

The action of ozone-depleting substances

When chlorofluorocarbons (CFCs or freons) were developed during the 1930s, they seemed to be a wonder product. Previously used refrigerants were very toxic and flammable, and were therefore quickly replaced by the non-toxic and non-flammable CFCs. Later on, CFCs were also used as propellants in spray cans and to expand plastic foam. CFCs are extremely stable and it took a long time before it was discovered that they were not so stable when exposed to UV radiation in the stratosphere. UV radiation releases chlorine atoms from CFCs. These chlorine atoms can react with ozone, which results in ozone destruction. They can also react with oxygen atoms, thereby preventing ozone formation. In both of these processes, the chlorine atoms are formed back and are again able to react with ozone or oxygen atoms. One chlorine atom can thus destroy many molecules of ozone in a chain reaction (an example of positive feedback).

While replacing CFCs in spray cans and as blowing agents for plastic foam is relatively easy, it is much more difficult to find a suitable refrigerant. The refrigerants used before the introduction of CFCs are not an option because of their dangerous properties. The most suitable CFC replacements are the so-called hydrochlorofluorocarbons (HCFCs). These substances are nearly as good refrigerants as CFCs and are also non-toxic and not inflammable. However, HCFCs also destroy ozone and also contribute to the greenhouse effect. Only their shorter lifetime in the atmosphere makes them less harmful to the ozone layer than CFCs.

CFCs are extremely stable, and will therefore persist in the atmosphere for up to 100 years after their release. Measures taken to prevent release of CFCs into the atmosphere will therefore take a long time to result in a thicker ozone layer.

Reducing ozone-depleting substances

Table 15.6 uses the "replace, regulate and restore" model of pollution management strategy to suggest some actions for reducing ozone-depleting substances (ODSs).

Strategy for reducing pollution	Example of action
Altering the human activity producing pollution	Replace gas-blown plastics
	Replace CFCs with carbon dioxide, propane or air as a propellant
	Replace aerosol propellants
	Replace methyl bromide pesticides but most of the gases that can be used to replace CFCs are greenhouse gases
Regulating and reducing the pollutants at the point of emission	Recover and recycle CFCs from refrigerators and air-conditioning units
	Legislate to have fridges returned to the manufacturer
	Capture CFCs from scrap cars
Clean-up and restoration	Add ozone to or remove chlorine from stratosphere – not practical

Table 15.6 "Replace, regulate and restore" model of pollution management for ozone depletion

Role of national and international organizations in reducing CFC emissions

The discovery of the ozone hole led to a fast response at national and international levels. But even before governments and international organizations took steps, the general public in many developed countries started to boycott CFC-containing products (mainly spray cans). The aerosol industry reacted quickly by changing to ozone-friendly spray cans and, even before CFCs were forbidden by law, hardly any CFC-containing spray cans were produced any more.

The United Nations organization involved in protecting the environment is UNEP (United Nations Environment Programme). UNEP forges international agreements, studies the effectiveness of these agreements and the difficulties in implementing and enforcing them. Apart from these activities, it also gives information to states, organizations and the public. One of the treaties signed under the direction of UNEP is the **Montreal Protocol** (1987). It is an international agreement on the reduction of the emission of ozone-depleting substances. The signatories agreed on freezing consumption and production of many CFCs and halons to 1986 levels by 1990 and on strongly reducing the consumption and production of these substances by 2000. Since 1987, the original Montreal Protocol has been strengthened in a series of amendments. In the protocol a distinction was made between MEDCs and LEDCs. The LEDCs got more time to implement the treaty.

Most countries followed the rules of the Montreal Protocol and made national laws and regulations accordingly. China and India, however, were still producing and using huge amounts of CFCs. These countries' need for refrigerators and air conditioners is quickly growing because of the fast economic growth they are experiencing. But China has now agreed to phase out ODS production two years before schedule and India also has a plan to do so.

The reduction of emissions of CFCs and other ozone-depleting substances has been one of the most successful international co-operative ventures to date.

Timeline for CFC reduction

Year	Events
1970s	Ozone-depleting properties of CFCs recognized. In 1974 USA and Sweden banned them from non-essential aerosol uses. Concerns continue to mount through 1980s
1985	British Antarctic Survey reports the ozone hole
1987	Montreal Protocol organized by UNEP. Over 30 countries agree to cut CFC emissions by half by 2000
1990	London Amendment to strengthen Montreal Protocol: Phase-out dates and rates inadequate so Montreal Protocol amended. Industrialized countries would eliminate CFC production by 2000 and developing countries by 2010
1992	Further measures to accelerate phasing out of ODS and replacement by substitute chemicals
1995	Nobel Prize for Chemistry awarded to Molina, Sherwood Rowland and Crutzen for their work on the ozone depletion issue
2006	NASA and NOAA (US National Oceanic and Atmospheric Administration) record the Antarctic ozone hole as the largest ever measured

Effectiveness of CFC reduction agreements

This has been a good example of international co-operation. However this is not the end of the story. Due to the long life of CFCs in the atmosphere, it was estimated that chlorine would not reach its peak in the stratosphere until 2005 nor return to normal levels much before 2050. LEDCs are still allowed to make and use CFCs and there is an illegal trade in ODSs across national borders.

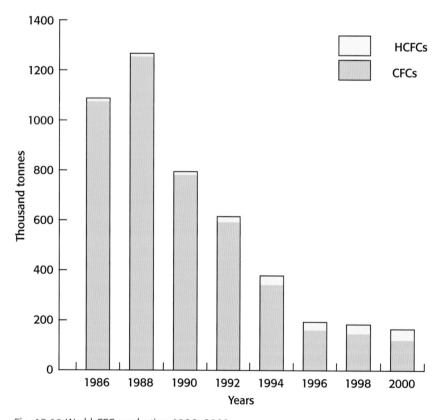

Fig. 15.10 World CFC production 1986–2000

To do

1 What steps is your national government taking to comply with the Montreal Protocol and subsequent agreements?

2 Name local or non-governmental organizations concerned with CFC reduction.

3 What actions are they taking to persuade governments to comply?

4 What can you do to help protect the ozone layer?

To do

1 Comment on the significance of the data in Fig. 15.11 in relation to MEDCs and LEDCs.

2 Discuss the success of UNEP and other international bodies in dealing with the CFC problem.

3 State one change in the data that the pie chart for the year 2006 might show.

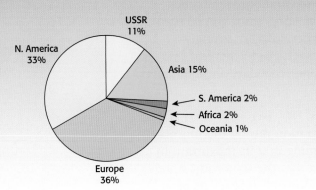

Fig. 15.11 Percentage use of CFCs in 1986 by region

Urban air pollution and the formation of tropospheric ozone

Ozone in the troposphere

Only 10% of ozone is in the troposphere (from Earth's surface to about 12 km in altitude). Here it is in concentrations of only 0.02–0.3 ppm. Ozone is also a greenhouse gas with a global warming potential 2000 times that of carbon dioxide.

Formation of tropospheric ozone

Among the pollutants emitted during the combustion of fossil fuels are hydrocarbons and nitrogen oxide (NO). Hydrocarbons are emitted because not all the fuel is combusted. Nitrogen oxide is formed when oxygen and nitrogen (both originating from air) react as a result of the high temperature during combustion reactions.

Nitrogen oxide contributes to the formation of tropospheric ozone (O_3). First nitrogen oxide reacts with oxygen to nitrogen dioxide (NO_2). This is a brown gas that contributes to urban haze. Other pollutants like hydrocarbons and carbon monoxide accelerate the formation of NO_2. When NO_2 absorbs sunlight, it breaks up into NO and oxygen atoms. These oxygen atoms subsequently react with oxygen molecules, forming ozone.

Tropospheric ozone causes serious problems. Ozone is a toxic gas and an oxidizing agent. It damages crops and forests, irritates eyes, can cause breathing difficulties in humans and may increase susceptibility to infection. It is highly reactive and can attack fabrics and rubber materials.

The formation of photochemical smog

On warm sunny days with lots of traffic, photochemical smog can be formed over cities. Because photochemical smog first caused problems in Los Angeles, is often called Los Angeles-type smog. Other cities that frequently suffer from this type of smog are Santiago, Mexico City, Rio de Janeiro, Sao Paulo, Beijing and Athens (see Fig. 15.11).

Photochemical smog is a formed when ozone, nitrogen oxides and gaseous hydrocarbons from vehicle exhausts interact with strong sunlight. Complex reactions create many chemicals in photochemical smog including VOCs (volatile organic compounds), PANs (peroxyacyl nitrates), ozone, aldehydes, carbon monoxide and nitrogen oxides. Because NO_2 is an important component of smog, smog can be seen as a brown haze above the city. All these chemicals are strongly oxidizing and affect materials and living things. Photochemical smog is a mixture of about 100 different primary and secondary air pollutants with ozone as the main pollutant.

Even though the main primary pollutants, NO_x (see page 295) and hydrocarbons, reach a maximum concentration during the morning and evening rush hours, photochemical smog is at its maximum in the early afternoon. This is due to the fact that the important smog-causing reaction is a photochemical reaction, so it reaches its peak during afternoon sunshine.

The occurrence of photochemical smog is governed by a large number of factors, including the local topography, climate, population density and fossil fuel use. Smog is most often formed over large cities lying in valleys. The hills or mountains surrounding these cities shelter them from most of the wind and on warm, calm days severe smog can occur. Thermal inversion makes things worse. Normally, air over cities is relatively warm and has a tendency to rise. On warm days, however, an even warmer layer of air on top of the warm polluted air can prevent this air rising, trapping the pollution at ground level. This occurs most often in warm dry climates. Weather plays an important role in the disappearance of smog: rain cleans the air of pollutants while winds can disperse the smog.

Fig. 15.11 Photochemical smog over Athens

Under the above conditions the concentration of pollutants can reach harmful or even lethal levels. Smog is not only affecting life in the city itself. Often, smog is blown out of the city by the wind and causes damage in the countryside, sometimes up to 150 km away from the city where the smog was formed.

London-type smog
London-type smog (smog = smoke + fog) is a completely different type of smog. It occurs at low temperatures. The main primary pollutants are also different: sulfur dioxide and smoke particles produced by burning coal. These smogs killed many people until clean air acts legislated to stop people burning coal and use smokeless fuels instead.

Possible effects of photochemical smog and ozone
Damage to plants: Tropospheric ozone is absorbed by plant leaves. In the leaves, ozone degrades chlorophyll so photosynthesis and productivity are reduced.

Damage to humans: At low concentrations, photochemical smog can cause reduced lung function and can cause eye, nose and throat irritation. At higher concentrations, smog can cause coughs and decreased ability to concentrate.

Damage to materials and products: Ozone attacks natural rubber, cellulose and some plastics. It reduces the lifetime of car tyres. It also bleaches fabrics.

Formation of particulates

Burning almost any organic material or fossil fuel releases small particles of carbon and other substances, referred to as particulates. Poorly maintained diesel engines, in particular, release large amounts of particulates (small solid particles) in exhaust fumes. They are called PM10 or particulate materials smaller than 10 micrometres in diameter. The problem with them is that our respiratory filters (hairs lining the nose and the passages of the bronchi and lungs) cannot filter them out so they enter our bodies and stay there causing asthma, lung cancer, respiratory problems and even premature death. Many particulates are carcinogenic and in areas close to industrial or dense urban areas and especially in developing regions crops become covered with particulates, which reduces their productivity.

Acid deposition

Acidity

Acids are chemicals that are able to donate a hydrogen ion (H^+) to other chemicals. The acidity of solutions is measured using the pH scale. On this scale, a pH value of 7 is neutral (pure water). Values below 7 indicate acidic solutions whereas values above 7 indicate basic (alkaline) solutions. The pH scale is not a linear scale. A solution with pH 2 is ten times more acidic than a solution with pH 3. Normal unpolluted rain is slightly acidic and has a pH of about 5.6. This is caused by the presence of carbon dioxide in the atmosphere. Precipitation is considered to be acidic when its pH is well below pH 5.6. Often precipitation with a pH of below 5 is said to be acidic. Certain pollutants make rain acidic, and its pH can sometimes fall lower than pH 2.

Acid deposition is the general term for acid coming down to the Earth's surface from the air. Often, the acid comes down in the form of rain (or snow); this is called **wet deposition**. However, sometimes the acid comes down as gas or dry particles, without water. This is called **dry deposition**.

Main pollutants and sources

Air pollutants can be divided into **primary** and **secondary air pollutants**. Primary pollutants are those pollutants that are emitted directly, e.g. those pollutants leaving the chimney of a factory or the exhaust pipe of a car. Primary pollutants can change to secondary pollutants by reaction with other substances in the atmosphere. So, secondary pollutants are made after the pollutants leave the chimney. The main primary pollutants leading to acid deposition are sulfur dioxide (SO_2) and nitrogen oxides (NO and NO_2, usually written as NO_x). The most important human activity that leads to the emission of these pollutants is the combustion of **fossil fuels** in motor cars, industry and electricity plants. Sulfur dioxide is formed when sulfur-containing fuels are combusted. Sulfur is common in

To do

1 List the major sources of urban air pollution.
2 What is a secondary pollutant?
3 Why is photochemical smog greater in afternoons than in mornings?
4 Outline the effects on humans and other species of urban air pollution.
5 What actions can be taken to reduce urban air pollution?

coal and oil, but is usually absent in natural gas. Nitrogen oxides are formed by reaction of oxygen and nitrogen from the air, which readily takes place at the high temperature during combustion of fossil fuel. So the nitrogen is not part of the fuel.

If the primary air pollutants remain in the atmosphere for a sufficiently long time, a variety of secondary air pollutants can be formed. Sulfur dioxide can react with oxygen from the atmosphere to form sulfur trioxide (SO_3). Both sulfur dioxide and sulfur trioxide can react with water and form sulfurous acid (H_2SO_3) and sulfuric acid (H_2SO_4) respectively. The nitrogen oxides can also react with water and form nitric acid (HNO_3). These secondary pollutants are very soluble in water, and are removed from the air by precipitation in the form of rain, hail and snow (wet deposition).

Effects of acid deposition on soil, water and living organisms
Acid depositions can have:

- direct effects, e.g. by weakening tree growth
- toxic effects, e.g. effect of aluminium ions on fish and plant roots
- nutrient effects, e.g. leaching of nutrients.

Effect of acid rain on coniferous forests
Acid rain damage became a focus of environmental attention in the early 1970s when Germany's Black Forest showed a dieback or *waldsterben* when trees of all ages, both coniferous and deciduous, showed signs of physical damage. Acid rain affects forest biology in several ways. Leaves and buds show yellowing (loss of chlorophyll) and damage in the form of lesions, thinning of wax cuticles. These and other changes reduce growth, allow nutrients to be leached out of the soil and washed away, and pathogens and insects gain entry to the trees. Symbiotic root microbes are killed and this greatly reduces the availability of nutrients, further reducing tree growth. Acid rain affects the soil by reducing the ability of soil particles to hold on to nutrients such as calcium, magnesium and potassium ions which are then leached out. Furthermore the acid dissolves toxic alumnium ions from soil particles which then damage root hairs. Trees can thus be weakened to the point of death.

Toxic effect of acid deposition

Effect on fish and other aquatic organisms
Aluminium is a common mineral element in the soil. Acid precipitation decreases the pH of the soil, making aluminium ions more soluble. The aluminium released from the soil eventually ends up in streams and rivers. Fish are particularly sensitive to aluminium ions in water. At low concentrations, aluminium disturbs the fish's ability to regulate the amount of salt and water in its body. This inhibits the normal intake of oxygen and salt. Fish gasp for breath and the salt content of their bodies is slowly lost, leading to death. At higher concentrations, a solid is formed on the fish's gills, leading to death by suffocation. As well as aluminium, other toxic metals can dissolve because of increased acidity.

Effect on lichens
Lichens which are a symbiotic pairing of an alga and a fungus are found growing on trees and buldings and are particularly sensitive to

Fig. 15.13 Forest death from acid deposition

gaseous pollutants like sulfur dioxide. They are used as indirect measures of pollution. Immediately downwind of a heavily polluting industrial region only a few tolerant lichen species are found. These are **indicator species** of high levels of air pollution. As the distance from the source of pollutants increases, more and more species are able to survive. Tables of lichen indicator species are used to estimate pollution levels.

Effect on soil fertility through nutrient removal

As described above, acid rain affects the soil by reducing the ability of soil particles to hold on to nutrients, such as calcium, magnesium and potassium ions which are then leached out. Acid rain also inhibits nitrogen-fixing bacteria and thus their ability to add nitrate ions to the soil.

Other effects

Acid deposition also affects human constructions. Limestone buildings and statues (including many with great archeological and historical value) react with acid and simply dissolve.

Recent research has also found that peat bogs affected by acid rain produce up to 40% less methane than before. This is because the bacteria that use the sulfates as a food source outcompete the ones that produce methane.

Regional nature of acid deposition

The effects of acid deposition are regional, in contrast to global warming or ozone depletion because, before the pollutants can spread over long distances, they return to the surface as dry or wet precipitation. It is therefore mainly the downwind areas of major industrial regions that are strongly affected. For example, Scandinavian forests and lakes were mainly affected by acid rain originating in Britain, brought over by prevailing southwestern winds. Industrial pollution from the USA is blown by prevailing winds towards Canadian forests. The acids seldom travel over longer distances than a few thousand kilometres. Dry deposition usually occurs quite close to the source of the acidic substances. It consists of sulfur dioxide, sulfur trioxide and the nitrogen oxides. Wet deposition occurs at slightly longer distances from the sources of the primary pollutants. It consists of sulfurous acid, sulfuric acid and nitric acid.

Soils or bodies of water are most often affected by acid rain. However, the impact of acid rain will depend very much on the geology of the area on which it falls. Acid rain does little harm to soils derived from calcium carbonate rocks, i.e. limestone and chalk. These are alkaline soils, which neutralize (or buffer) the acids. However, non-alkaline rocks produce soils which are very sensitive to acid rain. Acid rain leaches out minerals from these soils. This reduces biodiversity, and runoff affects nearby lakes.

Pollution management strategies for acid deposition

Table 15.7 uses the "replace, regulate and restore" model of pollution management strategy to list some actions for reducing acid deposition and their evaluations.

Strategy for reducing pollution	Example of action	Evaluation
Altering the human activity producing pollution	Replace fossil fuel use by using renewables	Also reduce carbon dioxide emissions but we live in a fossil-fuel reliant economy
	Reduce overall demand for electricity	Demand for power is ever increasing, particularly in India and China as they industrialize
	Use less private transport	
	Use low sulfur fuels, remove sulfur before burning, or burn mixed with limestone	
Regulating and reducing the pollutants at the point of emission	Clean-up technologies at "end of pipe" locations (points of emission), e.g. scrubbing in chimneys to remove the sulfur dioxide	Expensive and costs passed on to consumer Catalysers are cost effective if well maintained but are expensive to buy
	Catalytic converters convert nitrogen oxides back to nitrogen gas	
Clean-up and restoration	Liming acidified lakes and rivers (see below)	Effective in restoring pH but has to be repeated regularly. Costly
	Liming forestry plantations. Trees acidify soils as they remove nutrients	Affects biodiversity in other ways Treats symptoms and not the cause
	International agreements. See below	Agreements are difficult to establish and to monitor

Table 15.7 "Replace, regulate and restore" model of pollution management for acid deposition

The role of international agreements in effecting change

Timeline for political solutions

Year	Events
1970s	Evidence for acid rain accumulates including death of vast tracts of German forests, loss of lake biodiversity and accelerated weathering of buildings
1979	UN Convention on Long Range Transboundary Air Pollutants
1983	Convention modified and 15 European countries, USA and Canada agreed to cut sulfur emissions by 30% of 1980 levels by 1993
1993	Convention further modified to cut emissions by 80% of 1980 levels by 2003

Evaluation of success
An average of 50% reductions was achieved in Europe by 2000. However, LEDCs are rapidly industrializing and emissions are set to increase fast unless MEDCs can help LEDCs "leapfrog" to clean technologies and ecocentric lifestyles. Due to the regional character of acid deposition, international agreements concerning acid deposition tend to have a bilateral or regional character. One example is the EU Large Combustion Plant Directive. This directive regulates the emissions by large installations that have a thermal capacity of more than 50 MW, basically electricity plants and large industries.

Reduction of effects of acid deposition

Liming lakes to neutralize acidity
Starting in the 1950s, the loss of many fish and invertebrate species in large numbers of Scandinavian lakes was linked to high levels of lake acidity. By 1990, over 400 lakes were virtually lifeless. In the 1980s, Sweden experimented with adding powdered limestone to their lakes and rivers. However the results were mixed. The pH of treated lakes is quickly raised but short-lived because of the flow of water; liming only treats the symptoms and not the cause.

Biodiversity was not immediately restored; the lime seemed to affect nutrient balance as nutrients other than calcium were absent.

Reducing emissions

One way to reduce the emission of sulfur dioxide and nitrogen oxides is to reduce the combustion of fossil fuels. Reducing the need for electricity, reducing the use of cars, developing more efficient cars and switching to alternative energy sources like water power, wind and solar energy can do this. Also biological fuels and nuclear power will reduce SO_2 and NO_x emissions.

Precombustion techniques aim at reducing SO_2 emissions by removing the sulfur from the fuel before combustion. The sulfur removed from the fuel can be obtained in several useful forms: as the element sulfur which can be used in the chemical industry; as gypsum, which can be used in construction; or as dioxide which can be used in the production of sulfuric acid, one of the most used chemicals.

Acidification can also be reduced by **"end of pipe"** measures. These clean-up measures remove the sulfur dioxide and nitrogen oxides from the waste gases. Examples include waste gas scrubbers in electricity plants and the catalytic converter in cars. Waste gas scrubbers are intended to remove sulfur dioxide. The catalytic converter in cars removes nitrogen oxides, together with other pollutants. Catalytic-converter-equipped cars also produce no sulfur dioxide – the catalyst is destroyed by sulfur dioxide, so these cars use sulfur-free fuel. "End of pipe" is sometimes referred to as "points of emission".

Environmental impact assessments

An **environmental impact assessment** or **EIA** is a report prepared *before* a development project to change the use of land, e.g. to plant a forest or convert fields to a golf course. An EIA weighs up the relative advantages or disadvantages of the development. It is therefore necessary to establish how the abiotic environment and biotic community would change if a development scheme went ahead. An EIA will try to quantify changes to microclimate, biodiversity, scenic and amenity value resulting from the proposed development. These measurements represent the production of a **baseline study**.

EIAs look at what the environment is like now and forecast what may happen if the development occurs. Both negative and positive impacts are considered as well as other options to the proposed development. While often EIAs have to deal with questions about the effect on the natural environment, they can also consider the likely effects on human populations. This is especially true where a development might have an effect on human health or have an economic effect for a community.

What are EIAs used for?

EIAs are often, though not always, part of the planning process that governments set out in law when large developments are considered. They provide a documented way of examining environmental impacts that can be used as evidence in the

To do

1 What is acid deposition?
2 What causes acid deposition?
3 List the effects of acid deposition.
4 What actions can be taken to reduce acid deposition?
5 Describe one way to measure acid deposition directly and one way to measure it indirectly.

decision-making process of any new development. The developments that need EIAs differ from country to country, but the EIA process tends to be included in certain types of development in most parts of the world. These include major new road networks, airport and port developments, building power stations, building dams and reservoirs, quarrying and large-scale housing projects.

Where did EIAs come from?

In 1969, the US Federal government passed the National Environmental Policy Act (NEPA). NEPA made it a priority for Federal agencies to consider the natural environment in any land use planning. This gave the natural environment the same status as economic priorities. Within 20 years of NEPA becoming law in the US many other countries had also included EIAs as part of their planning policy. In the US, environmental assessments (EA) are carried out to determine if an EIA (called EIS – environmental assessment statement) needs to be undertaken and filed with the federal agencies.

What does an EIA need in it?

There is no set way of conducting an EIA, but various countries have minimum expectations of what should be included. It is possible to break an assessment down into three main tasks:

- identifying impacts (scoping)
- predicting the scale of potential impacts
- limiting the effect of impacts to acceptable limits (mitigation).

There is always a non-technical summary so that the general public can understand the issues.

To think about

EIAs are models of the system under study and allow us to predict the effects of the proposed change. A model is only as good as its parameters and asking the right questions is crucial. A change of land use will always have an effect but whether this is a net positive or negative one depends on the criteria used to measure it. Simplistically, if a factory blocks your view of the mountains that may be a loss to you but it may bring employment to the area, produce goods that would otherwise be imported and reduce the country's ecological footprint.

Cost-benefit analysis measures impacts of a development or change of land use translated into monetary values. In theory, this puts all costs into the same units of measure – money – so they can be assessed. Of course, how the assessment is made is critical to the values assigned and there are several ways to do this. For example, it may be based on the cost of restoring the environment to its previous state (e.g. after an open-cast mine operation) or it may ask people which of several options they would select or be prepared to pay for.

1 Strategic environmental assessment tries to measure the social and environmental costs of a development but this can be subjective or a not very accurate prediction. Does it also depend on the environmental worldview of those planning the assessment?

2 Imagine a development or change of land use in an area near to your school or home. Decide among your class what this will be (it may be an actual one that is about to happen or has happened) and discuss the following:

a What criteria would you use to select the factors you think will change (e.g. number of jobs provided, net profit, land degradation, habitat loss, pollution)?

b How would you value these? (Is there another way of measuring them apart from financial?)

c How would you weigh up the evidence to make a decision on whether the project should proceed, or proceed in a modified state?

Key words

pollution	CFCs
point source	halons
non-point source	stratospheric
pollutant	tropospheric
biochemical oxygen demand	Montreal Protocol
BOD	photochemical smog
indicator species	acid deposition
biotic index	wet deposition
replace, reuse, restore	dry deposition
solid domestic waste	primary air pollutant
reduce	secondary air pollutant
recycle	fossil fuel
landfill	sulfur dioxide
incinerator	sulfuric acid
compost	nitrogen oxides
eutrophication	regional
phosphate	liming
nitrate	EIA
ozone	baseline study
ozone-depleting substances	non-technical summary
ODS	

Practical work and ideas

Key points

- How to measure biotic and abiotic components of an ecosystem
- Measuring these along an environmental gradient and measuring these due to a specific human activity
- Constructing and using simple keys
- Estimating the biomass and abundance of organisms
- Estimating the biomass of trophic levels
- Applying Simpson's diversity index
- Environmental impact assessment (EIA) example
- Calculating population data
- Calculating sustainable yields
- Calculating ecological footprints
- Detection and monitoring of pollution by direct and indirect methods

"I hear and I forget,
I see and I remember,
I do and I understand."

Confucius

What is in this chapter?

This book is not an experimental or investigative guide to practical work but this chapter contains examples of practical work that is part of the course or that will help you understand the concepts. There is a basic outline of the type of practical work that you might see in a practical scheme of work (PSOW) and ideas for investigations. This chapter contains concepts that can best be taught through investigative hands-on work but which are part of the whole course.

Practical work

Hands-on field work or laboratory work is an essential part of this course. Internal Assessment (IA) comprises up to 20% of your final marks and the IB recommends that you spend at least 30 hours out of the 150 suggested for teaching the course on practical investigations. This is partly because you understand concepts better if you are an active learner and partly because you should have direct experience of field work and practical techniques. In carrying out investigations, you should always ensure that you work safely, do not damage the environment or inflict pain or suffering on other people or animals. Always work ethically.

Internal assessment (IA)

The IB course guide contains a section on internal assessment. We shall not repeat its contents here but will give you ideas for practical investigations as techniques are explained. The IB expects you to submit a list of the investigations you have carried out during the course (PSOW, practical scheme of work) which covers the breadth of the course. This work should be kept in an IA portfolio which may be a file or laboratory book. The investigations may be shorter or longer, field or laboratory-based, quantitative or qualitative observations and interpretation, computer simulations, questionnaire, data analysis or any other format that your teacher deems appropriate. Scientists and problem-solvers usually work in teams and you should also work together with your colleagues as well as individually. Of the four criteria that are assessed by the IB (see below), you have to submit two examples of your best work in three of these to an IA moderator. For the fourth, personal skills, your teacher gives you a mark at the end of the course based on your work during it. The IB moderates these teacher-awarded marks to make sure standards are similar everywhere.

At all times be academically honest. Do not plagiarize or collude with your colleagues, whatever the circumstances, in a piece of work that you submit to the IB for assessment. If you are discovered to have done so, the penalty is large and could mean that you do not get an IB Diploma or certificates. Everybody has to learn what being academically honest really means and your teachers will help you in this. One pitfall to avoid is submitting group work (e.g. data that you collected as a group) for assessment under the data collecting and processing (DCP) criterion. Of course, you can and should do group work, but only submit data you collected yourself for this criterion. For assessed work, your teacher is allowed by the IB to comment on first drafts but the next draft must be the final one.

The four criteria against which you are assessed are:
1 Planning (Pl)
2 Data collection and processing (DCP)
3 Discussion, evaluation and conclusion (DEC)
4 Personal skills (PS)

Each of the criteria have three aspects and you can achieve "complete" – 2 marks, "partial" – 1 mark, or "not at all" – 0 marks in each aspect.

The maximum mark for each of these criteria is therefore 6. You are assessed twice in each of the first three so this adds up to $6 \times 2 \times 3 = 36$ plus one Personal Skills mark out of 6 = a total of 42 marks. These are moderated then scaled to give a total out of 20%. To get full marks does not mean a faultless submission but the important thing is to make sure that you are submitting work that meets the criteria. Sometimes it can be too simplistic or your teacher gives you too much help and guidance (e.g. in planning) to allow you to score highly.

Not all investigations in your PSOW have to be assessed according to the IB criteria. In fact, many will not be and your teacher will follow your school or college policy in these. But the criteria are helpful in guiding your work. The IB does not make it mandatory for you to take part in the Group 4 project as part of this course but your school may require this. If it is on a suitable topic, you would be able to add it to your PSOW. It is essential that you read and understand the section on IA in the *Environmental Systems and Societies Guide*.

Use of ICT is encouraged and data-logging is a useful tool to be able to use and understand.

There are no requirements for previous subject-specific knowledge in this course although some basic mathematical requirements are listed on page 5 of the *Environmental Systems and Societies Guide*. A willingness to engage with the issues and to be curious about the world around us is important.

Criterion	Aspects			Level	
Planning **PI**	**State** a focused problem/ **research question** **Identify** the relevant variables – input (controlled) and output (managed) [Give a **hypothesis** where appropriate] **c p n**	**Design** a **method** for the effective control of variables **c p n**	[**List** equipment] **Describe** a method that allows for collection of **sufficient** relevant data **c p n**	ccc ccp cpp/ccn ppp/cpn ppn/cnn pnn nnn	6 5 4 3 2 1 0
Data collection **DCP**	**Collect** data (1^0 or 2^0) Give **quantitative** (numerical) and/or **qualitative** (observations) data Include **units** **c p n**	**Process** (1^0 or 2^0) data correctly **c p n**	**Present** processed data appropriately and effectively assist analysis **c p n**	ccc ccp cpp/ccn ppp/cpn ppn/cnn pnn nnn	6 5 4 3 2 1 0
Discussion Evaluation and Conclusion **DEC**	Clear, well-reasoned **discussion** **Review**, showing broad understanding of **context** And **implications** of results **c p n**	**Evaluate** procedure **Identify** weaknesses and limitations and Suggest realistic **improvements** **c p n**	State a **reasonable** conclusion Give a **correct explanation** based on the data **c p n**	ccc ccp cpp/ccn ppp/cpn ppn/cnn pnn nnn	6 5 4 3 2 1 0
Personal skills **PS**	**Carry out techniques** Fully **competent** and **methodical** in the **use of range of techniques and equipment** **c p n**	**Work in a team and consistently collaborates** and **communicates** well with the group and **incorporates other people's ideas** **c p n**	Always pay attention to **safety issues** Pay attention to **environmental consequences** of actions Maintains **academic integrity** **c p n**	ccc ccp cpp/ccn ppp/cpn ppn/cnn pnn nnn	6 5 4 3 2 1 0

Table 16.1 Summary of IA criteria for this course
c = complete, p = partial, n = not at all

Devising a practical programme

What your teacher and you decide to do depends very much on where you are and the resources available to you. Every institution will have a different practical scheme of work but all will commit at least 30 hours to the practical programme and often far more because you can learn and understand most of the course by doing it.

If you are able to go on a residential field course, you will benefit from seeing a different ecosystem to those around your institution and then be able to compare characteristics of these. Many schools teach you the techniques you need and then take you on a field course for you to carry out the investigations that you will submit for IB assessment.

There is no set number of investigations that you must do; some will be short, others longer. You may return to the same ecosystem several times to measure biotic and abiotic factors there with seasonal changes. You may have an ecosystem that you can get to during class time, e.g. the school grounds, a local pond, stream, public park, wood or grassland. If this is not possible, you can still set up mini-ecosystems in the classroom, e.g. an aquarium, terrarium, ant farm, closed and open systems, and carry out investigations on these.

If you are in the middle of a city, you can measure pollution levels in the air, water and soil. Visits to sites such as local farms, water treatment plants, dams, power stations, zoos, botanical gardens, parks and conservation projects should be possible in most locations. Visits from guest speakers on environmental issues, recycling, local conservation projects, endangered species, local community issues or waste management can be arranged. You could link these with your CAS activities if you are taking the Diploma and arrange a project within your institution to involve others.

In the rest of this chapter, we look at various methods to measure aspects of environmental systems and societies. This is not a practical handbook but there are ideas for your own investigations, some of which could also be extended essay topics. (But remember not to submit the same work twice for assessment to the IB.)

This chapter is organized broadly in this order:

Looking at systems
Ecological investigations
Investigating pollution
Human population calculations
Calculations on ecological footprints and sustainable yield.

To think about

There is a tension between the reductionist approach of science, in which we break down systems to look at the cellular, molecular and atomic levels of individual components, and the holistic and interrelated issues of the environment, in which we want to look at the complexity of cause and effect of parts of the system. Sometimes we miss the effects that components have on each other and the complexity of the climate change issue shows you how difficult this is to untangle. Social scientists have long been used to the complexity of systems that involve humans who may be unpredictable and not always rational in their actions. The natural sciences have tended to avoid considering the non-rational as the scientific method relies on predictability and replicable results. We formulate a hypothesis and test it to see if it can be falsified or not.

Discuss in class how these two approaches can benefit the study of the environment and human activities within it.

To do

- Consider different models of systems. This may be best done at the end of the course when you are revising and making links between the topics you have studied. These may be simple flow diagrams, three-dimensional models, graphical or computer models. Find some examples of models of systems, e.g. an energy flow diagram through an ecosystem, a graphical model of biome distribution, predator-prey graphical models, the demographic transition model, predictions of human population growth in various forms, AOGCM (atmosphere/ocean general circulation model) computer models of climate change, ENSO (El Niño southern oscillation) models, ozone depletion models, pollution management modelling, cost-benefit analysis modelling. You will be able to find others throughout this book. Evaluate the models' strengths and limitations.

- Select a particular issue covered in the course that involves human activities having an impact on natural systems. Can you construct a model (of any type) to show the results of changing a factor and the feedback mechanisms that may be involved?

- Create a board game (look at examples such as Monopoly, Risk, Trivial Pursuit, Snakes and Ladders or any others that you know) based on a system that you have studied. Try it out on students younger than yourself.

- Make a working or 3-D model of a system of your choice.

To do

Observe some simple systems such as a candle, a boiling kettle, a plant, yourself and draw flow diagrams to show their inputs, outputs, storages and flows. Consider the effects of increasing or decreasing each input and think about feedback mechanisms that regulate the flows.

Looking at systems

These activities may best be carried out at the end of the course to consolidate your understanding.

Ecological investigations

Ecology has simple questions which often have complex answers. The simple questions are:

> What is there?
> How many are there?
> Where are they?
> Why is it like this?

Measuring abiotic components of the system

Ecosystems can be roughly divided into marine, freshwater and terrestrial ecosystems. Each of these ecosystem types has a different set of important physical (abiotic) factors.

You should know methods for measuring any three significant abiotic factors and how these may vary in a given ecosystem with depth, time or distance.

Marine ecosystems

Abiotic factors: salinity, pH, temperature, dissolved oxygen, wave action

Salinity is the word used for the salt concentration. It is expressed in ‰ (parts of salt per thousand parts of water). Normal seawater has a salinity of about 35‰. The Baltic Sea has brackish water with a salinity between 1‰ and 10‰. The salinity can be determined by measuring the electrical conductivity or the density of the water.

Seawater usually has a **pH** of above 7 (basic). The pH can be measured using a pH meter.

Temperature is an important variable. Most organisms in marine ecosystems are ectotherms (their body has about the same temperature as the water), so their metabolism is strongly affected by water temperature. The temperature determines the solubility of oxygen in water, and thereby the availability of oxygen to organisms. Water temperature can also affect the mixing of water layers.

Many marine organisms rely on **dissolved oxygen** for respiration. Low oxygen concentrations can be an indication of water pollution. As stated above, the amount of dissolved oxygen strongly depends on water temperature: the higher the temperature, the lower the concentration of dissolved oxygen becomes. In measuring the amount of dissolved oxygen, care must be taken to avoid oxygen from the air affecting the measurements. Dissolved oxygen can be measured using an oxygen-selective electrode connected to an electronic meter, datalogging, or by a Winkler titration. (A series of chemicals is added to the water sample and dissolved oxygen in the water reacts with iodide ions to form a golden-brown precipitate. Acid is then added to release iodine which can be measured, and is proportional to the amount of dissolved oxygen, which can then be calculated.) Oxygen-selective electrodes give quick results, but need to be well maintained and calibrated in order to give accurate results. The Winkler titration is more labour intensive.

Wave action is important in coastal zones where organisms live close to the water surface. Areas with high wave activity usually have high concentrations of dissolved oxygen. Typical examples are coral reefs and rocky coasts.

Freshwater ecosystems
Abiotic factors: turbidity, flow velocity, pH, temperature, dissolved oxygen

Turbidity is the cloudiness of a body of fresh water (high turbidity = cloudy water, low turbidity = clear water). The turbidity is important because it limits the penetration of sunlight and thereby the depth at which photosynthesis can occur. Turbidity can be measured with optical instruments or by using a Secchi disk. The latter method gives a clear indication of how far light penetrates. A Secchi disk is a white or black-and-white disk attached to a graduated rope. The disk is heavy to ensure that the rope goes vertically down. The procedure is:

1 Slowly lower the disk until it disappears from view.
2 Read the depth from the graduated rope.
3 Slowly raise the disk until it is just visible again.
4 Read the depth from the graduated rope.
5 Calculate the average depth. This depth is known as the Secchi depth.

For reliable results a standard procedure should be followed (standing/sitting in the boat, with/without glasses, always work on the shady side of the boat).

Flow velocity is an important factor in determining which species can live in a certain area. Flow velocity can be very variable in time (e.g. melt water in the spring) and it also varies with depth. The easiest way to measure flow velocity is to measure the time a floating object takes to travel a certain distance. The floating object should preferably be partly submerged to reduce the effect of the wind. Oranges and grapefruits make suitable floats. This method gives the surface flow velocity only. The average flow velocity of a river can be estimated from the surface flow velocity by dividing the surface velocity by 1.25. Flow velocity can also be measured with a calibrated propeller attached to a stick. With this type of equipment you can measure the flow velocity at any depth.

pH values of fresh water range from moderately acidic to slightly basic, depending on surrounding soil, rock and vegetation. It can be measured with a pH meter or datalogging pH probe.

For temperature and dissolved oxygen see *marine ecosystems* (page 307).

Terrestrial ecosystems
Abiotic factors: temperature, light intensity, wind speed, particle size, slope, soil moisture, drainage, mineral content

Air temperature can be measured using simple liquid thermometers, min-max thermometers, more complex (electronic) thermometers. The latter equipment can be used to measure temperature continuously during a longer time as can a datalogging temperature probe.

Light intensity can be measured with electronic meters. The fact that light intensity varies with time (sunny period, clouds, time of the day, season) should be taken into account.

Wind speed can be measured with different techniques:

- A revolving cup anemometer consists of three cups that rotate in the wind. The number of rotations per time period is counted and converted to a wind speed. Revolving cup anemometers can be mounted permanently or hand-held.
- A ventimeter is a calibrated tube over which the wind passes. This reduces the pressure in the tube, which makes a pointer move. It is easy to use and inexpensive.
- By observation of the effect of the wind on objects. The observations are then related to the Beaufort Scale (a scale of wind speed from 0 to 12).

Particle size determines a soil's drainage and water-holding capacity. Large particles can be measured individually and subsequently the average particle size and the particle size distribution can be calculated: a simple, but time-consuming procedure. Smaller particles (medium size pebbles to fine sand) can be measured by sieving using a series of sieves with different mesh sizes. The smallest (silt and clay) particles can be measured by sedimentation or optical techniques. Sedimentation techniques are based on the fact that large particles sink faster than small particles. Optical techniques use light scattering by the particles (light scattering is what makes suspensions of soil particles in water look cloudy). Both sedimentation and light scattering can nowadays be done using automated instruments but are expensive for secondary school use.

Slope determines surface runoff and can give rise to erosion. Slope can be measured using a water level consisting of a plastic hose with two clear cylinders on its ends and a tape measure. It can also be measured with a simple field level and sticks marked in 20-cm bands in black and white.

To do

Abiotic factors of a temperate or tropical forest ecosystem

Using a table like the one alongside, list the abiotic factors you would measure if you were investigating the microclimate of a forest, the sensors that you would use and the typical units. Evaluate the effectiveness of the technique.

Then carry out the investigation.

Abiotic factor	Apparatus	Units	Effective use
temperature	thermometer	°C	Simple to use; accurate to nearest whole degree;

Soil moisture can be measured by drying soil samples at elevated temperature but not hot enough to burn off organic matter. The sample is weighed first and then dried until its mass becomes constant. This takes several days.

Mineral content – the ratio between mineral and organic material is important for a soil's water-holding capacity and fertility. The mineral content can be determined by the loss on ignition (LOI) method. Soil samples are heated at high temperatures of 500 to 1000°C for several hours and the mass loss is measured. There are no standard conditions for this method and the choice of temperature and time depends on the sample material. Preferably, the same conditions should be used when comparing samples.

Measuring biotic components of the system

To measure biotic components we need to observe and question. Why is it as it is? What has changed recently? Why does this grow here and not there? What impact do more people walking here have?

So walk around your institution's grounds or the local area. Is there a playing field? Is there a footpath on soil rather than concrete? Does the ground slope? Is it more shady or more moist in one area than another and what difference does that make to the type and number of species living there?

Measuring biomass

A simple way to measure biomass is to take a sample of a species from an area of known size and dry it at about 60–70°C until it reaches a constant weight. All the water is then removed and the dry biomass is left. The result can be extrapolated to the total biomass of that species in the ecosystem.

Sampling animals that can move away

Investigate soil organisms by collecting samples of different soils or soils from different areas of your campus or local park, e.g. soil from under the playing fields or a footpath and soil from a natural, untrampled area; soil in the shade and exposed to more light; soil at the top of a slope and the bottom. As well as investigating the abiotic factors (particle size, drainage rates, air and water content), investigate the biotic factors (organic matter content and living organisms).

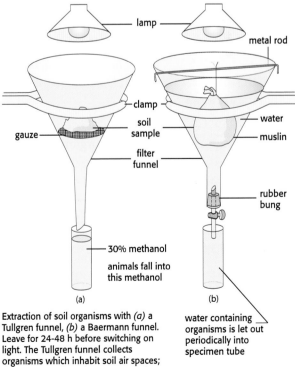

Extraction of soil organisms with (a) a Tullgren funnel, (b) a Baermann funnel. Leave for 24-48 h before switching on light. The Tullgren funnel collects organisms which inhabit soil air spaces; the Baermann funnel extracts organisms which live in soil water films.

Fig. 16.1 How to collect soil organisms

Microscopic soil organisms (nematodes, mites, other arthropods) can be extracted from soil by suspending soil samples under a heat source for a few days (see Fig. 16.1). The animals move down, away from the heat and can be collected in water or alcohol.

Sampling to estimate population size of motile animals can be done in various ways.

Small insects and invertebrates can be caught with a pooter – a small jar with two tubes attached (see Fig. 16.2). You suck gently on one tube and the animal is pulled into the jar. You cannot swallow it as there is gauze at the end of the mouthpiece tube!

Small crawling animals that cannot fly away can be caught by pitfall traps (see Fig. 16.3). Insects can be attracted by decaying meat or sweet sugar solution and will fall into the trap. The traps need checking frequently in case of predation.

Small mammal traps (e.g. the Longworth trap) are baited with food to catch small mammals humanely. Nets of various types can catch flying insects, butterflies and moths. Sticky traps hung on trees attract insects. Beating a bush with a stick and collecting the animals that fall off on a white sheet under the bush is effective; so is holding an umbrella under the bush. Night-flying moths will be attracted to a light behind which a white sheet is hung and the moths settle on this for you to observe.

In aquatic systems, nets of various mesh sizes and net sizes can be used to catch plankton, small invertebrates or larger fish. These can be towed behind boats or held in running water. Simple plastic sieves are effective. Kick sampling – standing in a stream holding a net downstream and gently kicking the substrate for a set time – loosens invertebrates, which drift into the net. Turning stones over is also effective.

Some of these methods are destructive and kill the organisms. You may think this is not acceptable but most do not harm the organisms caught and you should always return them to their habitats if at all possible.

Keys
Once you have collected the organisms, you may want to find out what they are called. A (biological) key is used to identify species. Ecologists make keys to specific groups of organisms, e.g. soil invertebrates in specific ecosystems, to help other interested people identify species. Keys come in two formats, a diagrammatic, dichotomous or "spider" key and a paired statement key.

Look at examples of published keys. Diagram keys are useful but professionals use paired statement keys because printed descriptions are more exact than pictures. Both keys are used by starting at the top each time and following the lines or "go to" numbers.

A list of pond organisms in a temperate ecosystem is shown in Fig. 16.4. It is from the UK Environmental Education Centre in Canterbury, Kent.

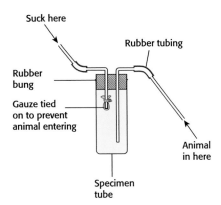

Fig. 16.2 A pooter

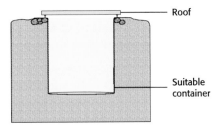

Fig. 16.3 A pitfall trap

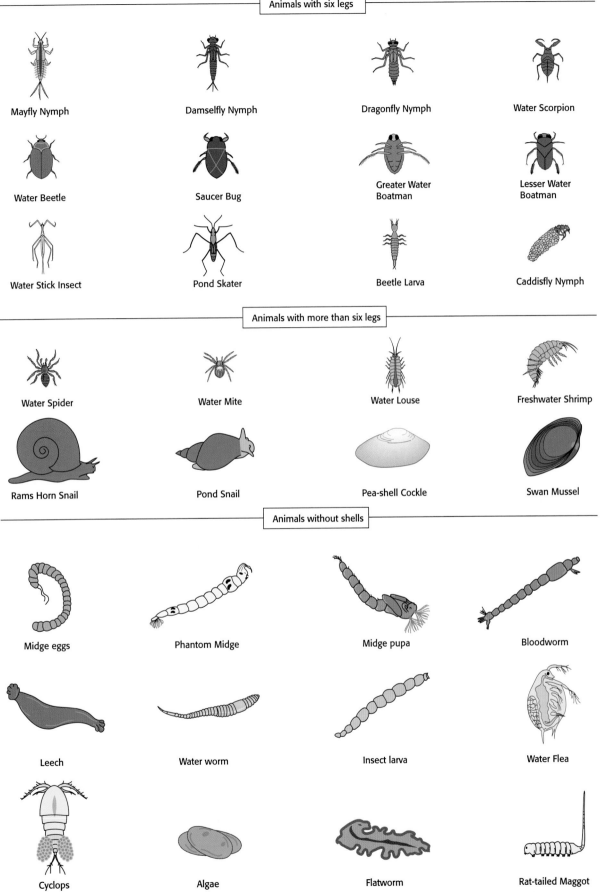

Animals with six legs

Mayfly Nymph
Damselfly Nymph
Dragonfly Nymph
Water Scorpion

Water Beetle
Saucer Bug
Greater Water Boatman
Lesser Water Boatman

Water Stick Insect
Pond Skater
Beetle Larva
Caddisfly Nymph

Animals with more than six legs

Water Spider
Water Mite
Water Louse
Freshwater Shrimp

Rams Horn Snail
Pond Snail
Pea-shell Cockle
Swan Mussel

Animals without shells

Midge eggs
Phantom Midge
Midge pupa
Bloodworm

Leech
Water worm
Insect larva
Water Flea

Cyclops
Algae
Flatworm
Rat-tailed Maggot

Fig. 16.4 A picture key of pond animals

A paired statement key has no pictures but is more accurate (see box).

Key to the species of mangroves in southeast Queensland

1 a Leaves alternate . . . . go to 2

 b Leaves opposite . . . . go to 4

2 a Sap milky; a pair of glands at junction of leaf blade and petiole . . . . *Excoecaria agallocha*

 b Sap not milky . . . . go to 3

3 a Leaf broadest towards the tip; a gland on underside near the leaf apex; salt crystals absent from leaf surface . . . . *Lummitzera racemosa*

 b Leaf broadest about the middle; no gland on underside; salt crystals usually present on leaf surface . . . . *Aegiceras corniculatum*

4 a Plants with conspicuous stilt roots; leaves with numerous specks on under surface; bud cover has 4 lobes . . . . *Rhizophora stylosa*

 b Plants lacking stilt roots . . . . go to 5

5 a Plants with erect, cylindrical aerial roots, projecting from the mud; leaves grey on the underside . . . . *Avicennia marina*

 b Plants with knee-like aerial roots projecting from the mud . . . . go to 6

6 a Stems distinctly buttressed; bud with cover of 4-6 lobes . . . . *Ceriops tagel*

 b Stems lacking buttresses; bud cover with 10 or more lobe . . . . *Bruguiera gymnorrhiza*

Source: *Currumbin Wildlife Sanctuary, Australia*

To do

1 Make a key to identify any of these:

 a Your classmates

 b A selection of office stationery, e.g. paper clip, treasury tag, pen, stapler, pencil sharpener, ruler, pen, hole punch, etc.

 c A selection of leaves

 d Animals that you find in a local pond or stream. When you have identified these, research their trophic levels and draw a food web.

2 Look at the arthropods in Fig. 16.5 and list body features that could be used to distinguish between species. Then make a paired statement key to identify them, naming them A–J.

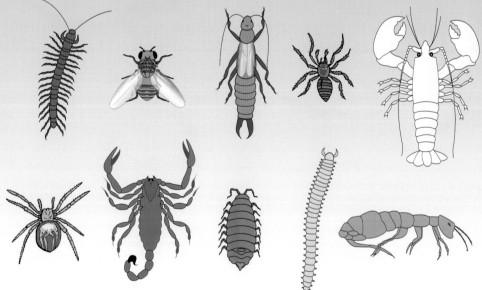

Fig. 16.5 A selection of arthropods

Measuring changes in the system

Changes along an environmental gradient

An environmental gradient is a trend in one or more abiotic and/or biotic components of an ecosystem. Examples can be found at shores of lakes, streams and seas and at forest edges. Measuring changes along an environmental gradient is most easily done on vegetation or on immobile animals (like corals) or very slow-moving animals (like mussels).

The technique used is called **transect**. In the simplest version, a **line transect**, a string or measuring tape is laid out in the direction of the environmental gradient and all species touching the string or tape are recorded and counted. Normally many line transects need to be combined to obtain sufficient data. A **belt transect** is a strip of chosen width through the ecosystem. It is made by laying two parallel line transects, usually 0.5 or 1 metre apart, between which all individuals are sampled.

A **continuous transect** is a line or belt transect in which the whole line or belt is sampled. In an **interrupted transect**, samples are taken at points along the line or belt transect. These points are usually taken at regular horizontal or vertical intervals.

Changes over time can occur over vastly different time spans, e.g. day-night cycles, seasonal changes or long-term (succession). Measuring these changes requires sampling/measuring on a regular basis over a suitable period. "Before" and "after" measurements are not enough. When measuring long-term changes you should do all measurements at the same time of the day to prevent interference with day-night cycles. For example, measuring temperature variation in a wood and a field will only give you meaningful results if you measure both at the same time on the same days.

Assessing the effects of human activities

It is not difficult to measure the effect of human activities in an ecosystem. Here are two examples but you will be able to find many others.

Effects of sewage treatment on water quality

Sewage treatment removes organic material before the water is discharged. Primary treatment takes out the larger particles; secondary treatment involves breaking down the organic matter using bacteria, often in a filter bed system where bacteria live on the surface of stones so the large surface area means plenty of oxygen is available. Other organisms live here too and eat the bacteria. It is not safe for your health for you to investigate this practically but the box below contains data which you can analyse.

To do

Sewage works filter bed
Refer to Table 16.2. SS are suspended solids – small organic and inorganic particles.

1 Describe how the sewage is changed by primary settlement.
2 What percentage of the original SS is removed by (a) primary settlement (b) action of bacteria in the filter bed?
3 Ammonia concentrations decrease after treatment. Explain what happens to the ammonia.
4 Why does the BOD fall?
5 What impact may nitrate have when the treated sewage is released into local rivers?

	SS (mg l^{-1})	BOD (mg l^{-1})	Compounds of nitrogen	
			Ammonia (mg l^{-1})	Nitrate (mg l^{-1})
Crude sewage	250	180	20–30	0
After primary settlement	100	100	18–25	0
After trickle filter bed (and resettlement to remove bacteria)	25	19	9	10

Table 16.2 Sewage works data

Assessing the effects of intensive agriculture on a stream
Intensive farming is high-input farming. High yields are obtained by use of machinery, pesticides and fertilizer. Machinery uses energy in the form of diesel fuel. Pesticides spread by wind during crop spraying may end up in groundwater and in streams. Pesticides in streams are likely to result in decreased biodiversity.

Fertilizers usually contain nitrate, ammonium and phosphate ions. If too much fertilizer ends up in groundwater and streams this can lead to eutrophication. In severe cases algal blooms can occur. Fertilizer in streams stimulates a strong increase in plant growth. Floating algae are increasing very fast in number, and strongly increase turbidity. Plants in deeper water layers do not get enough light, and die. Decomposition sets in, and can result in such a strong decrease in oxygen level that species living on dissolved oxygen like crustaceans and fish die.

All measurements need to be done both upstream and downstream of the field.

Techniques for measuring pesticides do exist but are usually too complicated for school use.

Biodiversity can be determined by sampling and using Simpson's diversity index.

Fertilizers (nitrate, ammonium and phosphate ions) can be measured using test kits.

Estimation of the abundance of organisms

Sampling

Unless your time is unlimited, you need to sample an area or a population rather than counting every organism within it. In doing this, we assume the sample is representative of the whole and so need to plan our sampling by using a standard sampling unit and thinking about how to count, what to count, how many samples and where they will be taken from.

The abundance of plants and sessile animals (animals that cannot move in their adult stage, e.g. limpets, barnacles, sea anemones) can be estimated by sampling using a **quadrat**, which is a frame of specific size which may be divided into subsections. Often, a quadrat area is $0.25\,m^2$ (the frame measuring $0.5\,m$ along each side), but the size of the quadrat chosen is dependent on the size of the organisms being sampled. An area of $0.25\,m^2$ is of no use in sampling mature tree distribution, when you may need a $25\,m^2$ quadrat. For sampling algae or lichens on tree trunks or walls, a quadrat of side $10\,cm$ would be large enough. There is a balance to strike between increasing accuracy with increasing size and time available and the number of times a quadrat is placed.

How many quadrat samples, and of what size?

These will vary depending on the ecosystem, size of organisms and their distribution. But you can work out how many samples to take and what size the quadrats should be, quite simply.

As you increase the number of samples, plot the number of species found. When this number is stable, you have found all species in the area, so in Fig. 16.6, eight samples are enough.

If you increase the size of the quadrat (e.g. from side length 10 cm, 15 cm, 20 cm and so on) and plot the number of species found, when this number reaches a constant, that is the quadrat size to use.

How to place quadrats?

Quadrats can be placed randomly or continuously or systematically (according to a pattern). **Random quadrats** may be by throwing the quadrat over your shoulder but we do not recommend this as it could be both dangerous and not random – you may decide where to throw. The conventional method is to use random number tables and place the quadrats on a grid marked on the ground following the random numbers generated.

Stratified random sampling is used when there is an obvious difference within an area to be sampled and two sets of samples are taken.

Continuous and systematic sampling is along a transect line. This is a tape or length of rope marked at intervals and laid carefully across the area to be sampled. You might use this to look at changes in organisms as a result of changes along an environmental gradient, e.g. zonation along a slope, a rocky shore or grassland to woodland, or to measure the change in species composition with increasing distance from a source of pollution.

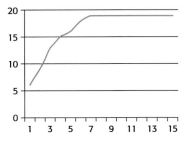
Fig. 16.6 Number of species and quadrat size

Continuous sampling records every species on the line but, more often, systematic sampling at set distances is used. Often this is widened out to a belt transect instead of a line and quadrats are placed at intervals along the belt. This is an **interrupted belt transect**.

Abundance

What is counted in the quadrats depends on what is there. Animals that do not move away can be counted as individuals, e.g. limpets and barnacles. Plants spread out and grow, so **percentage cover** is often measured instead of individual numbers. This is an estimate of the coverage by each species and it sometimes helps if the quadrat is divided up for this. As they overlap or lie in different storeys in a forest, the percentage cover within a quadrat may be well over 100% or much less if there is bare ground. **Density** (mean number of plants per m²) or **frequency** (the percentage of the total quadrat number that the species was present in) can also be measured within the quadrat.

The percentage cover can be estimated by comparing the sample area with Fig. 16.7. The percentage can then be graded on a scale from 0 to 5 or the ACFOR scale by using Table 16.3.

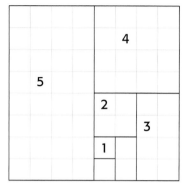

Fig. 16.7 Percentage cover estimation with a quadrat

Percentage cover (%)	ACFOR scale	Score
>50	Abundant	5
25–50	Common	4
12–25	Frequent	3
6–12	Occasional	2
<6, or single individual	Rare	1
absent		0

Table 16.3 Percentage cover scales

Sampling can be done in other ways, depending on the organisms you want to sample. Holly leaf miners may be sampled by picking leaves on a holly bush; fish parasites by capturing the fish and counting the parasites on each.

A **capture, mark, release and recapture** technique called the **Lincoln Index** is used to estimate the population size of animals which move about or do not appear during the day. The actual method of capture will depend on the size of animals.

In an area, a sample of the population is captured and marked in some way that is non-harmful and does not expose them to higher predation levels than non-marked individuals. For example, dog whelks on a rocky shore or woodlice in a woodland can be marked with a spot of non-toxic paint. They are then released and allowed to remix with the population. Once they are mixed, a second sample is taken in the same way as the first and the proportion of marked and unmarked individuals recorded. At least 10% of the marked sample should be recaptured if this estimate is going to be fairly accurate.

Assumptions made are that:

- Mixing is complete, i.e. the marked individuals have spread throughout the population.
- Marks do not disappear.
- Marks are not harmful nor increase predation by making the individual more easily seen.
- It is equally easy to catch every individual.
- There are no immigration, emigration, births or deaths in the population between the times of sampling.

This can be expressed as:

n_1 = number of animals first marked and released
n_2 = number of animals captured in the second sample
m_2 = number of marked animals in the second sample
N = total population

then $\dfrac{m_2}{n_2} = \dfrac{n_1}{N}$ or

$$N = \dfrac{n_1 \times n_2}{m_2}$$

To do

Lincoln Index calculations

Use the formula to calculate the following populations:

1 Woodlice are terrestrial crustaceans that live under logs and stones in damp soil. To assess the population of woodlice in an area, students collected as many of the animals as they could find, and marked each with a drop of fluorescent paint. A total of 303 were marked. 24 hours later, woodlice were collected again in the same place. This time 297 were found, of which 99 were marked from the first time. What is the estimated population of woodlice in this area?

2 While studying field voles, an ecologist caught 500 and ringed one foot on each vole before releasing them. Every day for the next two weeks he examined the waste material found in the nests of their predators. He collected a total of 300 field vole skulls and 15 rings. How many field voles were probably in the area examined?

3 100 water snails were captured from a pond and marked with a small spot of paint on the undersides of their shells; they were then returned to the pond. One week later, a second sample of 100 was captured, of which 25 had paint spots on their shells. What is the estimated number of snails in the pond?

4 2000 beetles of a particular species were caught in a given area, marked and liberated. Later, 200 individuals were recaptured and, of these, 20% had the mark. Based on this information, what is the estimated size of the beetle population in that area?

5 In a woodland, the undergrowth was sampled for snails and 430 were found and marked. They were then released and the population similarly sampled after a two-week period. This second sampling produced 410 snails, 100 of which were marked. What is the estimated snail population of the woodland?

6 A tropical rainforest is inhabited by a species of large carnivorous lizard. 119 of these lizards were counted in 240 hectares of the forest. What is the population density of the lizards?

Measuring species diversity

Species diversity is the number of different species and the relative numbers of individuals of each species, i.e. species richness and species evenness.

Ecologists try to express diversity in a number. This makes it possible to compare ecosystems or to see whether ecosystems are changing in time. The most common way to turn diversity into a number is **Simpson's diversity index**.

But be careful. The name "Simpson's diversity index" actually describes three related indices (Simpson's Index, Simpson's index of diversity and Simpson's reciprocal index). Here we are using **Simpson's reciprocal index** in which 1 is the lowest value (when there would be just one species) and a higher value means more diversity. The highest value is equal to the number of species in the sample. In the other indices, the value ranges from 0 to 1.

$$D = \frac{N(N-1)}{\Sigma n(n-1)}$$

where D = diversity index

N = total number of organisms of all species found

n = number of individuals of a particular species

Example

	Number of individuals of species		
	A	**B**	**C**
Ecosystem 1	25	24	21
Ecosystem 2	65	3	4

Number of individuals of three species in two different ecosystems

The diversity index of ecosystem 1 can be calculated like this:

$N = 25 + 24 + 21 = 70$

$$D = \frac{70 \times 69}{(25 \times 24) + (24 \times 23) + (21 \times 20)} = 3.07$$

In ecosystem 2, the diversity index is 1.22.

Both ecosystems have the same species richness (3) but in system 1 these are more evenly distributed so species diversity, by this measure, is higher.

A high value for D indicates a highly diverse ecosystem and often a stable and ancient site. In contrast, low values for D are found in disturbed ecosystems like logged forests. Pollution also results in low values for D. Agricultural land has extremely low values for D, as farmers try to prevent competition between their crops and other species (weeds).

Simpson's diversity index is most often used for vegetation, but it can also be applied to animal populations.

To do

1 What is the lowest possible value for D in Simpson's diversity index?

2 A certain ecosystem contains 79 organisms of species A, 15 of species B and 7 of species C. Calculate the diversity index.

3 The ecosystem in question 2 is sprayed with herbicide. Afterwards, 75 organisms of species A, 1 of species B and 0 of species C are found. Calculate the diversity index.

Measuring productivity of the system

Primary productivity

In **aquatic ecosystems** (both marine and freshwater ecosystems) the **light and dark bottle technique** can be used to measure both the gross and net productivity of aquatic plants (including phytoplankton). This is simple but has given us a good idea of the productivity of the oceans and of many lakes. In this technique, two bottles are filled with water from the ecosystem and equal amounts of aquatic plants of the same species. One of the bottles is made of clear glass, whereas the other is of dark glass or is covered. The productivity is usually calculated from the oxygen concentrations in the bottles. The procedure is:

1 The oxygen concentration of the water is measured by chemical titration (Winkler method) or oxygen probe, and recorded as mg oxygen per litre of water.

2 Equal amounts of plants of the same species are put into a clear glass bottle and a dark bottle.

3 Both bottles are completely filled with water and capped. (No air should be present.)

4 After incubation for several hours, the oxygen levels in both bottles are measured and compared with the original oxygen level of the water. The incubation can take place in the laboratory or outdoors in the ecosystem of investigation.

In the light bottle, photosynthesis and respiration have been occurring. In the dark, only respiration occurs. (See page 320 (To do) and pages 31–32.)

In **terrestrial ecosystems**, you can do a similar experiment with square "patches":

1 Three equally sized patches with similar vegetation (e.g. grass) are selected.

2 One of these is harvested, i.e. all vegetation is removed, the soil is washed out of the roots and the material is dried (patch A).

3 The second patch (B) is covered with black plastic whereas the third patch (C) is just left as it is.

4 After a suitable time period, patches B and C are harvested as described for patch A.

5 Now GPP, NPP and R can be calculated (usually per m^2) (see page 32).

Secondary productivity

In a typical experiment, a herbivore is fed with a known amount of food. The procedure is that the food and the herbivore(s) are weighed. After a suitable time period, the remaining food, the herbivore(s) and the faeces are weighed.

To do

Productivity calculations

GPP − R = NPP

Students measured the productivity of two ecosystems (see page 319); ecosystem 1 was a marine ecosystem, and ecosystem 2 was a terrestrial grassland. Answer the following questions.

1 Assume that the incubation period was 1 hour. Results:

Initial bottle = 8 mg O_2 L^{-1}

Light bottle = 10 mg O_2 L^{-1}

Dark bottle = 5 mg O_2 L^{-1}

The oxygen increased in the light bottle compared with the initial bottle due to photosynthesis, and the oxygen decreased in the dark bottle due to respiration.

Calculate GPP, NPP and R.

2 After 14 days, biomass is as follows:

patch A: 73 g m^{-2}

patch B: 61 g m^{-2}

patch C: 107 g m^{-2}

Calculate GPP, NPP and R.

Evaluation

This method is difficult if plants are large (trees).

Soil is hard to remove. Too careful washing will not remove all the soil, too rigorous washing will lead to root breakage.

Care must be taken that the three patches are identical at the beginning of the experiment (this includes the amount of vegetation, soil composition, temperature).

The dark patch cannot be covered for too long a time. This will change the moisture level compared to the light patch and can also lead to the plants' death.

Productivity is usually expressed per year.

a Calculate the annual productivity from an experiment that lasts only two weeks, assuming growth rate is constant.

b Think about seasonal changes and explain how this experiment could be made more accurate.

3 The secondary productivity of a snail is measured by feeding the snail with salad leaves during a two-week period.

	Mass at beginning of experiment (g)	Mass at end of experiment (g)
Snail	3.7	4.1
Leaves	16.2	13.1
Faeces	–	0.4

Results of secondary productivity measurement

NSP = GSP − R

a Calculate the gross and net productivity and respiration of the snail.

Evaluation

The food needs to be selected carefully. During the experiment the food may lose weight because of evaporation. The experiment is usually carried out under laboratory conditions. If small animals like snails and caterpillars are used, the amount of faeces might be too small to be measured accurately.

b How do you calculate the productivity per year?

c How do you account for drinking water and urine?

d Is the food representative of the food the animals will eat in nature?

e How can the results be translated to an ecosystem?

Ideas for practical work in ecology

This is far from exhaustive but here are some ideas:

1 What factors affect distribution of different animal or plant species across an environmental gradient? The gradient may be slope, water content, light intensity, human activity or any other.

2 Compare abiotic and biotic factors in two ecosystems, e.g. two ponds of different sizes or one nearer to human habitation; a grazed and ungrazed area of land; a footpath and no footpath; a wood and an open area.

3 What is the distribution of a species and why? Does it show a pattern of distribution? What abiotic factors change?

4 Does earthworm population (or that of any measurable soil or aquatic organism) change with different land use?

5 What factors influence decomposition rates of (any of) leaf litter, organic material, biodegradable plastic bags? Place the substance in mesh bags in the soil and observe over time by digging up

samples at intervals and recording changes. Or do this with bags of different mesh size. Or put under coniferous and deciduous trees or under trees and under grass.

6 Compare soil organisms in two habitats.

7 Investigate the effect of pollution levels on a biotic indicator. Look at lichen growth on walls or tree trunks with distance from a source of sulfur dioxide or from an industrial area into the countryside. Or look at freshwater species as biotic indicators in clean and more polluted waters.

8 Measure the oxygen concentration of a polluted and an unpolluted water sample by oxygen probe and/or the Winkler titration method.

9 Investigate succession on a bare surface over time. This may be a bare lot of soil, burnt area, microscope slides suspended in a pond or the sea. Remove a slide at intervals and look at it under a microscope.

10 Simulate the effect of acid rain on seed germination by measuring germination rates of seeds in different concentrations of sulfur dioxide (using sodium bisulfite tablets is a simple way of doing this and breaking them into a half, a quarter, an eighth).

11 Measure productivity rates of algae (light and dark bottles) or grassland. Compare different samples from different habitats.

12 Estimate population sizes of mobile animals by the capture-mark-release-recapture method. This could be a simulation with a tank full of seeds, beads, mealworms or in the field, with woodlice, dog whelks or other suitable animals.

13 Compare species diversity of plants in two different areas.

14 Simulate predator-prey interactions with your colleagues using a blindfolded "predator" who finds prey of different densities. The prey can be any small objects that can be picked up, e.g. beads, washers, marbles. Modify this in various ways by changing density of prey, territory to search, remove the blindfold and camouflage one type of prey (green and red beans scattered on grass).

15 Estimate the energy budget of a suitable organism, e.g. stick insect, silkworm, locust, by recording initial and final weight of population of organisms, food in, food wasted, waste out and oxygen uptake in a respirometer to calculate the efficiency of conversion.

16 Investigate heavy metal or salt tolerance in different grass species by planting seeds in different concentrations of a toxic agent and measuring germination and growth rates. Compare this with species growing next to a road/slag heap/industrial waste heap.

17 Sample an ecosystem for animals (e.g. use a pooter, pitfall traps, sweep nets) and identify the animals using published keys or make your own keys.

18 Make a key to identify a collection of organisms or objects. Draw up a paired statement key or a spider key.

19 Investigate a rotten log or fallen tree trunk. What is there in this community? Make a food web of the species. Monitor the process of decay.

20 Investigate where two ecosystems meet – the ecotone. What is here? Is species diversity different to each of the ecosystems?

21 If you live in a region where there are hedges separating fields, investigate species diversity in different hedges. This can give an estimate of the age of the hedge.

Calculations on human populations

As with any organism, we ask ourselves these questions with respect to human population:

How many people are there? Where are they?

What are the trends in population change?

Censuses record population statistics and researching these for your own country or others can give you a picture of population change.

A survey of obituaries in a local newspaper may provide ages and dates of birth of those who die.

To do

1 The USA has a population growth rate of 0.894%. How long would its doubling time be if this was maintained?

2 Why do CBR (crude birth rate) and CDR (crude death rate) decline? (Think about the demographic transition model.)

	EU	Afghanistan	Brazil	India	Japan	Germany	UK	USA
Population size (10^6)	491	32.7	192	1148	127	82	61	304
Median age	–	17.6	30	25	44	43	40	37
Natural increase rate	0.12	2.6	0.98	1.6	−0.14	−0.044	0.276	0.88
Crude birth rate	10	46	16	22	7.9	8.2	10.6	14.2
Crude death rate	10.2	19	6	6	9.3	10.8	10	8.3
Life expectancy	78	44	72	69	82	79	79	78
Literacy rate, male	–	13	88	73	99	99	99	99
Literacy rate, female	–	4.3	88	48	99	99	99	99
Total fertility rate	1.5	6.58	1.86	2.76	1.22	1.41	1.66	2.1

Table 16.3

3 Look at Table 16.3 taken from the *CIA World Fact Book*. The figures are estimates for July 2008. Literacy rate is the number who can read and write over the age of 15. The EU consists of 27 European countries.

a Calculate and compare the doubling rate of Afghanistan, Brazil, India and Germany.

b In which stage of the DTM are these countries and why?

c Comment on the median ages in each country and discuss the effects this will have on the society over the next 50 years.

d Why may a country's population size increase when its NIR (natural increase rate) is negative?

4 Look up your own country in the *CIA World Fact Book* and write a paragraph on the changes that you predict in the population over the next 50 years.

5 Describe and explain the consequences of the data in Figs 16.8 and 16.9.

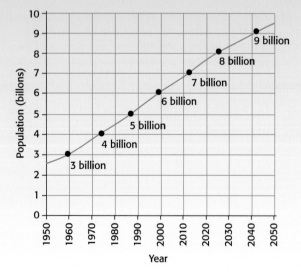

Fig. 16.8 World population 1950–2050, estimated

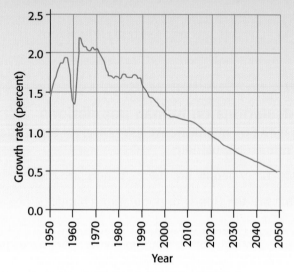

Fig. 16.9 World population growth rates 1950–2050

Calculating sustainable yields

Students should be able to calculate and explain sustainable yields from given data.

Sustainable yield (SY) may be calculated as the rate of increase in natural capital, i.e. that which can be exploited without depleting the original stock or its potential for replenishment. For example, the annual sustainable yield for a given crop may be estimated simply as the annual gain in biomass or energy through growth and recruitment. Thus,

SY = (total biomass or energy at time t + 1) − (total biomass or energy at time t)

SY = (annual growth and recruitment) − (annual death and emigration)

To do

1 In the Mkumazi game reserve in Tanzania where hunting is allowed the population of elephants is 750. During the course of a year researchers counted 60 births and 55 deaths. During the same time 10 young bulls left the reserve and 15 previously unknown elephants appeared.

  a What is the sustainable yield?

b During the year hunters killed 8 elephants for trophies. Is this sustainable utilization?

2 Researchers estimated the biomass of wood in a small temperate forest, which amounted to 500 000 tonnes. A year later an estimate amounted to 501 000 tonnes. What is the sustainable yield?

Calculating ecological footprints

Ecological footprint calculations convert various kinds of consumption and waste production into a land area needed to produce or service it. The four main areas to consider are fossil fuel and energy consumption, food land, forest products and land required for towns, roads and factories (consumed land).

Ecological footprints are stated as the land area in hectares needed by a particular population, e.g. a country or community, or as a per capita per annum of that population. Ecological footprints will always be conservative estimates. Why?

Ecological footprint formulae

Food land

This is the total crop and pasture for plant and animal food in the country (plus overseas land used for imported food items).

Calculation: area of cropland and pasture / population.
Units: w ha.cap^{-1}

Consumed land

This is land consumed for houses, factories, roads, sports grounds, etc.

Calculation: area of consumed land / population.
Units: x ha.cap^{-1}

Energy land

This is land to absorb carbon dioxide from the combustion of fossil fuels.

Calculation: fuel used in gigajoules / population × energy area.

Units: y ha.cap^{-1}

Forest land

This is the forest area needed to produce sawn wood, paper and other wood products.

Calculation: wood products (mass or volume) / population × productivity.

Units: z ha.cap^{-1}

To do

Ecological footprint of India

Calculate the ecological footprint of an average Indian person.

Use the data in Table 16.4 to calculate w, x, y and z. Add them together to find the total average ecological footprint for a citizen of India.

Population	1 148 000 000
Cultivated area	250 000 km^2
Consumed land	25 000 km^2
Fuel used	13 500 PJ (1 PJ = 10^6 GJ)
Energy area: fuel	6000 J = 1 ha yr^{-1}
Energy area: energy	200 GJ = 1 ha yr^{-1}
Wood products	300 000 m^3
Wood mass to volume	1 tonne = 1.8 m^3
Productivity (volume)	2.3 m^3 ha^{-1} yr^{-1}
Peta = 10^{15} ; giga = 10^9	1 PJ = 10^6 GJ
1 hectare = 10 000 m^2	1 km^2 = 100 ha

	Results
Cultivated area	
Covered land	
Fuel used	
Wood products	
Total	

Table 16.4

What else?

Obviously some areas of the course provide more opportunity for practical investigative work than others, but your portfolio of practical work should contain more than laboratory or field reports. You may make observational or qualitative notes on visits you make or workshops that you attend. You may include the results of questionnaires that you carry out, e.g. on recycling levels of different groups of people, or perceptions of environmental issues.

In topics other than those already covered, here are some suggestions for practical activities.

● Visit two farms – intensive and extensive, commercial and subsistence (or hobby), arable or livestock or mixed, large or small. They may be terrestrial or aquatic farms. Ask the farmer or manager to explain what they do in terms of systems.

● Visit a local dam or reservoir with a visitor or educational centre.

- Visit a local refuse collection area – landfill, incinerator, recycling plant. If this is not possible due to availability or health issues, ask a local council representative to come to you to explain their policy and strategy on waste management.
- Visit or hear from a local zoo/botanic garden/nature reserve.
- Develop your own nature reserve in the school grounds or nearby.
- Find an EIA (environmental impact assessment) on a local project, perhaps a new road or shopping centre.
- Do your own baseline study of an area and imagine a development that might be put there. What changes would there be? How would you monitor these?
- Survey SDW (solid domestic waste) production by your household/school/town.
- Research what happens locally to old fridge or air conditioners with ODSs (ozone-depleting substances).
- Ask a car dealer or manufacturer to talk to you about emissions from automobiles, diesel and petrol engines, engine efficiency.
- Arrange a question time for a panel of local politicians/teachers/ professionals and ask them questions about their policies on climate change, waste reductions, congestion, planning, loss of biodiversity, etc.
- Organize a debate on the issue of climate change, environmental worldviews or sustainability.
- Devise and carry out a survey of environmental viewpoints among different groups of people.

Key words

If you have worked through the preceding chapters, there should be no new terminology in this one. Perhaps this is the time to review the glossary in the *Environmental Systems and Societies Guide* and make sure you understand the terms.

17 Writing your extended essay

If you are studying for the full IB Diploma, then you must write and submit an extended essay as part of the requirement. This chapter gives you some advice on how to approach this task and to see it as something that will give you research skills that you need in tertiary education.

In brief, the extended essay is compulsory and marked externally, so it is sent to an IB examiner. It is up to 4000 words and *should* not take you longer than about 40 hours of work. You may choose any topic for your extended essay but it must be submitted under one of the IB Diploma Programme subject areas. You can be awarded up to 1.5 points (out of the maximum 45) for your extended essay and up to 1.5 points in Theory of Knowledge (essay and presentation). So the extended essay and Theory of Knowledge can gain you up to 3 core points. That may not seem like very many points for the amount of work involved but, if your extended essay is a good one, it may make the difference between getting into the university or college course of your choice or not.

Why the extended essay is a good thing

Researching and writing an extended essay may be the first time that you take real control of your academic study. Most of the time in education, you do what your teachers tell you to do to achieve the highest marks you can or to pass examinations. But you do not get much choice of the subjects which you study. Here is a chance, at last, to select a topic which you find of interest and which you can research yourself. In writing the extended essay, you will develop research skills that will be of benefit later on and develop skills of critical thinking and systematic research. We all hope that you also enjoy the process and gain satisfaction from intellectual discovery.

Tips 1

Ten useful tips

1 Keep it under control and do not allow your other work to suffer because of your extended essay. You are not likely to win a Nobel Prize for this piece of work, nor change the path of environmental research.

2 It usually makes more sense to be studying the subject and not just think that you can pick up the terminology of the subject as you go along.

3 Use your supervisor. They are allowed to direct you and help you but not to write your extended essay or do the practical work for you. If you have a meeting organized with them, be there. They are there to help.

4 Keep to interim deadlines. Your school may have these. If not, make your own. With a large project, you need markers so you can see your progress.

5 Select your research title with great care. Select subject area, then topic, then title. Make sure it is relevant, i.e. is about an environmental systems issue, do-able within the word count and ethically sound. You need a question to which you do not know the answer at first. That is why you are doing the extended essay. Be patient. It takes time to find the right question.

6 Aim to include at least some experimental or observational data that you have collected.

Analyse this and interpret it. Comparing it with secondary data is helpful.

7 Read and cite a range of sources from some of these: academic papers, journals, textbooks, personal conversations, reputable websites, books and reviews.

8 Read the extended essay guide. It is full of helpful information and instructions to you. There is no excuse for not reading this and following its advice.

9 Keep all your notes and your draft until you have your results. Your supervisor and the IB have to be sure this is your own work. These are your evidence.

10 Be safe and ethical. Do not do an extended essay in an area that inflicts damage on an organism or the environment, or may endanger you or anyone else.

Common errors
- Your research question is too broad or unfocused.
- Your research question cannot be answered in 4000 words.
- Your sources are not relevant to the research question.
- You cite websites uncritically.
- You have a one-sided argument and no counterclaims.

Your supervisor
How your school determines extended essay supervision is a matter for your school but you must have a supervisor who is a teacher within the school. Your supervisor may or may not be a subject specialist in your chosen subject but you should expect your supervisor to help you by:
- advising and guiding you on your research question and on research skills and resources
- ensuring that your research question is manageable and ethically sound
- giving you the assessment criteria
- reading and commenting on your first draft
- confirming that this is your own work
- submitting a predicted grade (A–E) to the IB.

Your supervisor will probably spend about 3–5 hours with you and conducts a short concluding interview (viva voce).

Your supervisor does not write your extended essay for you, edit your extended essay or do the practical work for you if there is any.

Essential documents
- Extended essay guide
- Academic honesty guide
- Recent extended essay subject reports
- Examples of extended essays in this subject or in Environmental Systems.

How to start
Decide if this subject area is the one in which you want to write your extended essay.

Ask yourself these questions: Do you enjoy it? Do you read around the subject anyway?

Then take some trips – to a local park or protected area, a local farm, reservoir, stream, pond, zoo, school grounds, your back yard or garden, a site that has construction going on, or has been built, a landfill site, etc. Go anywhere where organisms live and humans are having an impact.

Observe what is going on. Make some notes and ask yourself some questions. What is there? How many are there? Where are they? What are they doing? Why? The "they" may be plants, animals or human animals or all three. What has changed? Why has it changed? What impact have humans had here? If you can compare two similar sites or a "before and after" site – upstream and downstream of a pollutant, for example – so much the better.

Come up with a list of possible questions that you think you could investigate.

Then look again at this list and ask yourself:
- Can I find an answer to the question?
- Is the answer too obvious, e.g. Is the environment harmed by the new trunk road? (You would want to change the questions to ask how the environment is altered.)
- Does the answer need investigating by an experiment, questionnaire, literature review or a combination of these?
- Can I try to find the answer within the time I have for this? If it involves you comparing ecosystems that are on different continents because you are visiting the other one in the summer vacation, or growing plants that will take months, think carefully about this. You always need to build in extra time in case things to go wrong and you need to redo them.

Although you do not have to, an experimental or observational approach may be more rewarding to you than a pure literature review.

Keep it simple. Don't be overambitious in your choice of title or your need for equipment. You do not need complex equipment to undertake an excellent essay.

Read the extended essay guide, particularly the sections in it on Introduction Outline, Details, Assessment criteria and subject guidelines for Environmental Systems and Societies.

Read previous extended essays in this subject or in Environmental Systems.

What you must include

This subject looks at the systems of the natural world and the effect of humans upon these. So you need to include both in your extended essay. You do not need to give equal weight to environmental systems and societies but should include both to some extent. If you investigate an ecological principle such as zonation or succession, then include the interaction of humans with this. If you decide to research environmental philosophies, you also need, in your extended essay, to have an element of natural systems.

Take the systems approach in your extended essay. If you are only looking at one organism in isolation, the extended essay may be better submitted as a Biology extended essay.

You cannot just have a descriptive extended essay, it must have an analytical element. This means you will score poorly if you just describe a farm or a zoo or ecosystem. But you could compare it to another or to itself in the past or under different conditions and evaluate the relationship.

You need an argument to which there are counterarguments.

Keep the focus sharp. If it is too broad it can be superficial. See the extended essay guide for examples of focused and broad topics in the subject guidelines.

Tips 2

Advice for the best extended essay possible

1 Select your research question with great care. This makes or breaks the extended essay. It must be specific enough to address within 4000 words but also allow you to argue critically, not just describe. Comparisons are good for this.

2 Set the scene in an introduction by referring to the background and secondary sources. Explain what the implications of the research question may be.

3 Either collect primary data or secondary or both. Primary may be fieldwork, laboratory work, interviews, questionnaires or a combination of these. Secondary may be literature reviews or from other sources. If you plan to collect primary data, make sure your methodology is valid, reliable and repeatable and also reference some secondary data as well. If you are collecting secondary data, beware the website with no author or websites that are not reliable. You need to be sure that the information is from a reputable source, is cross-referenced if possible and is not trying to sell to you or persuade you of a cause and is not just plain wrong. Try to collect data from sources other than the Internet, however tempting it may be.

4 Manipulate and analyse the data. It is not enough just to copy someone else's or to put your results with no analysis and discussion. Present your analysis of results clearly. Draw graphs and tables if you can. Show your personal involvement and make arguments and draw conclusions.

5 Make links between the work you have carried out and the body of knowledge in the subject. You must use the terminology of the subject and a systems approach.

6 Give your own viewpoints on the evidence as well as those of others. Mention criticisms of your work and how you might address these if you had the time or equipment.

7 Summarize your conclusions at the end of the extended essay, mention work that you could have done and any new questions raised. The conclusion must address the research question or hypothesis directly.

8 Although you may write your extended essay using only internet sources or library shelves, you are more likely to score marks if you can show personal contact with the environment or people and initiative in selecting the research question that addresses the interrelatedness of components of a system.

Presentation

You can achieve up to 4 marks in presenting your extended essay correctly. This means simple things such as numbering the pages, having a title page and contents page and labelling graphs and tables clearly and correctly. Make sure you read the advice in the extended essay guide on references and bibliographies on pages 16–18, and follow it.

Abstract

The abstract goes at the front of your extended essay and is 300 words or fewer. It summarizes what the extended essay is about and contains the research question, investigation and conclusions in brief. Write it in the present tense and in the passive. An example is given in the box. Determine how many marks you would award this abstract.

Abstract

This essay addresses the environmental disasters of the Love Canal in the United States of America and Minamata Bay in Japan from which the question has arisen:

How did the companies involved cause the pollution and what was done to attempt to resolve the pollution?

The events in both disasters, and effects of pollutants are described and the common negligence of the companies and governments and action taken are compared. In brief, the chemical companies in question dumped waste chemicals into the neighbouring water systems without any consideration to the harm they may cause the natural and human environment. This practice continued for many years in the two nations and gradually more and more problems arose. The aquatic life became poisoned, the chemicals leaked into the environment and the communities of people and animals began to become ill from the slowly increasing toxicity of the environment. The communities attempted to alert the government through peaceful measures yet their concerns were not highly considered by the ministers. There was an obvious discrepancy of the degree of the problem in both cases and so little action was taken by those in the governing bodies to resolve the situation; the companies continued to freely dump the deadly waste chemicals. The Japanese government took eleven years from hearing of the problem to publicly charging the Chisso Company with the environmental damage. Similarly, the Hooker Chemical Company continued to pollute freely until the Love Canal community gained publicity by holding two agents hostage, prompting FBI involvement and eventual action from the government. The dispute of liability, in both cases, was a serious preventative as to the establishment of a clean-up program as it seems without lengthy legal action it was impossible to label the cause of pollution and so begin the restoration. The clean-up process in any environmental disaster is a slow procedure yet it is worth noting the painfully slow decontamination and dredging of Minamata Bay that, 65 years later, was still occurring.

The essay informs the reader of the similarities of negligence and lack of action in these two environmental disasters and highlights the necessity of regulation in the protection against pollution. Important criticisms have been made against the governments involved in representing the environments affected. In both cases, the officials failed to police the dumping of waste, protect the citizens and act effectively against those jeopardizing the environment. This essay considers the liability of the culprits but also draws attention to the responsibility of the government officials to act against pollution as well as react in the best interest of the public.

Academic honesty

The IB requires your school to have an academic honesty policy and you should have seen this and discussed what it means in practice. In the widest sense, being academically honest means that you do not use the ideas, creative works, words, photographs or diagrams of someone else without acknowledging that they came from that source. Of course, you build upon the ideas of others as we do not live isolated from the rest of human experience and knowledge, but it is only right that you do not imply that it is your authentic work if it is not. It is only too easy to cut and paste words from websites or to scan materials. There is nothing inherently wrong with that but attempting to persuade others that this is your own material, words or ideas without stating the original source is deemed to be plagiarism by the IB.

It is your responsibility to ensure that the extended essay is your own work and you have to sign a cover sheet when you hand it in to say that this is the case. Your supervisor also has to sign to say that they believe this to be your own work. Referencing and citations tell the reader where the information was found and show that you have acknowledged your sources. You must set out a reference so that the reader can find this information for themselves. You might reference a book or paper, magazine or newspaper, internet site or interview. Failure to acknowledge a source is plagiarism, whether intentional or by accident. Avoid it at all costs by giving citations to the facts and quotations you use unless they are your own or are common knowledge.

Collusion is another form of malpractice and this is when you support the malpractice of another candidate. It may be working together with someone else on a piece of work for assessment when it should be done individually, or taking your work from or giving it to another for the purpose of submitting to the IB. This carries the same penalties as plagiarism as do duplication of work for assessment and cheating in examinations.

While it would be easy if plagiarism fell into a black (yes you did) or white (no you did not) box, there is a grey area in the middle. This is when ideas or words may now be common knowledge (e.g. the Earth is not flat, gravity is a force pulling us into the Earth) yet were once the "intellectual property" of one person, or when you paraphrase someone else's words to the extent that they become your own. The best practice you can adopt is to keep track of all material that you read or note for your extended essay and other work and *if in doubt, cite*.

Keep your notes, drafts and copies of any papers or print-outs that you use until after the examination session closes (15 September for May examinations and 15 January for November examinations). Then you can be sure that you have a paper trail to show that you did the work. Your supervisor may ask to see these notes as they build up a picture of how you worked. They are your evidence.

The 10–15 minute long **viva voce examination** is a chance for you to talk about your extended essay with your supervisor and show what you have learned.

What exactly is plagiarism?

We emphasize the positive aspects of academic honesty – know how to reference correctly – but, sadly, plagiarism is the most common type of malpractice that the IB finds in work submitted for assessment. In many cases, this is due to negligence or ignorance and is not intentional but you are still guilty of malpractice and the penalties range from no mark being awarded for that component to no mark awarded for the subject. It is your responsibility to get this right.

This article is by Chris Willmott and published in the *Journal of Biological Education* (2003) volume 37 number 3 pp 139–140.

What exactly is plagiarism?

Taking someone else's words or ideas and presenting them as your own work is known as plagiarism. But how much do you need to change something before it becomes a legitimate re-working?

The paragraph below is taken from *Pharmacology* (4th edition, 1999) by Rang, Dale and Ritter. Study the essay extracts in the table and decide whether or not you consider the author of the work to be guilty of plagiarism – some may be more obvious than others!

During the last 60 years the development of effective and safe drugs to deal with bacterial infections has revolutionised medical treatment, and the morbidity and mortality from microbial disease have been dramatically reduced.

Essay extract	Plagiarism? (✓ or ✗)
During the last 60 years the development of effective and safe drugs to deal with bacterial infections has revolutionised medical treatment, and the morbidity and mortality from microbial disease have been dramatically reduced.	
During the last 60 years the development of effective and safe drugs to deal with bacterial infections has revolutionised medical treatment, and the morbidity and mortality from microbial disease have been dramatically reduced. (Rang et al., 1999)	
"During the last 60 years the development of effective and safe drugs to deal with bacterial infections has revolutionised medical treatment, and the morbidity and mortality from microbial disease have been dramatically reduced." (Rang et al., 1999)	
In the 4th edition of their textbook *Pharmacology* (1999), Rang, Dale and Ritter state that: "*During the last 60 years the development of effective and safe drugs to deal with bacterial infections has revolutionised medical treatment, and the morbidity and mortality from microbial disease have been dramatically reduced.*" Such a bold assertion understates the ongoing threat posed by microbial infection. It is estimated, for example, that worldwide there were over 8 million cases of tuberculosis in 1998 (WHO, 2000).	
The development of safe and effective drugs to deal with bacterial infections has dramatically reduced the death rate arising from microbial diseases.	
During the post-war years, the development of effective and safe drugs to deal with bacterial infections has transformed medical treatment, and death and illness resulting from microbial disease has been dramatically reduced.	
The availability of antimicrobial compounds has transformed healthcare in the period since the second world war. People are far less likely to die or even be seriously ill than they had been prior to the introduction of these drugs.	

Plagiarism – the answers

1 The first version listed is an "ice-breaker". It is clearly a verbatim account and is thus seriously guilty of plagiarism.

2 The second version is marginally better, but is still not acceptable. The original work has been acknowledged as a source of ideas and information, but no indication has been made that the text itself has actually been used.

3 In this case the addition of quote marks makes an important distinction from the previous versions. The author is clearly acknowledging that both the ideas and the word order have come from the textbook. It is not therefore guilty of plagiarism. We include this version to highlight a different weakness, namely that stringing together a series of quoted "chunks" of text is a poor way to construct an essay and work written in this way is therefore likely to score low marks.

4 This version of the essay is fine. The quotation is indicated and is used in an appropriate way; it is being critiqued by the author and contrasted with a view supported by a second reference. Not plagiarised.

5 Here we get to the crux of the matter. The fifth and sixth versions of the essay are illustrations of practice that undergraduate students early in their studies consider acceptable but we do not. They are derivatives of the original work with only cosmetic alterations. The wording and sentence construction of version 5 bears a very close relationship with the source and is guilty of plagiarism.

6 Similarly, this is a "thesaurus-ed" or word-swapping version of the same text. A few words have been replaced with synonyms but this is not sufficient to be considered new work.

7 The author of the final essay has made a serious attempt to produce a novel account of the subject. It is still not perfect—lined up as it is here with one original source document, there are still echoes of the thought processes within the work and we would ideally want the student to draw on a number of sources in order that the essay has genuine originality. Nevertheless, significant effort has gone into bringing freshness to the text and we would consider that this is not guilty of plagiarism.

Some extended essay titles

While your research title will be specific to your interests, some of these and a few comments about them may help you in clarifying your ideas. All these titles are real, past titles submitted over the last

10 years as Group 4 Environmental Systems extended essays. Those that address both the environment and societies to some extent would also be acceptable as transdisciplinary extended essays in Environmental Systems and Societies. Those that have an asterisk beside them may not be as suitable.

High scorers

- A comparison of a dairy farm in the Netherlands to one in Tanzania.
- Evaluation of water quality in Ramallah district springs.
- Is the decline in wild salmon in Vancouver due to the increase in number of farmed salmon?
- Which ecosystem within the UK suffered the worst ecological fallout as a result of the Chernobyl disaster?
- A comparison of the diversity and height of plant species on burnt and unburnt areas of Mediterranean coastal ecosystem near Nice, France.*
- An investigation in the soil erosion rates and effect on people and habitat in two areas of the lower slopes of Mt Kilimanjaro, Tanzania.
- Comparing inputs, outputs and efficiencies in an organic and non-organic sheep farm in New Zealand.
- What are the environmental and social impacts of building the Three Gorges Dam in China? (Although based on secondary data, some siltation experiments were carried out.)

Too general

All these are far too broad and global and cannot be addressed within the extended essay to meet the criteria. Avoid such general titles.

- Tourism in Antarctica.*
- Global dimming.* (But this could be a good idea if it were experimental and addressing light and/or heat transmission through different levels of atmospheric pollutants and their effect on the growth/germination of plants.)
- Oceans and their coral reefs.*(But investigating a specific coral reef bleaching or death over time and comparing primary and secondary data would work.)
- The greenhouse effect.** (Don't go there unless you do specific experiments. If thousands of scientists are working on this, what would you say in 4000 words?)
- The effects of oil spillage on marine life.* (Again, if you made this specific to one oil spill, or a comparison of two and their clean-up and restoration, you could have a good title.)
- The nuclear winter.*

Too one-sided or just not right

- The wisdom of Lynx reintroductions in Colorado. (The title suggests the conclusion – that they are not wise – and there is a danger in only putting one side of the argument. Better to keep the title open as Lynx reintroductions in Colorado, success rates, methodology and evaluation.)

- Temporary habitats for aspiring Martians.* (Perhaps a little too flippant as a title. It was about terraforming but, again, too general to score well.)
- The effects of the Chernobyl catastrophe 1986 on Germany and the Bavarian woods 19 years after the event. (How could you compare then and now?)

Too vague

- Do the positive consequences of destroying the Amazon rainforest to benefit humans outweigh the consequences or are the negative consequences so critical that there is nothing worth risking them?
- The study of one key aspect to determine success of a community based conservation project.
- Effect of fertilizer runoff on freshwater ecosystems.*
- The qualities of water and their impact on plants.*

Not addressing the subject

- Antimicrobial resistance: a growing ethical dilemma between economics and the environment.
- Heavy metal contamination in dietary supplements.*
- Does the frequency of noticeable earthquakes off the west coast of S California and the Gulf of Alaska correspond to Southern Oscillation Index Values?*
- An investigation of teenage noise pollution.*

Referencing and citations

As the IB extended essay guide states, academic referencing may follow one of several styles. It does not matter which one you select but you must be consistent and it does matter if there is a full stop or a comma in the right places. Be accurate. Find a guide to referencing on many university websites by putting "academic referencing" into your search engine. One from OUP Australia is at http://au.oup.com/content/general.asp?ContentID=52

About the examinations

Don't be afraid of examinations. If you have prepared well, they can be a satisfying experience and a culmination of your studies in secondary education. Worldwide, there is continuous debate about the purpose and value of examinations but they represent a system that society has evolved in which we are tested and sometimes ranked in order to select us, perhaps for particular careers or courses. They have been criticized as unfair, merely testing how good you are at revising and doing the tests, and a poor means of judging people. However, at the moment, you are in this system and should try to see the IB Diploma examinations as assisting you in your growth and development as a life-long learner and one of the "internationally minded people who, recognizing their common humanity and shared guardianship of the planet, help to create a better and more peaceful world", to quote the IB Mission Statement.

We hope that your studies have not focused almost exclusively on the final assessment (external assessment or EA in IB terminology) but that you have gained knowledge, thought critically and reflectively about that knowledge and enjoyed at least parts of the experience. If you reach the end of the course and think that we have degraded the planet to such an extent that you are pessimistic about its ability to recover or the human race's ability to live in balance with the resources on Earth, don't be. We are resourceful as a species and, being sentient beings, we are aware of our own existence and its impact on others. The Earth may be moving to a new equilibrium, civilization will change, but we are capable of managing our population size and resource exploitation and humans have never yet not used an innovation or discovery that we have found.

But, back to the examinations. By the time you start the examination in this subject, you will have up to 20% of your final marks as IA. You will not know how many until the results are published but you will have done nearly all the work by the time you go into the examination.

There are two papers in EA. Paper 1 is one hour long and worth 30% of the final marks. All questions must be attempted and they require short answers testing your knowledge, application of knowledge and responses to data-based questions. Paper 2 is two hours long and worth 50% of the final marks. It contains questions on a resource booklet in Section A and essay-style questions in Section B, of which you answer two out of four. For both papers you are allowed five minutes of reading time before you start writing. The resource booklet in Paper 2 can contain a lot of information but much of it is in the form of diagrams or tables so do not be daunted by its length. The information is fairly easy to find within it.

> "*I was thrown out of college for cheating on the metaphysics exam; I looked into the soul of the boy sitting next to me.*"
>
> Woody Allen

Tips 1

Read the course guide

Read the course guide carefully. There is more in the guide than just the syllabus or course content. Make sure you have your own copy of this or access to it from your institution. It not only tells you what the course covers but gives a great deal of guidance and advice. Look at these sections:

1 **Nature of the subject**, aims and assessment objectives – these tell you what the course covers, how to approach it and what we hope you will be able to do when you have completed it.

2 **Assessment statements** (AS) – these are expressed as outcomes – what you are expected to be able to do, e.g. Define the term biome or Evaluate the impacts of eutrophication. Each AS is classified as objective 1, 2 or 3 in terms of the command terms (on pages 68–69 of the course guide). These help you to know if the AS is a lower-order skill (1 and 2), e.g. define, list, describe, or a higher-order skill (3), e.g. evaluate, explain. These command terms are also used in the examination papers and you will not be expected to answer an examination question that is of a higher order than the AS. If the AS says "describe", the command term in the examination paper

for that AS would not be of a higher order, e.g. explain. Get into the habit of spotting the command terms and even highlighting them in your tests.

3 **Teacher notes** – under most AS are teacher notes, some shorter, some longer. If the topic of the AS is a well-known one with many resources available, the teacher's notes may be short. But, if the topic is less well-documented or needs additional explanation, the teacher's notes are longer. In Topic 3, they can be detailed and give you much guidance on the AS and the depth to which you should go for the assessment. Do read them.

4 **Command terms** – we have mentioned these already. Make sure you know the meaning of the terms and have practised using them.

5 **Glossary of terms** – the last appendix of the guide is a glossary. Make sure you have a copy of this as it is really useful throughout the course as a reference. The glossary includes words used in the AS and some others which are listed to reduce ambiguity in their meanings. You are not expected to learn the definitions of all these words by rote but you should know what the terms mean and be able to express the meanings in your own words.

Tips 2

What else should you read?

1 Past papers and markschemes
2 Subject reports

Past papers and markschemes are published and sold by the IB. Towards the end of the course, you should look at and do some of these. Practise getting your timing right so that you do not run out of time in the examination. Read the markschemes carefully so you can see how the examiners award marks.

Subject reports are written by the senior examiner after every examination session. They report on each question and how it was answered and on the IA. Here the grade boundaries are published as well. Included in

the subject report are recommendations to your teacher and to you for future examinations. Here are some of the recommendations from the May 2007 subject report for the pilot of this subject.

- A good understanding of terminology and key concepts means that candidates can work quickly in an examination, and are less likely to run out of time.
- Remind candidates to allocate time according to the number of marks given for a question and to use the number of lines as a guide for how much is expected.
- Encourage candidates to think creatively about designing methods for collecting ecological/environmental data.

- Remind candidates to "have a go" at answering questions, as they are often able to pick up marks.
- Clarify the difference between eutrophication and acidification of lakes when teaching these topics.
- Remind candidates that when two action verbs are present in a question, they will be required to do both, e.g. compare AND explain, or describe AND evaluate, to gain all the marks.
- In order to excel in this transdisciplinary subject students are required to think on a number of levels and planes. Therefore teachers should encourage students to think "out of the box". Environmental matters are never far from the news headlines and so teachers should nurture the skill of bringing to the table ideas from a range of sources and combine them with students' syllabus knowledge. These sources could include work covered in other topics, other subjects, background reading and personal experience. Great effort is made to build examination questions and papers that are fair yet challenging and, although great care is taken that syllabus knowledge alone is enough to gain full marks for a question, credit is often given to students who bring into an answer that which is beyond the syllabus; this is particularly relevant in section B of Paper 2.
- Students still need to spend more time considering the true context of the examination question asked and then structure an answer that targets the question. There would still appear to be the familiar "tell them everything I know about topic X", this approach rarely produces high marks.
- Clarity of handwriting is also very important; if answers are difficult to read it is likely that students will lose marks over the paper as a whole.

Tips 3

Grade descriptors

The IB Diploma examinations are graded according to criterion-based assessment. This means that your work is marked according to a markscheme and then the boundaries between each grade, e.g. between a 6 and a 7 or a 3 and a 4, are determined according to written criteria. These are given below for this subject. The first paragraph is for the EA, the second for IA. This means that everyone could, in theory, gain a grade 7 if their work meets this criterion. In practice, with a large number of students, some will do very well, some not so well and most are grouped around the average mark so plotting marks gives a normal distribution. But there is no set percentage each session that has to gain a particular grade. This means that your teacher cannot say how many marks a grade 7 equates to as it varies from examination session to examination session. This is fair for you as it removes any discrepancies if one examination is harder or easier than previous years.

Grade 7 *Excellent performance*

Demonstrates: comprehensive understanding of factual information, terminology, concepts, methodology and skills; high level of ability to apply and synthesis information, concepts and ideas; quantitative and/or qualitative analysis and evaluation of data thoroughly; fully reasoned balanced judgments based on holistic consideration of multiple viewpoints, methodologies or opinions; communication of a well reasoned and justified personal viewpoint, whilst critically appreciating the views, perceptions and cultural differences of others; a high level of proficiency in communicating logically and concisely using appropriate terminology and conventions; insight and originality.

Demonstrates personal skills, perseverance and responsibility in a wide variety of investigative activities in a very consistent manner. Works very well within a team and approaches investigations in an ethical manner, paying full attention to environmental impact. Displays competence in a wide range of investigative techniques, paying considerable attention to safety, and is fully capable of working independently.

Grade 6 *Very Good*

Demonstrates: broad detailed understanding of factual information, terminology, concepts, methodology and skills; good level of ability to apply and synthesis information, concepts and ideas; quantitative and/or qualitative analysis and evaluation of data; reasoned balanced judgments based on holistic consideration of multiple viewpoints, methodologies or opinions; communication of a reasoned and justified personal viewpoint, whilst critically appreciating the views or perceptions and cultural differences of others; a competent level of proficiency in communicating logically and concisely using appropriate terminology and conventions; occasional insight and originality.

Demonstrates personal skills, perseverance and responsibility in a wide variety of investigative activities in a very consistent manner. Works well within a team and approaches investigations in an ethical manner, paying due attention to environmental impact. Displays competence in a wide range of investigative techniques, paying due attention to safety, and is generally capable of working independently.

Grade 5 *Good*

Demonstrates: sound understanding of factual information, terminology, concepts, methodology and skills; competent level of ability to apply and synthesis information, concepts and ideas; some quantitative and/or qualitative analysis and evaluation of data; balanced judgments based on holistic consideration of multiple viewpoints, methodologies or opinions; communication of a reasoned personal viewpoint, whilst appreciating the views or perceptions and cultural differences of others; a basic level of proficiency in communicating logically and concisely using appropriate terminology and conventions;

Demonstrates personal skills, perseverance and responsibility in a variety of investigative activities in a fairly consistent manner. Generally works well within a team and approaches investigations in an ethical manner, paying attention to environmental impact. Displays competence in a range of investigative techniques, paying attention to safety, and is sometimes capable of working independently.

Grade 4 *Satisfactory*

Demonstrates: secure understanding of factual information, terminology, concepts, methodology and skills with some information gaps; some level of ability to apply and synthesis information, concepts and ideas; descriptive ability with quantitative and qualitative data, limited analytical skills and rudimentary evaluation consideration of some contrasting viewpoints or methodologies or opinions; communication of a personal viewpoint, some awareness of perceptions and cultural differences of others; a basic limited proficiency in communicating using appropriate terminology and conventions;

Demonstrates personal skills, perseverance and responsibility in a variety of investigative activities, although displays some inconsistency. Works within a team and generally approaches investigations in an ethical manner, with some attention to environmental impact. Displays competence in a range of investigative techniques, paying some attention to safety, although requiring some close supervision.

Grade 3 *Mediocre*

Demonstrates: limited understanding of factual information, terminology, concepts, methodology and skills with information gaps; basic level of ability to apply and synthesis information, concepts or ideas; descriptive ability with quantitative and qualitative data; vague consideration of some contrasting viewpoints or methodologies or opinions; basic communication of a personal viewpoint, weak awareness of perceptions and cultural differences of others; communication lacking in clarity, with some repetition or irrelevant material.

Demonstrates personal skills, perseverance and responsibility in some investigative activities in an inconsistent manner. Works within a team and sometimes approaches investigations in an ethical manner, with some attention to environmental impact. Displays competence in some investigative techniques, occasionally paying attention to safety, and requires close supervision.

Grade 2 *Poor*

Demonstrates: partial understanding of factual information, terminology, concepts, methodology and skills with some fundamental information gaps; little ability to apply and synthesis information, concepts or ideas; limited descriptive ability with quantitative and qualitative data; no consideration of contrasting viewpoints or methodologies or opinions; weak communication of a personal viewpoint, occasional awareness of perceptions and cultural differences of others; communication frequently lacking in clarity, repetitive, irrelevant or incomplete.

Rarely demonstrates personal skills, perseverance or responsibility in investigative activities. Works within a team occasionally but makes little or no contribution. Occasionally approaches investigations in an ethical manner, but shows very little awareness of the environmental impact. Displays competence in a very limited range of investigative techniques, showing little awareness of safety factors and needing continual and close supervision.

Grade 1 *Very poor*

Demonstrates: very little understanding of factual information, terminology, concepts, methodology and skills with many fundamental information gaps; inability to apply and synthesis information, concepts or ideas; inability with handling quantitative and qualitative data; no consideration of contrasting viewpoints or methodologies or opinions; lacks communication of a personal viewpoint, or awareness of perceptions and cultural differences of others; communication mostly lacking in clarity, repetitive, irrelevant or incomplete.

Rarely demonstrates personal skills, perseverance or responsibility in investigative activities. Does not work within a team. Rarely approaches investigations in an ethical manner, or shows an awareness of the environmental impact. Displays very little competence in investigative techniques, generally pays no attention to safety, and requires constant supervision.

In the examination

To state the obvious, be prepared. By this, we mean, have done your revision, reached the examination room in good time, have all the equipment you need (and some spares) with you. Although you can bring any revision notes to the door of the examination room, **do not** take them inside. You do not want to be found guilty of malpractice because you forgot you had some notes in your pocket. Always check.

Once at your desk, listen to the person starting the examination. Check that you can see the clock and have all that you need. Use the five minutes' reading time to first scan the question paper (and resource booklet in Paper 2) and then read it more thoroughly. Start to decide which essay questions you will attempt in Paper 2.

When you are allowed to start to write, do not be rushed if you see your colleagues writing furiously. You can score high marks with a few well-chosen words, or write a great deal and not answer the question.

Examination technique can be learned and you can get better at it. You will practise answering examination-style questions over the course and, in the end, do timed practice papers. Remember, you cannot get a mark for knowing something, only for writing it down. You cannot get a mark if you do not answer the question. You may write a brilliant answer but if you "describe" and you were asked to "explain", you do not get the marks. Always answer the question and answer the right number of questions. You must write on two of the four possible essays on Paper 2. Do not omit one essay, for example, or you immediately lose 20 marks out of the 65 for Paper 2. Do not write three essays or the last will not be considered and has wasted you time.

Check the time that you have. If it looks like you will run out of time, make a list, submit your essay plan, answer the easier parts – do anything to get the marks that you can. But, before you reach that point, practise timed essay writing to make sure you do have enough time in the examination.

Examiners are humans too and want you to do your best. But they have to mark according to the markscheme and they need to be able to read your writing. They will do their best to read what you have written and do not penalize you for spelling or grammar as long as the meaning is clear. You may word-process more than you write in longhand, but examinations are not yet at the stage where you can word-process all your answers. Most of you will have to write and the speed of your writing can make a difference between you getting the mark or running out of time. So practise writing clearly and quickly.

If it is not written down, you cannot get the mark. So, in calculations, show your working. Remember to put the units. Do not assume the examiner knows you know.

You can answer the questions in any order so you may do the ones you are most confident about first. Never leave a gap: have a guess and write something as you are not penalized for trying.

The command terms are vital to your understanding of what you should do. Check that you know which term it is in the question and that you note any other clues, e.g. "Justify, giving two reasons ..." – so give two, not just one; "State, giving a named example, ..." – always have named examples ready in your mind.

Sometimes you will be asked for a diagram or annotated diagram. Sometimes you will not be asked for this but you know it would help you answer the question. So do draw one if you think it will help you.

Use the terminology of the subject and technical language correctly.

In Paper 1 and Section A of Paper 2, there are spaces in which you write your answer and the mark allocated is also given. The number of lines for your answer are worked out carefully and most candidates do not need to write more than this. There are usually two lines per mark. You are not penalized if you do write more on an additional sheet but it is not often that the longer answer gains more marks than the concise one. Think before you write the first thing that you think of. The number of lines and mark allocation also give you a clue. If the question asks you to describe and evaluate and there are 4 marks allocated, you might guess that 2 of these are for describing and 2 for evaluating. It does not always work like that but get into the habit of thinking about this.

The resource booklet is to be read in conjunction with Paper 2, Section A. This tests both your ability to extract and interpret data and your knowledge of the subject. Do not be put off by the size of the resource booklet. Much of it will be diagrams or tables. Scan the booklet, then refer to the questions in Section A and go back to the booklet and read it in more detail. Some of the information in it may not be necessary to answer the questions but the skill of finding the relevant information is an important one.

In Section B of Paper 2, there are four essay-style questions of which you answer two of your choice. Most of these questions are divided into three subsections and mark allocations are given. Up to two of the 20 marks for each essay are for your expression of ideas. This means how you communicate clearly, use relevant examples, justify your viewpoint, make the counterarguments and write in continuous sentences.

Some students like to do an essay plan for both essays at the start of the examination. Then they add to this as new thoughts come to them. If you run out of time you can even just hand this in as it will probably gain some marks.

Some sample examination-style questions

These questions have been modified from past examination papers in the IB Group 4 Environmental Systems course (used with IB permission) and the pilot course Ecosystems and Societies preceding this transdisciplinary course (used with authors' permission). They are grouped under the following headings for your convenience but some questions overlap topics as they are designed to help you see the holistic nature of the course and to make connections between areas of your knowledge. We hope that by the time you enter the examination room, you will have made these connections and be confident that you can answer this type of question well. The answers are given in the next chapter.

Topics covered:
A Systems and environmental philosophies
B Ecology
C Biodiversity
D Pollution
E Resources
F Climate change
G Human population

A Systems and environmental philosophies

1 A desert with very low precipitation and little vegetation is an example of which type of system?

   A open   B closed
   C isolated   D closed and isolated

2 Which of the following is an example of negative feedback?

   A Loss of vegetation, leading to soil erosion, leading to further loss of vegetation.
   B Animals failing to reproduce when food is abundant.
   C More carbon dioxide favouring plant growth, so plants absorb more carbon dioxide.
   D A population of small mammals in a forest decreasing due to a fire.

3 A population of organisms the size of which remains approximately constant, but which fluctuates over the short term, may be described as being in a state of

  A static equilibrium   **B** stable equilibrium
  C positive feedback   **D** demographic transition

4 As disease spreads through a population, numbers fall. As the result of a reduction in contact between individuals, the rate of spread of the disease is reduced. This is followed by a recovery in numbers. This is an example of

  A positive feedback   **B** negative feedback
  C demographic transition   **D** entropy

5 What does the first law of thermodynamics tell us?

  A Doing work always creates heat.
  B Entropy tends to increase.
  C Energy cannot be recycled.
  D All energy comes from other energy.

6 Of what is the diagram below an example?

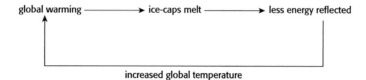

  A negative feedback
  B positive feedback
  C steady-state equilibrium
  D static equilibrium

7 Suggest how an environmental manager may reduce the impact of eutrophication.

8 **(a)** Outline what is meant by a *model*. *[2]*

  (b) Evaluate the models used:
    to predict the growth of human populations
    to predict climate change
    to assess demands human populations make on their environments. *[9]*

  (c) With reference to examples, explain the importance of understanding cultural factors when designing policies to control population growth. *[7] Expression of ideas [2]*

9 Cartoons which each express an environmental message.

  (a) Suggest what message cartoon (a) is trying to depict about attitudes to environmental problems. *[2]*

  (b) Cartoon (b) below suggests that a conflict exists between resource use and the needs of indigenous people. Explain why such a conflict might exist in tropical rainforests. *[3]*

Fig. (a)

Fig. (b)

B Ecology

1 Which of the following is most likely to be a parasite?

A A fungus in the living tissues of a tree.
B A fungus cultivated as a source of food by ants in their nest.
C A fungus growing on the decaying branch of a dead tree.
D A caterpillar consuming the tissue of a living fungus.

2 Which unit is the most appropriate to express the gross primary productivity of an ecosystem?

A kg
B kg y^{-1}
C kg m^{-2}
D kg m^{-2} y^{-1}

3 An *r*-strategist generally

A Gives considerable parental care to its offspring.
B Is small and short-lived.
C Lives in a stable environment.
D Produces small numbers of offspring.

4 In an area of grassland, a sudden severe frost reduces the population of butterflies. Of what is this frost an example?

A a density-dependent factor
B biological competition
C predation
D a density-independent factor

5 The frequency of bush fires increases as the abundance of an inflammable plant species increases. At the same time, the increased frequency of fires leads to a decrease in the population of certain animals. In what way(s) is the fire acting in this example?

A As a density-dependent factor for plants and a density-independent factor for animals.
B As a density-dependent factor for both plants and animals.
C As a density-independent factor for both plants and animals.
D As a density-dependent factor for animals and a density-independent factor for plants.

343

6 Which statement about productivity is correct?

A Net primary productivity is measured as energy fixed per unit area.

B Net productivity is the gain in energy or biomass per unit area per unit time after allowing for respiration losses.

C Net productivity is equivalent to the total solar energy fixed by photosynthesis in green plants.

D Net productivity is the total biomass gained by heterotrophic organisms per unit time.

7 Which statement is correct?

A A food chain can never have more than four members.

B Ecosystems that have a high productivity often also show high biodiversity.

C One ecosystem can never contain another within it.

D Pioneer communities usually contain more plant species than climax communities.

8 Which diagram best represents the change in the biomass of a forest ecosystem as it passes through a succession from a pioneer community to climax?

A
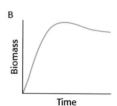

B
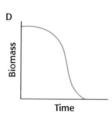

C
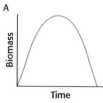

D
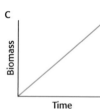

9 The table on the right shows invertebrates caught within a small woodland nature reserve using a series of pitfall traps.

(a) Draw and label a pitfall trap. *[2]*

(b) The pitfall trap technique was used over a four-year period to monitor invertebrate species diversity. Each year the value fell by approximately 12%. Suggest how this information may be used to manage the woodland nature reserve in the future. *[2]*

(c) Outline a method for estimating the population size of ground beetles in the woodland nature reserve. *[2]*

Invertebrates	Number
Springtails	520
Ground beetles	400
Greenflies	43
Spiders	33
Centipedes	34
Flies	24
Harvestmen	27
Woodlice	12
Ants	13
Wasps	17

Simpson's diversity index = 2.895

10 The figure below shows succession in a sand dune ecosystem.

(a) Define the term *succession*. *[2]*

(b) State what variable may be appropriate for the x-axis in the figure above. *[1]*

(c) Outline what will happen to soils as the ecosystem in the figure above changes from A to B. *[2]*

11

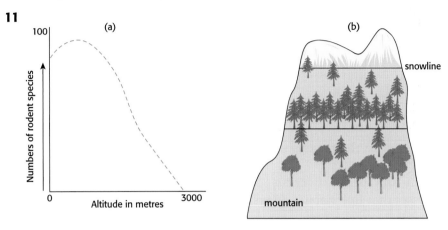

Fig. (a) Relationship between number of rodent species and altitude.
Fig. (b) Altitude habitat model.

(a) With reference to Figure (a) describe the relationship which appears to exist between altitude and the number of rodent species. *[1]*

(b) Predict **three** ways in which the altitude habitat model in Figure (b) might change as a result of global warming. *[3]*

12 The figure below shows energy flow through a food chain.

(a) Calculate the percentage energy loss to humans from the initial input of 900 kcal m^{-2} y^{-1}. *[2]*

(b) Explain why farming systems based on crop production are more energy efficient than harvesting from the sea. *[2]*

(c) Suggest **two** ways in which energy may be lost from the system. *[2]*

(d) Suggest **three** reasons why livestock (cattle, goats, sheep, etc.) form a part of most farming systems. *[3]*

C Biodiversity

1 The plants and animals of Australia had a unique evolution because the tectonic plate on which the continent is situated

A is the oldest

B is the youngest

C was never attached to any other plate

D has been isolated for a long period

2 For selected regions of the world, the bar chart below provides data for the total number of native plant species (species that occur naturally in the region), and the number of introduced species (species that have been brought in through human activities).

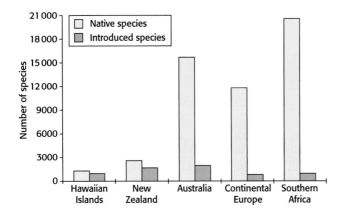

(a) Suggest reasons for differences between the regions using the data. *[4]*

(b) Define the terms habitat diversity and species diversity. *[2]*

(c) Outline the evolutionary processes that link habitat diversity to species diversity. *[5]*

(d) Name a species of plant **or** animal that has become extinct since 1600, and list **two** factors that help to explain why that species became extinct. *[2]*

3 The figure below shows wolf (predator) and moose (prey) population trends from 1959 to 1983.

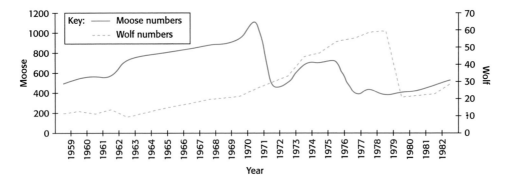

(a) Using data from the figure above, describe the trend between wolf and moose numbers between 1972 and 1979. *[2]*

(b) Suggest **two** possible explanations for the dramatic fall in wolf numbers in 1979. *[2]*

(c) Predict what might happen to the moose population and their habitat if wolves become extinct. *[2]*

(d) Wolves are often described as a pest species. Outline the arguments for protecting wolves. *[3]*

4 (a) Discuss the strengths and weaknesses of a species-based approach to conservation. *[3]*

(b) Discuss the strengths **or** weaknesses of the shapes of the nature reserves in the figure below. *[3]*

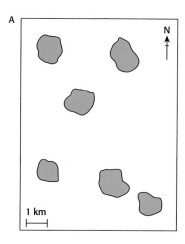
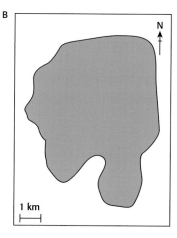
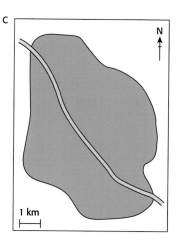

Key: nature reserve road

D Pollution

1 Which of the following is most likely to lead to increased mutation rates in phytoplankton?

   A increase in stratospheric ozone
   B decrease in stratospheric ozone
   C increase in tropospheric ozone
   D decrease in tropospheric ozone

2 Which of the following has been most effective in reducing world emissions of CFCs?

   A Improved technologies in the development of electric cars.
   B Aid programmes to developing countries.
   C Policies to minimize the combustion of fossil fuels.
   D International agreements, including the Montreal Protocol.

3 What happens when sunlight interacts with pollutants in the lower atmosphere?

   A Destruction of the ozone layer occurs.
   B Stratospheric ozone is produced.
   C Thermal inversion occurs.
   D Photochemical smog is produced.

4 Which statement is correct?

   A Ozone is destroyed by carbon dioxide released by burning of fossil fuels.
   B During recent decades the ozone hole over the Antarctic has become larger, producing global warming.
   C Ultraviolet radiation is involved in the production of stratospheric ozone.
   D Ozone is essential for human respiration.

5 Which statement about acid rain is correct?

   A Acid rain is a problem of developed countries because it is always deposited onto industrial areas.
   B Rain with pH above 7 is considered to be acid rain.
   C Lime can be used to restore acidified lakes.
   D Corrosion of buildings and rising sea levels are two of the main effects of acid rain.

6 Which statement describes a recycling process?

 A Using old newspapers for the manufacture of cardboard.
 B Control of water consumption by allowing watering of gardens only between certain times.
 C Using plastics instead of timber in furniture manufacture.
 D Re-conditioning old car (automobile) engines for resale.

7

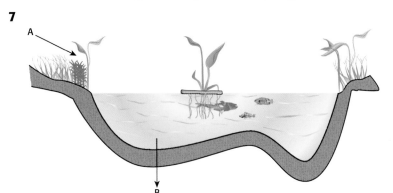

Diagrammatic representation of a eutrophic lake

(a) Define the term *eutrophication*. *[2]*

(b) Identify process A and process B. *[2]*

(c) Suggest **one** agricultural source and **one** non-agricultural source that may account for high phosphate levels. *[2]*

8 The figure below shows the world distribution of primary production in different biomes.

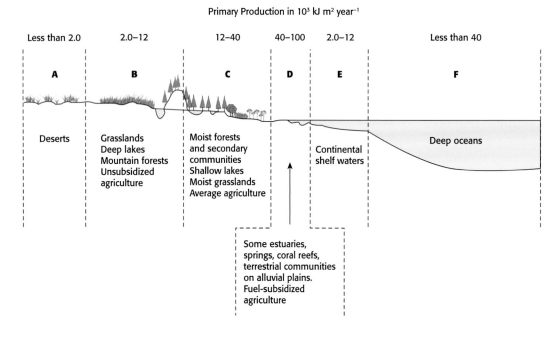

(a) State which of the zones A to F is responsible for the largest proportion of primary production. *[1]*

(b) Distinguish between *primary productivity* and *secondary productivity*. [2]

(c) With reference to the figure above, give **two** reasons why some biomes are more productive than others. *[2]*

E Resources

1 Which is an example of replenishable natural capital?

A a forest
B a hydroelectric power station
C groundwater
D a herd of cattle

2 For a particular year, for the fishing grounds of an island nation, assume:

R = the biomass of young fish reaching harvestable size
G = the growth in biomass of fish already of harvestable size
M = the loss of fish through death and emigration
B = the biomass of the total fish population at the start of the year.

The sustainable yield of the fishing grounds could be calculated using which of the following?

A $R + G - M$ B $B - M$
C $B + R + G - M$ D $R + G$

3 Which statement is correct?

A More than 10% (by volume) of the Earth's water is freshwater.
B The Earth's hydrological cycle is considered an open system because it exchanges matter and energy.
C Major cold ocean currents flow from polar regions towards the equator.
D The major reservoir of the Earth's water is in ice caps and glaciers.

4 Which statement is correct?

A Only resources that have a commercial value are considered to be natural capital.
B Nuclear power is a replenishable natural resource.
C Natural capital is a resource that can always produce an income.
D A catch of tuna fish is a form of natural income.

5 Which of the following best represents an example of sustainable yield management of resources?

A Reducing the amount of oil consumed by encouraging the use of smaller vehicles.
B Switching from fossil fuel power stations to nuclear power stations.
C Harvesting the timber of a forest at the rate that the trees produce it.
D Banning the catching of rare fish species.

6 (a) Outline the concept of sustainability. *[3]*
(b) Evaluate the importance of global summits in shaping attitudes towards sustainability. Refer to specific summits in your answer. *[5]*
(c) Discuss the factors which affect the choice of contrasting energy sources adopted in **two** societies you have studied.
[10] Expression of ideas [2]

349

7

Alternative renewable energy source	How the energy is produced	Major limitation
Tidal power	Energy is produced by using the ebbing and/or flooding tide to turn turbines and produce electricity	....
Wind power	Wind turbines are driven by available wind energy. The wind energy is turned into electrical energy via a generator. The electrical energy is supplied to an electrical grid to do work.	Dependent on the wind: no wind equals no energy.
Biofuel	....	Produces emissions and requires large areas to grow biofuel crop.

(a) State **one** other form of alternative renewable energy source not listed above. *[1]*

(b) Complete the table above for tidal power and biofuel. *[2]*

(c) Most MEDCs are still dependent on non-renewable forms of energy. Suggest reasons why MEDCs have not adopted renewable energy sources. *[3]*

(d) With reference to a **named** food production system you have studied, describe two ways in which food supply per unit area can be increased. *[2]*

(e) State the difference between **carrying capacity** and ecological footprint. *[3]*

F Climate change

1 Which list contains only greenhouse gases?

A Carbon dioxide, water and methane

B Methane, CFCs and sulfur dioxide

C Carbon dioxide, lead and methane

D Nitrogen, water and CFCs

2 If a tax were to be placed on fossil fuels proportional to the quantity of carbon dioxide released into the atmosphere when they were burnt, which would be the most likely effect?

A Global warming would increase.

B Generation of nuclear power would decrease.

C The occurrence of smog in cities would increase.

D Consumption of fossil fuels would decrease.

3 The figure below shows mean global climate change from
1850 to 1990.

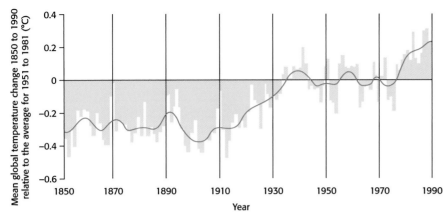

(a) With reference to the figure above, describe the trend in
mean global temperature between 1870 and 1990. *[2]*

(b) Suggest **two** explanations for this trend. *[2]*

(c) Suggest **one** method of preventing further increases in mean
global temperature. *[1]*

The table below shows different aspects of the environment and
society which might be affected by climate change.

(d) Complete the table below to state likely impacts of global
warming on the environment and society. *[4]*

Impact on the environment		Impact on society	
Impact			*Impact*
Ice and snow		Water resources	
Ocean and coast		Food and agriculture	
Hydrology	Increased flooding	Coastal living	
Ecosystems		Human health	

(e) Predict, giving reasons, whether the impacts outlined in part
(d) are likely to have a greater effect on less economically
developed countries (LEDCs) **or** more economically
developed countries (MEDCs). *[2]*

4 (a) Describe some of the contrasting responses to the issue of
global warming. *[5]*

(b) Outline the reasons why people have such different opinions
on the issue of global warming. *[5]*

(c) Describe what is meant by *carrying capacity* and evaluate the
role that technology could play in ensuring that the earth's
carrying capacity is not exceeded by human populations. *[8]*
Expression of ideas [2]

G Human population

1 China's average family size has fallen from 5.8 to 2.4 in the last
20 years. Which of the following could have produced this
change?

A High levels of emigration.

B Increase in the death rate in rural areas.

C Education about contraception methods.

D Genetic changes in the population.

2 Which of the following would be most likely to reduce the birth rate in a country?

A A decrease in the average age of marriage.

B An increase in educational opportunities for women.

C An increase in the carrying capacity of the country.

D A reduction in the rate of tax paid by families with several children.

3 The data below relate to the world population at a particular time.

Crude birth rate = 35 per thousand

Crude death rate = 20 per thousand

What was the natural increase rate at the time?

A 1.5   **B** −1.5   **C** 15   **D** −15

4 Why is estimating the carrying capacity for human populations difficult?

I Because of the high level of technological development.

II Because populations generally import resources from outside their immediate environment.

III Because birth rate is very low.

A I, II and III   **B** I and II only

C II and III only   **D** III only

5 If the crude birth rate of a country in a particular year is 16 per 1000, and the crude death rate is 8 per 1000, what is the annual population growth by natural increase?

A 0.8%   **B** 8.0%   **C** 2.0%   **D** 0.2%

6

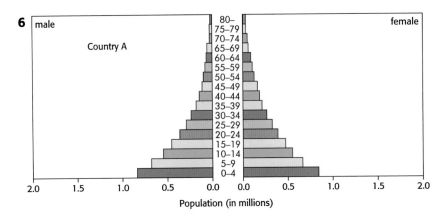

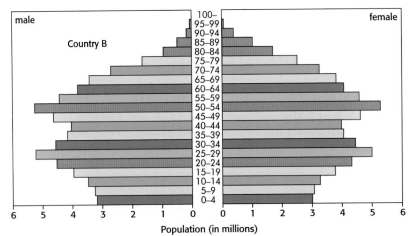

The figures above show the population pyramids for an MEDC and LEDC in the year 2000. The table (right) shows two ecological footprints for 2001.

(a) State **two** differences between the population pyramids shown in the figures above. *[2]*

(b) (i) With reference to the figures and the table, deduce which footprint belongs to country A and which footprint belongs to country B. *[1]*

(ii) Justify your answer to (b)(i). *[1]*

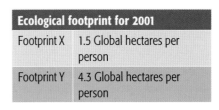

Ecological footprint for 2001	
Footprint X	1.5 Global hectares per person
Footprint Y	4.3 Global hectares per person

7 The table below shows population change and water consumption for 1971 and 2001 in an MEDC.

	1971	2001	Growth rate per year
Population / 10^4	48.6	57	0.53%
Total water consumption / 10^5 m^3 day^{-1}	42.7	83.3	2.3%

(a) Compare relative growth rates for population and water consumption between 1971 and 2001. *[1]*

(b) Suggest **two** factors which may explain the difference you have identified in (a). *[2]*

8 The figure below shows factors which affect human population density, distribution and changes.

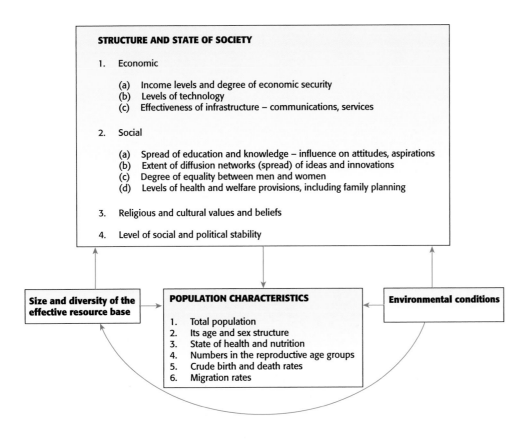

(a) Suggest what is meant by the *age and sex structure* of a population. *[2]*

(b) With reference to specific case studies, discuss how the "structure and state of society" can affect human population dynamics and growth. *[9]*

(c) Explain how environmental conditions can affect human population characteristics. In your answer you should refer to specific examples. *[6] Expression of ideas [3]*

In the markschemes, each marking point is signified by means of a semicolon (;). An alternative answer or wording is indicated by a "/" and either wording can be accepted. Words in (brackets) are not necessary to gain the mark. The order of points does not have to be as written and similar wording or meaning are acceptable.

A: 1 A, **2** C, **3** B, **4** B, **5** D, **6** B.

7 Identify potential sources of eutrifying materials (detergents, fertilizers, effluents); identify if source is point or non-point pollution; put in strategies for removing or reducing these pollutants, e.g. reduce fertilizers use; use organic fertilizers / use detergents with no phosphates / prevent run-off, etc.; remove contaminated sediment from water body / dredge it; reoxygenate water body; nets to reduce organic input; educate water source users about eutrophication problems; [3 max]

8 (a) A simplified description; designed to show the structure or workings of an object, system or concept; require approximations to be made; [2 max] (b) Predicting the growth of human populations: [3 max] growth of human population depends (at a simple level) on birth rates and death rates; from this rates of natural increase can be calculated and population total predicted; population pyramids enable policy-makers to chart what proportion of the population are in the fertile age bracket, helping to predict likely birth rates; demographic transition model shows how population growth is linked to economic development; enables the reasons for population growth to be understood; but not all countries conform to the stages identified; models are hugely simplified, and may not reflect the complex and unpredictable factors which affect growth rates, e.g. war/disease; Predicting climate change: [3 max] models can demonstrate anticipated changes to climate based on carbon emissions; model only as good as the data that goes in and it may be suspect; conflicting models can show different effects in same place; hugely complex in terms of numbers of factors involved in atmospheric systems so in process of oversimplification accuracy is lost; e.g. role of feedback or ocean systems not fully understood; *Accept other examples of feedback.* Assessing demands human populations make on their environments: [3 max] ecological footprints can be effective for comparing environmental impacts of different societies; able to provide a quantitative estimate of human carrying capacity; a quantification of what can be a very complex set of factors; can be useful tools for getting people to think about their impact; stresses the systems approach and interconnectedness of eco- and social systems; very difficult to calculate figures, e.g. per capita emissions; [9 max] *To receive full marks answers must have a balance of strengths and weaknesses. Award credit if other relevant models are evaluated.* (c) Strategies for controlling growth include availability of contraception, financial incentives, public information and legislative changes (e.g. making abortion illegal); often the reasons for family size can be attributed to cultural factors so for policies to be effective they need to understand the underlying reasons why people decide to have a certain number of children; the need for male children in some cultures is linked to the traditional practices and structures, e.g. inheritance by male heirs and dowries for females; sometimes cultural factors indirectly play a role in fertility rates, e.g. education and employment opportunities for women lead to delayed marriages and lower birth rates; provision of contraception in e.g. remote, rural communities may not be enough – programmes to educate males to be willing to use the contraception are also needed; cultural norms may be ingrained/deeply felt and policies need to address these at the deepest level to change attitudes, e.g. religious beliefs in catholic countries; culture and tradition evolve over time / cultural change can occur and governments can be a part of this; education and economic development are important factors in bringing about cultural change; [7 max] *Award [4 max] if no examples are used. Examples can be of cultural practices and do not need to be located in named geographical contexts. Expression of ideas:* [2 max] *Total:* [20]

9 (a) Perhaps cartoonist is suggesting that politicians/society refuse to act because they claim that more research needs to be done first; despite the fact that evidence (falling birds) is in front of their eyes; [2] *Accept similar interpretations of cartoon, no need to mention acid rain.* (b) Conflict might exist because different groups see the resource differently; economic value of timber/land is incompatible with leaving forest standing for other uses (indigenous cultures); indigenous tribes need large amounts of space in which to live sustainably; reserves left for indigenous people may be too small to sustain them; forest is cut down by outsiders ignoring the needs of indigenous people; intrinsic value of forest (biorights) is ignored by exploitative users only interested in economic use; difference between sustainable use of forest (natural income) and users who exploit natural capital; conflict between short-term and long-term perspective (indigenous people); [3 max]

B: 1 D, **2** D, **3** B, **4** D, **5** A, **6** B, **7** B, **8** B.

9 (a)

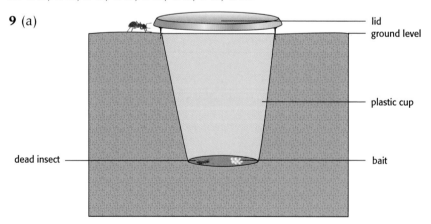

(b) A falling value would suggest a loss of diversity / in this instance a loss in the invertebrate biodiversity; this data could be used to identify a biodiversity decay problem / kick-start a management strategy to assess the cause and reverse the trend; [2] *Award credit if specific management strategies to address loss in biodiversity are suggested.*

(c) Use Lincoln index / capture-mark-release-recapture method; capture ground beetles, mark and release; after a period of time recapture and count those with and without marks; [2 max]

10 (a) The orderly process of change over time in a community; changes in the community of organisms cause changes in the physical environment; this allows another community to become established and replace the former through competition; leading often to greater complexity; [2 max] (b) Time; distance (from sea); [1 max] (c) Soils will become more mature; soils will be deeper; contain more organic material; become more complex; develop distinct horizons; [2 max]

11 (a) As altitude increases rodent species number decreases; there are fewer species of rodent at high altitude; there are higher species numbers at low altitude; inverse relationship; [1 max] (b) Migration of species up the mountain; melting/loss/reduction of snow cap / migration of snowline up mountain; appearance of new species; loss of species / species unable to adapt to rapid changes / lower diversity; increased growth rates; [3 max]

12 (a) 900 − 6.6 = 893.4 kcal lost, 893.4//900 × 100 = 99.3% loss; [2] (b) There is energy lost from respiration and waste production at each level within a food web; crop production harvests food from lower down in the food web than harvesting fish from the top, therefore it is more efficient; crops capture energy directly from primary source; fish harvesting utilizes a resource that is several steps away from primary production; [2 max] (c) Heat; respiration; faeces; [2 max] (d) Animals provide a source of protein (essential for the human diet); animals convert vegetation to food that would not be available to humans directly; produce diverse products (milk/meat/blood/wool); taste and culture affect demand; additional benefit that they are working animals; [3 max]

C: 1 D.

2 (a) Number of native species in Continental Europe/Africa/Australia is high because of large area; number of native species is high in Continental Europe/Africa/Australia because of wider range of habitats; number of native species is high in Continental Europe/Africa/Australia because of wider range of climate; (converse: NZ and Hawaii – smaller range of species because of smaller area/range of habitats/climate) number of native species high in Australia because of isolation throughout evolutionary history; number of native species low in NZ/Hawaii as few species can colonize remote islands; islands more liable to be colonized by introduced species; high proportion/percentage of total species in NZ and Hawaii are introduced; because native species may be adapted to specialized environments and are less resistant to competition; [4 max] *Any other reasonable points.* (b) Habitat diversity = the number of habitats/ecological niches per unit area/in a certain area; species diversity = the number of species of organism per unit area/in a certain area; [2] (c) Habitat diversity is an indication of the ecological variety of an ecosystem; and therefore the number of species it can accommodate; complex habitats provide more ecological niches for organisms; habitat diversity may cause isolation of populations;

natural selection ensures organisms are adapted to environment and way of life; as environmental pressures influence frequency of genetic traits in populations; reproductive barriers may arise through (divergent) evolution; so the more environments an ecosystem represents, the greater the possibility of speciation; e.g. an ecosystem with several layers such as tropical forest is likely to have a higher diversity than single layered ecosystems such as temperate scrub/heathland; [5 max] *Any other reasonable points.* (d) Name of species, e.g. dodo; factors will depend on example selected, e.g. for the dodo: confined to small island/limited distribution (Mauritius); small population; useful source of food for visiting sailors; extreme tameness; large and conspicuous; slow rate of reproduction; habitat destruction; competition with introduced organisms (e.g. pigs); [2 max] *Any other reasonable points. Award [1] for two factors.*

3 (a) From 1959 to 1970 both species show gradual increase trend; from 1970 to 1972 moose numbers crash/fall; wolf numbers continue to rise; 1978 to 1979 wolf numbers crash; [2 max] (b) Poor climate conditions, winter deaths due to cold; disturbance; hunting/trapping; disease; loss of prey species / decline of moose earlier; [2 max] *Accept other appropriate answers.* (c) Moose numbers may increase further; leading to greater densities; greater grazing and a change in habitat type; vegetation change would occur as moose preferentially graze; moose may ultimately outstrip their own food supply; may cause a habitat shift; decline in moose numbers; [2 max] (d) Ethical arguments – wolves have a right to exist; esthetic arguments – beautiful creatures; genetic arguments – loss of diversity, once it is gone it cannot be regained; ecological arguments – role in food web, maintaining numbers of prey species; commercial arguments – pelts / wildlife tourism / trophy hunting; [3 max] *Accept other reasonable arguments.*

4 (a) High profile/charismatic species catch public attention both nationally and internationally (e.g. tiger – India); however, species-based conservation favours charismatic organisms and is less successful in saving "non-cuddly" species; saving a named species requires preserving the animal's habitat, this benefits all other organisms in that habitat; however a species can be artificially preserved (e.g. in a zoo) while its natural habitat is destroyed (e.g. Giant Panda); [3 max] (b) Area A: [1 max] fragmented and small with a large perimeter area ratio / large edge effect so lots of disturbance; fragmented so difficult migration between fragments; small size may limit species contained / limit population sizes; *Accept other reasonable responses.* Area B: [1 max] large perimeter area ratio / relatively small edge effect so less disturbance; large size promotes high biodiversity; large size so good for large vertebrates/top carnivores/large species populations; *Accept other reasonable responses.* Area C: [1 max] as large as B but bisected by a road which acts as a barrier to species migration; road increases edge effect, some more disturbance; road allows easier access to the interior of reserve for monitoring; road gives easier access for poachers; some evaluative element is required (i.e. how the characteristic relates to the ecosystem in a positive or negative way); [3] *Accept other reasonable responses.*

D: 1 B, **2** D, **3** D, **4** C, **5** C, **6** A.

7 (a) The enrichment of a water body (lake, stream, etc.) by increased nutrient inputs (primarily phosphates and nitrates); depletion of oxygen content of water; leads to the development of algal blooms; the enrichment may be natural/artificial; accelerated by human activity; [2 max] (b) Process A: (agricultural) run-off / surface flow / nutrients entering lake; process B: leaching/ sedimentation/seepage / dead remains sinking / infiltration; [2] *Accept other reasonable responses.* (c) Agricultural source: fertilizers (containing phosphates and nitrates) / animal waste / manure; *Do not accept pesticides.* Non-agricultural source: detergents (domestic) / industrial effluent / sewage; [2]

8 (a) Zone D; [1] (b) Primary productivity is the gain in energy/ biomass by producers/autotrophs whereas secondary is gain by heterotrophic organisms; primary productivity is the conversion of solar energy whereas secondary involves feeding/absorption; [1 max] (c) Availability of light, e.g. deep oceans dark below surface, limits productivity of plants; availability of water, e.g. tropical rainforests receive lots of rainfall each year whereas deserts have little rain which is limiting to plant growth; temperature, e.g. rainforests warm throughout the year so have a constant growing season and higher productivity; nutrient availability, e.g. estuaries receive lots of sediment from rivers; [2 max] *Award [1 max] for no reference to the biomes.*

E: 1 C, **2** A, **3** C, **4** D, **5** C.

6 (a) Use of resources at a rate that allows natural regeneration; and minimizes damage to the environment; e.g. a system of harvesting renewable resources at a rate that will be replaced by natural growth might be considered to demonstrate sustainability; any society that supports itself in part by depleting essential forms of natural capital is unsustainable; if human well-being is dependent on the goods and services provided by certain forms of natural capital, then long-term harvest (or pollution) rates should not exceed rates of capital renewal; sustainability means living within the means of nature, on the "interest" or sustainable income generated by natural capital; [3 max] (b) Global summits can play a leading role in shaping attitudes to sustainability, e.g. UN Conference on Human Environment / Stockholm, 1972 was the first time that the international community met to consider global environment and development needs; and can play a pivotal role in setting targets and shaping action at both an international and local level, e.g. Rio Earth summit (in 1992) led to Agenda 21 and Rio declaration, which set out key policies; and to legally binding conventions, e.g. on climate change / Montreal (1987); however, countries can break these agreements and there is little the international community can do; in terms of shaping public opinion media can also be important, e.g. *Silent Spring* by Rachel Carson was pivotal; attitude change may occur without summits, e.g. UN commissioned the Brundtland report, which established initial definition for sustainable development; summits may not achieve their initial goals, however, they may act as a catalyst in changing the attitudes of governments,

organizations and individuals; [5 max] *Accept any other reasonable answers. Award [3 max] if summits are described but not evaluated.* (c) Factors could include availability, economic, cultural, environmental and technological factors: e.g. fossil fuels in UK availability – large oil, coal and gas reserves in UK have historically meant they were an obvious choice for exploitation; as the most easily accessible reserves have been used up the costs of exploitation have increased and alternative sources have been sought; changing awareness of environmental implications of fossil fuel exploitation has increased demand for renewable, non-polluting sources; cultural fears based on perception of nuclear accidents/waste have made this an unpopular choice politically; leading to greater investment and research into alternatives, e.g. wind and tidal; as public awareness of threats of global warming has increased there has been a shift in attitudes towards, say, wind power, despite the esthetic and environmental implications; e.g. firewood in India where a huge proportion of population rely on local sources of firewood for energy because it is most readily available/cheap; it is the traditional source of energy, which has always been used; and technology such as solar-powered stoves is not available or affordable; in a drive to develop economically the Indian government has sought to harness other sources of cheap energy to stimulate industrial development; specifically hydroelectric power, which has sometimes been extremely controversial for social/environmental reasons, e.g. Narmada dam; [10 max] *Award credit if figures are used. Award [5 max] if no societies are referred to. Societies do not need to be contrasting, but energy sources should be. Accept other reasonable responses. Expression of ideas:* [2 max] Total: [20]

7 (a) Wave power / solar radiation / heat pumps / water wheels; [1] *Accept other suitable answers if appropriate.* (b) Tidal power: good tidal range required / right shape of coastline / interferes with navigation / impact on wildlife / expensive; (only one limitation required); biofuel: plant material burned directly to produce heat / transformed into ethanol (used as fuel) / converted to methane (methane digestion); (only one method required) (c) MEDCs traditionally/culturally dependent on fossil fuels; fossil fuels are energy-efficient / easy to transport / relatively cheap; changing to renewable energy on a large scale requires massive capital investment / cultural inertia against change to renewables / many renewables depend on environmental conditions that are not constant (e.g. wind, sunshine, waves); [3] (d) Named food production system: e.g. rice paddies application of fertilizer; using herbicides; insecticides; irrigation; changing crop type/variety; using GM crops; [2 max] *Award [1 max] if no named food production system.* (e) Carrying capacity is the number of individuals/species/load an area of land / an environment can support (providing resources and absorbing waste); ecological footprint is area of land (and water) required to support an individual/population (providing all resources and absorbing waste); ecological footprint is a theoretical area whereas carrying capacity refers to a real area; they are the opposite/inverse of each other; carrying capacity involves sustainable support of a population, whereas footprints are not necessarily sustainable; [3]

F: 1 A, **2** D.

3 (a) The trend shows temperature with high years and low years with an overall upward trend; accelerating in the last quarter of the graph; temperatures prior to about 1935 were below the average (for 1951–1980); but from 1980 they were above the average; [2 max] (b) The onset of global industrialization; and the subsequent production of fossil fuel derived pollution; deforestation, particularly of rainforest; perhaps volcanic activity; sunspot activity; [2 max] (c) Control the amount of atmospheric pollution; reduce atmospheric pollution; stop forest clearance; increase forest cover; develop alternative renewable energy sources; improve public transport; set national limits on carbon emissions; carbon dioxide capture; [1 max] *Accept other reasonable answers.* (d) Impact on the Environment Impact on Society Impact Ice and Snow deglaciation /glacier retreat; Water resources may cause drought; Ocean and Coast sea level rise / coastal flooding; Food and Agriculture agriculture moving north / south; Hydrology increased flooding Coastal living flooding / storms; Ecosystems biome shifts / species change; Human health increased disease; [4 max] (e) LEDC because technologically/economically less able to cope; greater percentage of population already vulnerable; weak infrastructure/communications/emergency services; [2 max] *Accept other reasonable suggestions.*

4 (a) Some politicians believe action should be taken immediately by all nations to curb emissions of CO_2; to change lifestyles and plan to reduce fossil fuel dependence; whereas others argue that it is unreasonable to expect LEDCs to curb emissions until they have developed economically like the MEDCs have done; most scientists are now convinced that there is a causal link between CO_2 levels and global temperature change; whereas some scientists argue that relationships are more complex and that the effects of global warming are unclear; even that recent temperature changes are merely parts of natural fluctuations in the Earth's temperature; some ordinary citizens feel they have a responsibility to change the way in which they live to reduce their personal contribution to the problem; others do not believe that actions at an individual level can make much difference; others do not prioritize environmental issues including global warming; responses by organisms rather than people, migration/extinction/adoption; [5 max] (b) Opinions will depend to a large extent on what scientific evidence they find most convincing; this will depend on their specialized knowledge and their level of education; overall awareness of the issue, which can also depend on the profile of environmental issues in the media; environmental paradigms can shape how they read scientific literature; their attitudes to our relationship with the environment (e.g. whether we should live in harmony with it or control it using technology); environmental paradigms will stem from cultural context including prevailing religious attitudes (e.g. whether we have any moral obligation to future generations); the growth of the environmental movement (which has grown exponentially in profile and influence) has played a large role in raising awareness of the issue; cultural/religious group; where people live might affect their views, e.g. near the sea; socioeconomic status, e.g. extreme poverty leads to short-term view / wealth leads to faith in money to solve problem; age, e.g. young more concerned than old; [5 max] *Accept other reasonable responses.* (c) Carrying capacity is the maximum number of species that can be sustainably supported by a given environment; it is determined by availability of resources (e.g. food, water, space) to the population in an area; a country is said to be overpopulated if the carrying capacity is exceeded; it is a problematic term for human populations because technology has a huge influence on the resources that are available to human populations; our tastes and demand for particular resources changes at such a rapid rate; at a country level technology can help to ensure carrying capacity is not exceeded; e.g. by importing new resources with transport technology; but at a global level technology can be used to intensify the way in which we use resources, e.g. increased agricultural production on the same plot of land by using HYV rice; substitutions of resources, e.g. developing alternative energy technologies to fossil fuels; technology can also play a part in reducing human population size, e.g. through contraception/ medicines (reducing infant mortality and thereby reducing the incentive for high birth rates in many poor countries); technology alone may not be the full solution, attitudes to resource use may need to be altered; [8 max] *Award [2–3 max] for describing carrying capacity and [5–6 max] for role of technology. Expression of ideas:* [2 max] Total: [20]

G: 1 C, **2** B, **3** A, **4** B, **5** A.

6 (a) Country A is an expanding population whereas country B is a declining population; country A has a high proportion of young people / wide base whereas country B has a low proportion of young people / narrowing base; country A has low proportion of elderly / narrow top whereas country B has a higher proportion of elderly people / wider top; country A has a larger population than country B; [2 max] (b) (i) Country A: footprint X country B: footprint Y [1] *Both answers needed to receive* [1]. (ii) Country A is an LEDC and therefore people use fewer resources / more local resources / generate less pollution whereas country B is an MEDC and therefore people use more resources / more imported goods / generate more pollution; [1 max]

7 (a) Water consumption has increased at a faster rate than population growth; [1] *Figures are not needed.* (b) Increased demand for domestic goods/luxury items, e.g. washing machines and swimming pools; increased economic development so more water used in industry; agricultural development so greater use of water in irrigation (for intensive) farming; cultural change towards greater personal hygiene; [2]

8 (a) Age structure is the proportion of the population in different age groups (typically young dependents under 15, economically active aged 15–65 and elderly dependents aged 65+); sex structure is the proportion of the population divided into gender groups; [2] [9 max] (b) *Credit should be given for discussion of specific case studies.* In LEDCs such as Bangladesh many people who do not feel economically secure will have lots of children to help them on the

family farm; in MEDCs financial insecurity may be given as a reason for limiting family size (e.g. Japanese birth rate is very low); effectiveness of communications and availability of health services may affect how easily people can find out about / access contraception; attitudes to the family and expectations for lifestyles will be affected by mass media and communications; might lead to people wanting to limit family size and pursue other goals, e.g. travel; if women are educated they have alternatives to child bearing and may decide to reduce their family size; e.g. Bangladesh, one in four women is illiterate and birth rates are high; spread of ideas and new innovations in contraception are a high priority for the government in China where people are only allowed to have one child; empowerment of women is seen as a major step by many population agencies for reducing birth rates; religious attitudes may cause high birth rates if abortion or contraception, for example, are not allowed (e.g. Catholic countries such as Ireland or Mexico); birth rates tend to drop during war time or increase during periods of prosperity; e.g. baby boom generation after the second world war;
(c) natural hazards such as drought or floods; can be beneficial, e.g. volcanic eruptions and floods bringing fertile silt; improving the agricultural resource base; improving health and nutrition; or can be devastating, e.g. leading to crop failure and soil erosion / inundation of land with sea water increasing death rates; global warming / climate change can lead to inundation and migrations as people leave; these factors will affect birth rates and in turn numbers surviving into adulthood; biological infestation, e.g. locusts can lead to loss of crops; disease epidemics, e.g. malaria, can increase death rates; [6 max] *Expression of ideas max* [3 marks] *Total* [20 marks]

These pages contain resources and website addresses for further reading. It is not exhaustive and is a personal choice of the author. But it is a start. We cannot be sure that the addresses will not change but most are of established organizations and should be there for some time. As things change so rapidly on the Earth, the data in this book may well date quickly and we recommend that you look at more recent data if necessary. Many university websites contain lecture notes and presentations on topics in this book. Happy reading.

Websites

General

http://about.greenfacts.org	Belgian non-profit organization, aim to bring complex scientific consensus reports on health and the environment to the reach of non-specialists
http://earthtrends.wri.org/	Online database maintained by the World Resources Institute
http://info-pollution.com/myths.htm	Debunking anti-environmental myths pages – a good read
http://ue.eu.int/index.asp	Council of the EU
www.cia.gov/library/publications/the-world-factbook/	The CIA World Fact Book with detailed facts about the world and countries
http://news.bbc.co.uk/	Links to updates on environmental issues as do daily newspaper and other media channel websites
www.earthportal.org/	Comprehensive resource for objective, science-based information about the environment. Includes the Encyclopaedia of the Earth
www.nasa.gov	A huge site from the US space agency, Earth sections are full of data, photos are fantastic
www.newscientist.com	UK weekly science magazine with reputable articles and reviews, e.g. on climate change, seas, biodiversity
www.noaa.gov	A US federal agency, full of info in climate, oceans and coasts
www.peopleandplanet.net/	Global review and internet gateway with very readable articles on many topics
www.unep.org	UN Environment Programme
www.unesco.org/mab/mabProg.shtml	UNESCO's Man and the Biosphere programme

www.worldwatch.org/	A 35-year-old research organization that reports on global environmental events
www.wri.org	A 25-year-old environmental think-tank with great articles and publications about the state of the Earth
www.iiasa.ac.at/	International research organization based in Austria
www.oxfam.org/	Oxfam International – charity and NGO that works around the world
www.foei.org/	Friends of the Earth International – a large grassroots organization with 2 million members
www.greenpeace.org/international/	Greenpeace International – a global campaigning NGO
www.unc.edu/~rowlett/units/index.html	A dictionary of units of measurement
www.oneworld.net	International civil society network with the aim of building a more just society

Worldviews and environmental philosophies

http://plato.stanford.edu/entries/ethics-environmental/	Stanford University
www.worldchanging.com	Ideas for a better future

Ecology

www.envirolink.org/	Links to hundreds of environmental sites
www.globe.gov	The GLOBE programme coordinates hands-on science projects between schools around the world
www.epa.gov/	US Environment Protection Agency
www.epa.gov/highschool/	Has some good links
www.enviroliteracy.org/	All things environmental
	Put ecology lecture notes into your search engine to find university course notes.

Biodiversity

www.panda.org	Worldwide Fund for Nature main site
http://cms.iucn.org/	IUCN main site
www.iucnredlist.org	The entry point to the IUCN Red Lists of endangered species
www.eol.com	Encyclopaedia of life

Climate change

www.ipcc.ch	IPCC website with their reports and views
www.epa.gov/climatechange/	US Environment Protection Agency site
www.ieagreen.org.uk/	International Energy Agency offshoot

Human populations

www.unfpa.org/	UN Population Fund, international development agency that promotes the right of every woman, man and child to enjoy a life of health and equal opportunity
http://esa.un.org/unpp/	UN Economic and Social Affairs Department searchable population database
www.census.gov/ipc/www/idb/	US government census site with dynamic population pyramids
www.census.gov/ipc/www/popclockworld.html	World population clock
www.undp.org/mdg/	UN Development Programme Millennium Development goals

Resources

www.earth-policy.org/	Dedicated to building a sustainable future
www.iisd.org/	International Institute for Sustainable Development

Ecological footprints

www.footprintnetwork.org	Ecological footprint calculator
www.carbonfootprint.com/	Carbon footprint calculator
www.ecologicalfootprint.com/	Simple calculator

Energy resources

www.bp.com	British Petroleum site, one of the world's largest energy providers. BP reports are full of information
www.shell.com	Another big energy company with some renewables information on the site

Water resources

www.peopleandplanet.net/doc.php?id=671&section=14	
www.wateraid.org	Find practical solutions to water issues

www.ifpri.org/media/water2025.htm	Water resources booklets
www.worldwatercouncil.org/i	International, multi-stakeholder platform for water issues

Soil resources

http://soil.gsfc.nasa.gov/	Soil made exciting by NASA again
http://soilerosion.net/	Just what it says

Food resources

http://faostat.fao.org/	UN Food and Agriculture Organization database
www.iwcoffice.org	If you care about whales
www.ifpri.org/	International Food Policy Research Institute – good on food as well

Pollution

www.atm.ch.cam.ac.uk/tour/index.html	Cambridge University ozone hole tour: very informative
www.eia.doe.gov/kids/energyfacts/saving/recycling/solidwaste/plastics.html	Everything you need to know about recycling plastics
www.howproductsimpact.net/impacts/photochemicalsmog.htm	Photochemical smog
www.umich.edu/~gs265/society/waterpollution.htm	Water pollution from Michigan University
www.wasteonline.org.uk/	UK solid waste site full of facts on waste disposal

Textbooks

There are many books that you might dip into to increase your understanding of parts of this course and there is just a small selection here.

Byrne, K. (2001) *Environmental Science. Bath Advanced Science.* 3rd ed. Cheltenham: Nelson Thornes.

Chapman, J.L & M.J. Reiss (1998) *Ecology:Principles and Applications.* Cambridge: CUP.

Cunningham, W., M.A. Cunningham and B.W. Saigo (2005) *Environmental Science: A Global Concern.* 9th ed. McGraw-Hill.

Miller, G. Tyler (2006) *Living in the Environment: Principles, Connections and Solutions.* 14th ed. Brooks Cole.
The 13th edition has a companion website at www.brookscole.com which is full of information.

www.biozone.co.uk/ Biozone office in UK, Australia and New Zealand. IB texts for Biology relate to this course in some areas.

James Lovelock – read any of his books. *Gaia* is the first in which he talks about the Gaia hypothesis. *The Revenge of Gaia* (2007) is where he pictures a very different and warmer world.

Flannery, T. (2007) *The Weather Makers: Our Changing Climate and What It Means for Life on Earth*. London: Penguin.

Myers, N. & J. Kent (2005) *The New Atlas of Planet Management*. University of California Press.

Wilson, E.O. (1993) *The Diversity of Life*. London: Penguin.

Future prospects for study and careers in sustainable development

by Tejas Ewing

Now that you have read this Course Companion, and worked through the issues involved in the interaction between humans and the natural environment, you may find yourselves interested in the prospects of further study and work in a field that offers you the chance to make a positive change to the world around you, and also offers excellent career prospects.

Study and work related to environmental issues is no longer the preserve of hard scientists with a penchant for long days in the field. Careers in conservation, climatology and resource management still exist, and always will, but what is interesting about the field of sustainable development is that its focus on the interaction of human behaviour and the resultant effects on our environment are creating a multitude of new and innovative study and work options.

For example, climate change has come to be material to many fields of life, from business to politics to finance to media. There are growing opportunities in the fields of corporate social responsibility and climate change consulting for business. In London, there are a number of specialized firms that work purely on public relations and communications consulting related to environmental issues. And in the field of finance, emissions trading is one of the fastest growing fields in the global financial sector.

What this all means is that traditional concepts of working and studying environmental issues are changing. Because of high oil prices, food shortages, droughts related to climate change and increased political pressure, the environment is now so much more relevant to the way the world operates. This means that whatever your core interests – be they business, economics, media, politics, development – you can find a way to combine that interest with a desire to engage with sustainable development. When it comes to further study, universities are actively looking for multidisciplinary students who have ideas about bridging the gaps between environmental awareness and standard practice. And, further along, when it comes to job searching, it means that an environmental degree is no longer a specialized degree that limits you to a particular field. For example, every major multinational company, financial institution, media outlet and political party in the UK now has experts in environmental issues or sustainable development on their staff. This is now a mainstream issue.

What this course, the concept of sustainable development and the increased interest in climate change should make clear is that the management of human behaviour and our natural environment pervades almost every aspect of our lives, meaning that opportunities are diverse, growing and exciting. If you are interested in an exciting future, please check out the links below.

Further study resources
www.greenconsumerguide.com/university.php
www.enviroeducation.com/
www.learningforsustainability.net

Career resources
www.environmentjob.co.uk
www.ecojobs.com/
www.environmentalcareer.com/
www.acre-resources.co.uk/
www.sdgateway.net/jobs
www.learningforsustainability.net/jobs/

Tejas Ewing is a graduate of the IB Diploma Programme, having completed it at the United World College of South East Asia in 1998. He went on to study human geography, with a specialization in sustainable development, at the University of British Columbia in Vancouver. He recently completed a Masters degree in leadership for sustainable development, run by Forum for the Future in London. He now works full time in the field of sustainable development, currently as a carbon offset analyst.

Index